The vvhole art of navigation in five books ... by Captain Daniel New-House. (1685)

Daniel Newhouse

The vvhole art of navigation in five books ... by Captain Daniel New-House.
Newhouse, Daniel.
Added t.p.
[23], 311 [i.e. 315], 128, 113-131 p., [2] folded plates :
London : Printed for the author, 1685.
Wing / N920
English
Reproduction of the original in the Henry E. Huntington Library and Art Gallery

Early English Books Online (EEBO) Editions

Imagine holding history in your hands.

Now you can. Digitally preserved and previously accessible only through libraries as Early English Books Online, this rare material is now available in single print editions. Thousands of books written between 1475 and 1700 and ranging from religion to astronomy, medicine to music, can be delivered to your doorstep in individual volumes of high-quality historical reproductions.

We have been compiling these historic treasures for more than 70 years. Long before such a thing as "digital" even existed, ProQuest founder Eugene Power began the noble task of preserving the British Museum's collection on microfilm. He then sought out other rare and endangered titles, providing unparalleled access to these works and collaborating with the world's top academic institutions to make them widely available for the first time. This project furthers that original vision.

These texts have now made the full journey -- from their original printing-press versions available only in rare-book rooms to online library access to new single volumes made possible by the partnership between artifact preservation and modern printing technology. A portion of the proceeds from every book sold supports the libraries and institutions that made this collection possible, and that still work to preserve these invaluable treasures passed down through time.

This is history, traveling through time since the dawn of printing to your own personal library.

Initial Proquest EEBO Print Editions collections include:

Early Literature

This comprehensive collection begins with the famous Elizabethan Era that saw such literary giants as Chaucer, Shakespeare and Marlowe, as well as the introduction of the sonnet. Traveling through Jacobean and Restoration literature, the highlight of this series is the Pollard and Redgrave 1475-1640 selection of the rarest works from the English Renaissance.

Early Documents of World History

This collection combines early English perspectives on world history with documentation of Parliament records, royal decrees and military documents that reveal the delicate balance of Church and State in early English government. For social historians, almanacs and calendars offer insight into daily life of common citizens. This exhaustively complete series presents a thorough picture of history through the English Civil War.

Historical Almanacs

Historically, almanacs served a variety of purposes from the more practical, such as planting and harvesting crops and plotting nautical routes, to predicting the future through the movements of the stars. This collection provides a wide range of consecutive years of "almanacks" and calendars that depict a vast array of everyday life as it was several hundred years ago.

Early History of Astronomy & Space

Humankind has studied the skies for centuries, seeking to find our place in the universe. Some of the most important discoveries in the field of astronomy were made in these texts recorded by ancient stargazers, but almost as impactful were the perspectives of those who considered their discoveries to be heresy. Any independent astronomer will find this an invaluable collection of titles arguing the truth of the cosmic system.

Early History of Industry & Science

Acting as a kind of historical Wall Street, this collection of industry manuals and records explores the thriving industries of construction; textile, especially wool and linen; salt; livestock; and many more.

Early English Wit, Poetry & Satire

The power of literary device was never more in its prime than during this period of history, where a wide array of political and religious satire mocked the status quo and poetry called humankind to transcend the rigors of daily life through love, God or principle. This series comments on historical patterns of the human condition that are still visible today.

Early English Drama & Theatre

This collection needs no introduction, combining the works of some of the greatest canonical writers of all time, including many plays composed for royalty such as Queen Elizabeth I and King Edward VI. In addition, this series includes history and criticism of drama, as well as examinations of technique.

Early History of Travel & Geography

Offering a fascinating view into the perception of the world during the sixteenth and seventeenth centuries, this collection includes accounts of Columbus's discovery of the Americas and encompasses most of the Age of Discovery, during which Europeans and their descendants intensively explored and mapped the world. This series is a wealth of information from some the most groundbreaking explorers.

Early Fables & Fairy Tales

This series includes many translations, some illustrated, of some of the most well-known mythologies of today, including Aesop's Fables and English fairy tales, as well as many Greek, Latin and even Oriental parables and criticism and interpretation on the subject.

Early Documents of Language & Linguistics

The evolution of English and foreign languages is documented in these original texts studying and recording early philology from the study of a variety of languages including Greek, Latin and Chinese, as well as multilingual volumes, to current slang and obscure words. Translations from Latin, Hebrew and Aramaic, grammar treatises and even dictionaries and guides to translation make this collection rich in cultures from around the world.

Early History of the Law

With extensive collections of land tenure and business law "forms" in Great Britain, this is a comprehensive resource for all kinds of early English legal precedents from feudal to constitutional law, Jewish and Jesuit law, laws about public finance to food supply and forestry, and even "immoral conditions." An abundance of law dictionaries, philosophy and history and criticism completes this series.

Early History of Kings, Queens and Royalty

This collection includes debates on the divine right of kings, royal statutes and proclamations, and political ballads and songs as related to a number of English kings and queens, with notable concentrations on foreign rulers King Louis IX and King Louis XIV of France, and King Philip II of Spain. Writings on ancient rulers and royal tradition focus on Scottish and Roman kings, Cleopatra and the Biblical kings Nebuchadnezzar and Solomon.

Early History of Love, Marriage & Sex

Human relationships intrigued and baffled thinkers and writers well before the postmodern age of psychology and self-help. Now readers can access the insights and intricacies of Anglo-Saxon interactions in sex and love, marriage and politics, and the truth that lies somewhere in between action and thought.

Early History of Medicine, Health & Disease

This series includes fascinating studies on the human brain from as early as the 16th century, as well as early studies on the physiological effects of tobacco use. Anatomy texts, medical treatises and wound treatment are also discussed, revealing the exponential development of medical theory and practice over more than two hundred years.

Early History of Logic, Science and Math

The "hard sciences" developed exponentially during the 16th and 17th centuries, both relying upon centuries of tradition and adding to the foundation of modern application, as is evidenced by this extensive collection. This is a rich collection of practical mathematics as applied to business, carpentry and geography as well as explorations of mathematical instruments and arithmetic; logic and logicians such as Aristotle and Socrates; and a number of scientific disciplines from natural history to physics.

Early History of Military, War and Weaponry

Any professional or amateur student of war will thrill at the untold riches in this collection of war theory and practice in the early Western World. The Age of Discovery and Enlightenment was also a time of great political and religious unrest, revealed in accounts of conflicts such as the Wars of the Roses.

Early History of Food

This collection combines the commercial aspects of food handling, preservation and supply to the more specific aspects of canning and preserving, meat carving, brewing beer and even candy-making with fruits and flowers, with a large resource of cookery and recipe books. Not to be forgotten is a "the great eater of Kent," a study in food habits.

Early History of Religion

From the beginning of recorded history we have looked to the heavens for inspiration and guidance. In these early religious documents, sermons, and pamphlets, we see the spiritual impact on the lives of both royalty and the commoner. We also get insights into a clergy that was growing ever more powerful as a political force. This is one of the world's largest collections of religious works of this type, revealing much about our interpretation of the modern church and spirituality.

Early Social Customs

Social customs, human interaction and leisure are the driving force of any culture. These unique and quirky works give us a glimpse of interesting aspects of day-to-day life as it existed in an earlier time. With books on games, sports, traditions, festivals, and hobbies it is one of the most fascinating collections in the series.

JAMES R.

JAMES the Second, by the Grace of God, King of *England, Scotland, France,* and *Ireland, Defender of the Faith,* &c. To all to whom these Presents shall come, Greeting : Whereas We are humbly informed, That our Trusty and Well-beloved Captain *Daniel Newhouse* hath with great Art, and at the Expence of much Time and Money, composed a Treatise, Intituled, *The whole Art of Navigation,* in Five Books, which hath been perused by several the most Eminent Professors of the *Mathematicks,* and received their Approbation ; and the said Captain *Daniel New-house* having humbly besought Us to Grant him Our Royal License for the sole Printing and Publishing the said Book ; We have thought fit to condescend unto that his Request : And we do accordingly hereby grant our Royal License and Priviledge unto the said Captain *Daniel New-house,* his Executors, Administrators, and Assigns, for the sole Printing and Publishing the aforesaid Treatise or Book, under the Name or Title aforesaid, for and during the term of Fourteen Years, to be computed from the day of the first setting forth of the same. And Our Royal Will and Pleasure is, and We do hereby require and command, That during the said term of Fourteen Years, no *Printer, Publisher,* or other Person whatsoever, being Our Subjects, do presume to Imprint, or cause to be Imprinted, without the Knowledge and Consent of him the said Captain *Daniel New-house,* his Executors, Administrators, or Assigns, the aforesaid Treatise or Book, or any part thereof, under the Title aforesaid, or under any other Name or Title, or to sell the same, or to import into Our Kingdom of *England* any Copies thereof, Imprinted in any parts beyond the Seas, upon pain of the Loss and Forfeiture of all Copies so Imprinted, Sold, or Imported, contrary to the Tenor of this Our Royal License, and of such other Penalties as the Laws and Statutes of this Our Realm will inflict. And of this Our Pleasure, the Master, Wardens, and Assistants of the Company of *Stationers* are to take notice, that the same may be entred in their Register, and due Obedience be yielded thereunto.

Given at Our Court at White-hall *the Sixteenth Day of October,* 1685. *in the First Year of Our Reign.*

By His MAJESTY's Command.

SUNDERLAND.

RADIUM

NON

EXCUTIENT

THE WHOLE
ART of NAVIGATION
By Captain Newhouse.

THE
VVHOLE ART
OF
NAVIGATION;
IN FIVE BOOKS.

CONTAINING

I. The Principles of *NAVIGATION* and *GEOMETRY*.

II. The Principles of *ASTRONOMY*.

III. The Practical Part of *NAVIGATION*.

IV. The Description and Use of such Instruments, as are useful in taking Observations at Sea, and therein, the Use of a large new *Sinical Quadrant*, performing with more exactness than any yet extant, all Questions relating to *Navigation*; rendered so easie as to be understood by the meanest Capacity.

V. Useful **TABLES** in *NAVIGATION*, wherein those of the *Suns* and *Stars* Declination and Right Ascension, &c. are newly Calculated. The whole delivered in a very easie and familiar Stile, by way of Dialogue between a *Tutor* and his *Scholar*; approved by the ablest *Mathematicians*.

By Captain *DANIEL NEW-HOUSE.*

LONDON, Printed for the Author, 1685.

TO HIS MOST
EXCELLENT MAJESTY

JAMES II.

By the Grace of GOD
KING *of* England, Scotland, France,
and Ireland, *Defender of the Faith,* &c.

DREAD SOVEREIGN,

F Malice *and* Envy, *those
two great Enemies of* Vertue
and Glory, *do not often spare*
Persons *of the* first Rank, *but
endeavour to blast what is most
admired by the* just *and* wise;
a Person *so inconsiderable as I am, may well fear
their attack; and at the same time hope to be
excused*

excuſed for my *Ambitious Addreſs*, ſince I cannot ſecure my ſelf from it, but by the *Support of a Patron* endowed with all the *Eminent Qualities* that are ſufficient to ſhelter this *Work* from all the *Storms* that would deſtroy the advantages that its *Subject* deſerveth: And in this, I find not any whoſe *Protection* can be ſo advantageous to it, as that of *Your* MAJESTY's, to make it gain upon the *Lovers* of *Navigation*, who cannot be ignorant how perfect a *Maſter* Your MAJESTY is in this *Noble Art* (and other *Parts* of the *Mathematicks*) after ſo many *Proofs* You have given in the greateſt *Dangers*, when you did expoſe Your *Life* for the defence and ſafety of this *Kingdom*; and that with ſo much *Affection* and *Courage*, that this *Nation* cannot forget it without the greateſt *Ingratitude* imaginable; eſpecially, ſince nothing but *Malice* it ſelf can attribute the *Bravery* of Your *Exploits* and good *Succeſs*, to *Fortune*, but muſt call them by their

proper

proper *Names*, the effect of *ſteady Prudence*, and *ſkilful Conduct*, the reſult of *wiſe Counſels*, and *generous Reſolutions*: For without this, it would have been impoſſible to *Conquer* and *Reduce* ſo many *Potent*, and *Inveterate Enemies*, (as *you* have done:) And therefore I am very ſecure of the general approbation of this *Work*, if Your M A J E S T Y be pleaſed to receive it with the ſame favour with which you receive all thoſe that wiſh you a long and happy *Reign*. I am troubled, Great S I R, that there is nothing in it proportionable to the *Nobleneſs* of Your *Birth*, and *Greatneſs* of Your *Soul*; but what can I do in that particular, if the moſt ſublime *Wits* cannot attempt it, without being taxed with *Preſumption*? Without doubt, the boldneſs that I take to preſent to Your M A J E S T Y a thing ſo mean and inconſiderable, would be inexcuſable from any leſs *Generous* than Your ſelf. But the great condeſcention with which You receive the offerings of

obedient

obedient Hearts, makes me presume to give this Testimony of my Duty, and of the Zeal I have to be known in the World, by the Title of,

Your Sacred Majesty's

Most Loyal and most

Dutiful Subject,

DANIEL NEW-HOUSE.

To the Right Honourable

JOHN,

EARL of *Bridgwater:*

Viscount *Brackley*; Baron of *Ellesmere*;
One of the Lords of His MAJESTIES
Honourable Privy-Council; and Lord
Lieutenant of the Counties of *Bucking-*
ham and *Hertford:*

THE

Whole ART of Practical

NAVIGATION

IS HUMBLY PRESENTED

BY

NEW-HOUSE.

TO THE

READER.

HIS is not to beg your excuse for the plainneſs of this Work , for, as it is chiefly deſigned for Beginners, (although many Pilots may want it,) I have endeavoured to make it ſo, to render *Navigation* as eaſie and intelligible as poſſible I could: that is the reaſon that I chuſe to make it by Dialogues, and that I have paſſed over ſome few *Mathematical Demonſtrations*, which I have thought too hard for them to underſtand, and have explained the Principles of *Aſtronomy* by the *Ptolemaick Syſteme*, rather than by the *Copernican* ; however

ever, there is enough in it to learn to carry a Ship safe to any part of the World. Take only what pleases you best, and leave the rest for those whom Nature hath not endowed with your parts: And if you are for higher things, have a little Patience, and I shall endeavour to satisfie you in the next, which I design principally for Artists. In the mean time, accept of this as a Token only of my Ambition to serve those Gentlemen that incline to *Navigation*; and if you be of the Number, give me leave to subscribe my self,

Your affectionate

and humble Servant,

NEW-HOUSE.

To

To Captain NEW-HOUSE, upon his BOOK of NAVIGATION.

WHen Mortals first, (steel'd with the hopes of gain)
    Made an attempt upon the dangerous Main,
Rude of all Art, in shapeless Vessels, they
Commit themselves unto the faithless Sea:
Strange Neighbouring Lands discover, and explore
Nigh Continents, they n're had seen before.
In time far distant Realms, each other knew,
Whilst still by use the Sailors knowledge grew;
At last to Chart, and Compass they arriv'd,
The utmost thought, yet still the Science thriv'd.
How many Schools are Founded, to instruct
The Youth design'd, proud Vessels to Conduct?
What Volumes writ, to teach the safest way
To guide these floating Castles on the Sea?
And amongst these my Friend, may I devine;
None can be worthyer than this of Thine.
'Tis true, I ignorant my self must own,
In all the Rules of *NAVIGATION.*
But 'tis sufficient, that I know your Wit,
And can the surest judgment give from it;
For how can Plants, that good and generous are
But like themselves, fruit good and generous bear?

W. M.

A

TABLE

OF THE

Definitions, Propositions, *and* Problems, *contain'd in the* First Book.

GEOMETRICAL PROBLEMS:

S E R V I N G

For the Construction of the Figures contain'd in the following Books.

The TABLE of the Second Book.

Prop. 11.

Prop. 29.

The TABLE of the Third Book.

Prop. 19.

c   Prop. 39.

The TABLE of the Fourth Book.

The

The TABLE of the Fifth Book.

THE

THE
Compleat ART
OF
NAVIGATION.

THE FIRST BOOK.

The Principles of Navigation.

Begin with the Principles of Navigation, it being necessary to explain them here in very intelligible terms, to the end that they be Perfectly understood even by those of the meanest Capacity, who are desirous to Understand or Learn the Art of Navigation; most of them being necessary for Practice, and to make it yet more plain, I shall lay down the whole Art of Navigation by way of Dialogue, between a Young Scholar and his Tutor.

The Definition of Navigation, and what a Pilot should know.

S. WHAT is *Navigation?*

T. Navigation is a Science or Art which contains certain Rules, absolutely necessary for every man to know that undertakes the Conduct of Ships from one Country or Harbour to another.

B   *S.* Why

S. Why do you name it a *Science* or *Art* ?

T. Becaufe Navigation is confidered two ways, to wit, in *Theorie* and in *Practice*, that being properly an Art which puts into practice the Precepts Invented or Taught by Theorie, which is a Science.

S. How many *Sorts* of Navigations are there?

T. There are *Two Sorts* of Navigation, the one *Common* and *Short*, fuch as is Practifed by thofe that only go from Harbour to Harbour along the Coaft, and feldom lofe the fight of the Land. And the other *Artificial* and *Great*, becaufe it undertakes great Voyages, and requires greater knowledge than the firft, fince it carries Ships to all the moft diftant Parts of the World.

S. What is it neceffary to *know* for Practicing thefe two forts of Navigation?

T. For the *Common* or *Coafting* Navigation, befides the Ufe of the Sea-mens *Compafs* and *Card* for Soundings, it is alfo neceffary to know well how to judge of the *Ships-way*: The Ufe of the *Calender* for diftinguifhing the Common year from the Biffext, and for underftanding the Ufe of the *Golden Number* and *Epact*, to find by it the New Moons and her Ages, the time of full Sea, or low Water, the Spring and Ebb or Neap Tides: The force of the *Currents*, and how they fet, and *what time* you may Enter or put into any Harbour, whofe *Situation* muft be perfectly known, (that is to fay, what Point of the Compafs the New or Full Moon makes High Water in it) as well as the *Depth* and *Quality* of the Waters. The *variety* of Grounds; the *Rocks*, *Sands*, *Shoalds*, *Capes*, (or Points of Land) with their diftance, and how they Lye: *Clifts*, *Light-houfes*, or any other Land or Sea Marks; the knowledge whereof confifteth chiefly in Experience, which you may the fooner attain by help of the *Great Wagoner*, which Book, I recommend to you for that purpofe.

As for the *Artificial* and *Great* Navigation, befides all this which is *Common* to both; you ought alfo to underftand the *Definitions* and *Ufes* of the *Sphere*, the *Sun* and *Stars Declination* for finding the Latitudes; with the Ufe of the *Crofs-Staff*, *Quadrant*, *Aftronomical Ring*, (or *Aftrolabe*,) and for that purpofe the *Right Afcenfion* of the Sun and Stars, (for to know what time they will come to the Meridian, and the hours of the Night) the *Ufe* of the *Tables of Amplitude*, with the perfect Practice of the *Sinical Quadrant*, *Gunters Scale*, or the like, to reduce their feveral Courfes and to Correct them: And befides all this, you muft needs know very well how to obferve or find out the *Variation* of the Compafs, to Correct it, and Rectifie your Courfes, a thing fo neceffary to a Pilot, that without it he cannot be capable to undertake the Conduct of a Ship.

S. Is this *all* that a Pilot fhould know?

T. All good Pilots are obliged at leaft to know well *what I have named*, and to anfwer to it when Examined about it. But thofe that aim at a greater perfection in that Art are not fatisfied with the Practice of the Sinical Quadrant or Scale, but apply themfelves to other more exact

Geometrical

Geometrical Practices, as *Trigonometry*, and the Ufe of the Tables of *Logarithms*, *Sines*, and *Tangents*, which is the moft perfect and exact Navigation or way of Working a Traverfe, fince it agrees both to the *Triangles*, *Right-lines* and *Sphericks*: Befides every Pilot or Captain that will perfect himfelf, ought to learn at leaft how to *Calculate* the Declination of the Sun and Stars, (and to *Correct* the Tables of the Sun's Declination when neceffary) their Right Afcentions, and Amplitudes, Ortive and Occafive; their Azimuth, and other Aftronomical Practices which concern Navigation.

S. Is it neceffary for a Pilot to carry to Sea a *Crofs-Staff*, a *Quadrant* and an *Aftronomical Ring* or *Aftrolabe ?*

T. Yes, in long Voyages; fince many good Obfervers have found by experience, that the *Quadrant* is *beft* when the Sun is near the *Zenith*, but that *otherwife* the *Crofs-Staff* is *better*. The *Aftronomical Ring* and *Aftrolabe* is alfo neceffary to take obfervation, when there is no clear Horizon, (a thing that often happens to thofe that Sail to *New-found-land*, and fifh upon the grand Bank) but of thefe *two*, the *Ring* is the beft, the degrees being greater by half, and in every refpect more eafie and fitter for the Sea than the Aftrolabe.

PROPOSITION I.

Of the Year *Biffext*.

S. WHAT do you call a *Year ?*

    T. A Year is that *Interval* or *Space* of time, that the Sun takes to go about the *Twelve Signs* of the Zodiack.

    S. Are there *Several kinds* of Years ?

T. Yes, but I fhall treat or fpeak here only of *three* kinds, to wit, of the *Aftronomick*, which is Compofed of Hours, Minutes and Seconds, over and above 365 Days. And of the *Politick* or *Civil* Year, which contains only Days, and is divided into the *Common* which is of 365 Days, and the *Biffext* which is of 366, and this laft happens only once in Four Years.

S. Why do you call the *Fourth* Year *Biffext ?*

T. Becaufe the *Sixth* Day next before the Calends of *March* is *twice* repeated, a Day being added to the *Twenty-fifth* of *February* on which it falls, and therefore they count or repeat twice the *Sixth* of the Calends, in Latin *Bif-fexto Calendas*, from which two firft words *Biffext* is derived, *February* being that Year of 29 Days.

S. What do you mean by *Calends ?*

T. Calends are the *Firft* Days of every Month, from which the *Romans* counted the Days of the Month, they are derived from the *Greek* Verb *Calo*, which fignifies to Call, becaufe their Crier that Day (ftanding in a high place) after several Calls, made known to the people *how many* Days in that Month, the Fairs and Markets should last.

S. Why did they institute or set up *this Custom* of Bissext?

T. Because that the *Common* Year contains almost six Hours *more* than 365 Days, which *Hours* make a *Day* in *Four* Years time; and so by that means the Common and Astronomical Year almost agree.

S. How shall I know *when* a Year proposed is *Bissext* or *Common*?

T. *Divide* the proposed Year (since *Christ*) by 4, and the Remainder of the Division will show you what Year it is. But if there Remains Nothing, the proposed Year is Bissext.

Example.

I would know what the Year 1685 *will be*?   4) 1685 (421
Divide 1685 by 4, the Remainder is 1, thus,   001

Which shews, that the Year 1685. will be the *First* Year after Bissext.

Another way more easie.

Is to cut off, from the proposed Years, the Thousands, Hundreds, Scores, and all the Fours and the Remainder will show you what Year it is, but if there Remains nothing the proposed Year is Bissext. As for Example: If you would know what the Year 1684 is, first cut off the Thousand and Hundreds, and there will Remain 84, which is Four Scores and Four, which being likewise taken off, there Remains Nothing, by which you know that the Year 1684 is Bissext, and by the same directions you will find the Year 1686 to be the second Year, and the Year 1687 the third after Bissext.

S. Is not that *Institution* of Bissext Year *Interrupted*?

T. No not in *England*, but it will be interrupted in *Holland*, *France*, *Spain*, and other Countries that follow the *Gregorian* or New Calender, to wit, every Hundred Years, except the Fourth Centurie, to reckon from the Year 1600, and therefore the Year 1700, 1800, 1900, 2100, 2200, 2300, shall be Common Years amongst those Nations that observe the Constitutions of the Popes, but the Year 2000, 2400, &c. will be Bissext according to the decree of Pope *Gregory* the XIII. in the Year 1582. from which Year in the same time he cut off 10 Days to reform the *Julian* Calender which made the Year too long, and is the cause that we do and shall count our Month 10 Days later than the *Dutch* and *French*, 'till the Year 1700, that we shall differ 11 Days, because they make it a *Common* Year, and we a *Bissext*, by which means this difference will increase a Day, and will do so every Hundred Years, (unless our Calender be reformed) except the Fourth Century or Hundred, because then their Year shall be Bissext as well as ours.

P R O P.

PROP. II.

Of the Roman Indiction.

S. WHAT is the *Roman* Indiction?

T. It is a Revolution or Number confifting of 15 Years, which is now of no Ufe to *Navigation*, nor any thing elfe, altho it is moft commonly fet down in the Calenders, and in all the Charters and Writings of the Bifhops and Prothonotaries of *Rome*.

S. Since this Indiction is of *no Ufe* to us, why do they put it then in the Calenders?

T. 'Tis put in only to follow the Cuftom of the Antient *Romans*, which did ufe the like Indiction of Years, and therefore it is called the *Roman Indiction*.

S. Since it is put into our Calenders, I fhould be glad to know the firft occafion of it.

T. Then you muft be informed, that it was a Cuftom amongft thofe *Foreign* Nations that were *Tributary* to the *Roman* Empire (and dwelt a-far off) to pay their Tributes to the *Romans* as followeth.

The firft Five Years they paid Gold, in token of their obedience to the Empire. In the fecond Five Years they paid Silver, for the Soldiers pay. But in the laft Five Years they paid only Brafs, towards the Reparation of Armour and Munition: So this Cuftom of three different Payments in 15 Years, hath been the Caufe of this Indiction, the thing you defired to know.

S. How do you find out this Number of Indiction?

T. 'Tis found out thus, add 3 to the Year given, and divide the Sum by 15, the Remainder of the Divifion fhall be the Number of Indiction Required, which is to be counted from *September*; (and not from *March*, as the *Foaft*) and if nothing remains after your Divifion, then 15 is the Number of Indiction.

PROP. III.

Of the Golden Number.

S. I Have often Read of this *Golden Number*, but do not underftand it well, pray be fo kind to explain it to me in intelligible Terms.

T. To underftand well what is meant by the *Golden Number*, you muft know, that the *Conjunction* of the Sun and Moon (which we commonly call *New-Moon*) doth not happen every Month at the *fame* time, but only *once in* 19 Years, and yet not *exactly* then neither, for the Moon finifheth finisheth her Period an Hour and almoft Twenty Eight Minutes fooner than fhe begun it 19 Years before; however *this* being the *Number* of Years on which the Sun and Moon comes *nearest* to end their Revolutions together, it was chofen before *any other*, to find out by it the *New Moons* and *Eafter Days*, for which it was in fuch efteem amongft the Antient Aftronomers that they kept an exact Account of it, and did mark it in their Calenders every Year in Gölden Letters: Thus, I. II. III. IV. V. VI. and fo forth, untill XIX. which is the *reafon* it is called the *Golden Number*, to fignifie that as *Gold* furpaffeth *other* Metals, fo *this Number* furpaffeth *all others*, for this particular Ufe.

S. What is the *Ufe* of it now?

T. Its Ufe is only to find out the *Epact*, and not the *New Moons* and *Eafter* as formerly, becaufe it is found *Defective*.

S. How is this (*Golden*) Number to be found out?

T. There are feveral ways to find the Golden Number, but the eafieft is to cut off the Thoufands and Hundreds of the propofed Years, and to add 5 to the reft, from which take all the Scores, and for as many Scores as you cut off, add fo many times 1 to the Remainder, and that will make up the Golden Number required, as you will better under-ftand by this

Example.

I would know the Golden Number for the Year 1685.

Therefore, I cut off the Thoufand and Hundreds of the propofed Years, and there remains 85, to which I add 5 and it makes 90, from which I take or cut off the 4 Scores, and there remains 10, to which I add 4 (for the four Scores I took) and that makes 14, the Golden Number required.

S. Altho this is fufficient, I fhould be glad to learn *fome* of the *other* ways to find this *Golden Number*.

T. I do not think it neceffary; however to fatisfie you in it, I will fhew you *Three* other ways, (to the end that ycu may underftand as much of it as any;) the *Firft* of which is to cut off 1500 of the propofed Years, and to Divide the reft by 19, and the Remainder of the Divifion will be the Golden Number required, but if there remains Nothing, the Golden Number will be 19.

The *Second* is to cut off 1600 of the Years propofed, and to add 5 to the reft; and then to Divide it by 19, and the Remainder of the Divifion fhall be the Golden Number required, but if there remaineth Nothing, the Golden Number will be 19, as before.

And the *Third* Practice is to add 1 to the propofed Year, and to Divide it by 19, and fo forth, according to the precedent directions; by all which you will find the Golden Number of any propofed Year, this is fo plain that it needs no Example.

S. Why

S. Why do you add one to the Year of our Lord?

T. 'Tis becaufe the Golden Number was 2 the Year that *Chrift* was Born.

S. What will the Quotient of this laft Practice (or Divifion) fhow?

T. The Quotient will fhow how many *Lunary Cycles* (of 19 Years) have paft, fince the Year of *Jefus Chrift*, and that is the reafon why the propofed years are Divided by 19.

S. What time of the Year is it, that this Golden Number begins?

T. It begins always the *Firft* of *January*.

PROP. IV.

Of the Epact.

S. WHAT do you call the *Epact*?

T. The Epact is nothing elfe, but that difference of 11 Days, which is between the two Common Years of the Sun and the Moon; otherwife called Year *Solary* and Year *Lunary*.

S. How fhall I underftand this *Difference*?

T. You may eafily underftand it, if you confider that the Common Lunary Year is but of 12 Moons, and every Moon but of 29 Days and a Half, which makes but 354 Days; and therefore is fhorter by 11 Days than the Solar Year, which as you have read, contains 365 Days.

S. I underftand now very well that the Epact takes its Original from this Difference of 11 Days, but muft thefe 11 Days be added every Year to the former Epact?

T. Yes, to wit, the next Day after the laft of *February*, for the *Epact* never begins fooner than the *Firft* of *March*.

S. Is this order never interrupted?

T. No never, with us, except the Epact happens to be 29, as it will infallibly happen in the Year 1690. or every time that the Golden Number is 19, for then you muft add 12 for the following Year only; to the end that the Epact may keep the fame order as before, with the Golden Number. You are alfo to take notice, that the Epact never paffeth 30, and therefore when by adding 11, you find that it paffeth 30, you muft Subtract the 30, and the Remainder fhall be the Epact required.

S. How fhall I know the Epact?

T. The Epact may be known feveral ways, but the eafieft after you know the Golden Number, is to fuppofe three Numbers placed upon your left Thumb, as the following Figure fhews you.

And

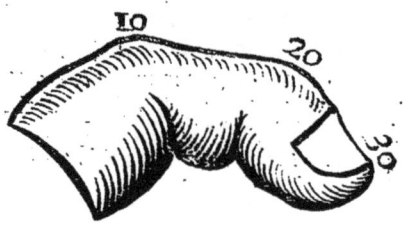

And then count your Golden Number on your Thumb: Thus, firſt begin on the lower Joint where the 10 ſtands, and reckon 1, then 2 on the middle where the 20 ſtands, and 3 upon the end or Number 30: Then begin again at the lower Joint, and there ſay 4; and ſo continue in the ſame order as before, untill you have counted the Golden Number of the propoſed Year. For that Number on which it falleth being added to it, will be the Epact for that Year: So that the Sum do not exceed 30, for if it doth, you muſt Subtract 30, and the Remainder ſhall be the Epact required.

Example.

I deſire to know the Epact for the Year 1685.

Therefore, I look firſt for the Golden Number of 1685, (as before taught) and find it to be 14, which being counted upon my Thumb, (in the ſame order as before) I find that it falls on the middle of it, on Number 20, which being added to 14, there comes 34, from which I Subtract 30, and there remaineth 4 for the Epact of the Year 1685.

S. This is plain and eaſie enough, however I deſire to know ſome other way or method.

T. The Second way to find the Epact, is to Multiply the Golden Number of the propoſed Year by 11, and the Product will be the Epact, (if it be under 30) but if the Product be above 30, then you muſt Subtract 30, and the Remainder ſhall be the Epact required.

Example.

I would fain know the Epact for the Year 1685. *whoſe Golden Number is* 14.

The

The Golden Number is 14
I Multiply it by 11
$$\underline{}$$
14
14
$$\underline{}$$
The Product . . . 154

Which divided by 30, the Remainder is 4;   30) 154 (5
the Epact required.   4

S. What is the Epact good for?

T. The Epact is good for *Three* things: First, it serveth to find the *Change* or *New Moon.* Secondly, it serveth to find the *Age* of the *Moon.* And Thirdly, the *Days* of the Month. (The Age of the Moon being known.)

PROP. V.

How by the Epact to find the New Moon and its Age.

S. WHAT do you call *New Moon?*

T. The New Moon is the *Conjunction* of the Sun and Moon, that is to say the Position of them both, under the same degree of the same Sign of the Zodiack.

S. And what is the *Full Moon?*

T. The Full Moon is the *opposition* of the Moon to the Sun, which happens when the Sun and Moon, are in the same degree of the opposed signs.

S. What is the *increase* and *decrease?*

T. The increase is when the Moon *increaseth* her *Light,* that is to say, from her *Change* or the New Moon, to her *Full,* untill which time she is on the *East* side of the Sun, and goeth down after him. The decrease is, when the Moon's Light *Diminisheth,* to wit, from her *Full* untill she be *New* again, and then she is *West* of the Sun and goeth down *before* him.

S. How is the *Change* or New Moon to be *known?*

T. Add the Epact of the proposed Year to the Months past since the first of *March,* and that Sum Subtract from 30, the Remainder shall be the Day of the Change or New Moon: But if the Epact and Month being added together exceed 30, they must be deducted out of 60.

Remember that in *January* you must add nothing to the Epact for the Month: in *February* you must add 2, in *September* 8, and in *November* 10, and by so doing you will come nearest to the true time of Change, or New Moon, the same is to be observed for finding its Age.

C   *Example.*

Example. 1.

I would fain know the Day of the Change or New Moon, in February, 1685.

The Epact of the proposed Year is 4
For the Month of *February* add 2
                                          The Sum 6

Which being Subtracted from . . . 30
                                                  6
                   There Remaineth . . . 24

Which is the Day that it will be New Moon in *February*, 1685.

Example 2.

I desire to know the Day of Change or New Moon, in October, 1684.

The Epact of 1684 is 23
The Months from the first of *March* 8
                              The Sum 31

Which being Subtracted from . . . 60
                                                  31
                   Remaineth . . . 29

By which I know that in *October* 1684. the Day of the New Moon will be the 29*th.* as was required.

S. There is enough of that, but how shall I know the *Age* of the Moon?

T. You shall know the Age of the Moon, if you add the Epact, and Month from *March*, to the Day of the Month proposed, for those Three Numbers, being added together, will be the Age of the Moon: But if it exceed 30, you must Subtract the 30, and the Remainder shall be the Age of the Moon.

Example. 1.

I desire to know the Age of the Moon, the 17th. of November, 1684.

The

The Epact of the Year 1684, is 23
For the Month of *November*, add . . . 10
The Day of the Month, add 17
                  The Sum . . . 50
From which I Subtract 30
               Remaineth . . 20

For the Age of the Moon, the 17th. of *November*, 1684.

Example 2.

The 7th. of *June* 1685. *I would know the Age of the Moon?*

The Epact of the Year 1685, is 4
For the Month, add 4
The Day of the Month, add 7
               The Sum . . . 15

For the Age of the Moon the 7th. of *June* 1685, and by it I know that it is Full Moon.

PROP. VI.

By the Age of the Moon, how to find the Day of the Month.

S. WHAT must I do, to find the *Day* of the *Month* by the *Age* of the *Moon?*

T. You must add the Epact of the proposed Year, to the number of Months past from *March*, (as before) and Subtract the Sum from the Days of the Moons Age, the Remainder shall be the Day of the Month required: But if the Epact and Months past from *March*, being added make more than the Days of the Moons Age, you must add 30 to the Age of the Moon, and Subtract the Sum (of your addition) as before.

Example 1.

In June 1685, *the Moon being the* 15 *Days Old, I desire to know what Day of the Month it is?*

C 2   The

The Epact of the proposed Year 4
The Month from *March*, add 4

        The Sum 8

Which being Subtracted from 1 5
                                8

There Remaineth 7

For the Day of the Month required.

Example 2.

In November 1684, *the Moon being then* 20 *Days Old*, *I desire to know what Day of the Month it is?*

The Epact of the Year 1684, is . . . 23
For the Month of *November*, add . . . 10

        The Sum . . . 33

Which I should Subtract from the Age of the Moon, but cannot; therefore I add 30 to the 20 Days of the Moons Age, which makes 50; from which I Subtract the Sum 33, Remaineth 17, for the Day of the Month in *November* 1684.

S. Is it necessary to know the *Age* of the Moon, for to find the *Day* of the Month?

T. Yes, and therefore if you don't know it, you must find it out, by observing when the Moon is upon your Meridian: You must also know the just hour of the Day or Night at that same time; which hours being Multiplied by 15, and the Product divided by 12, the Quotient will be the Age of the Moon: But take notice, that when the Moon comes to your Meridian from Noon to Midnight it is increase; but when from Midnight to Noon it is decrease, and then you must add 15 to the Moons Age, (found by your Division) and the whole will be the Age of the Moon required.

Example 1.

The Moon being on the Meridian at 9 *of the Clock at Night.* (*And by consequence in her increase.*) *I demand her Age?*

The hour is : 9
I Multiply it by 15

        The Product . . . 135

                                             Which

Which I divide by 12. 12) 135 (11 Days, 6 Hours,
                            3

And the Quotient showeth that the Moon is 11 Days 6 Hours Old; for the remainer of the Division being doubled, shows the Hours.

Example 2.

The Moon being upon the Meridian at 6 of the Clock in the Morning. (And by consequence in her decrease.) I demand her Age?

The Hour is 6
I Multiply it by 15
                The Product 90

Which I divide by 12.   12) 90 (7 d. 12 h.
                            6 15 added
              The Sum . . . . 22 Days, 12 Hours,

For the Age of the Moon.

S. Why do you Multiply the Hours by 15?

T. It is, to reduce them into Degrees; for an hour is equal to 15 degrees; for the Sun by his Diurnal Motion (from East to West) maketh 360 degrees in 24 hours, which is 15 degrees an hour.

S. Why do you divide it by 12?

T. I divide it by 12 to reduce it into Days, (of the Moons Age,) for the Moon moves 12 degrees more than the Sun in a Day, (by her own proper Motion) and therefore if you divide the degrees of her distance from the Sun, by 12, you will reduce it into Days and hours Lunar, as in the precedent Example.

S. Is there no other way to find the Moons Age?

T. Yes, you may also find it by the Cross-staff, observing therewith the distance between the Sun and the Moon, and the degrees being divided by 12 will show you the Moons Age.

You may also give a near guess at the Sun and Moons distance, by the Sea-man's Compass, setting by it the Sun and Moon, and allowing 11 degrees 15 minutes for every Point (of the Compass) contained between them both: But this way being not so exact as the first, I shall not recommend it to you.

S. Can you not find the Moons Age by common Arithmetick?

T. Yes, it may be found out by the Rule of Three; if you can tell how many hours the Moon comes later to the Meridian than the Sun; which you may easily do by a Dyal or by the Watch, or the Glasses run since Noon. *Example.*

Admit that I obferve the Moon on the Meridian 5 hours after the Sun, and would know the Age of the Moon ?

I make a Rule of Proportion, and fay: If 24 give 30: What will 5 hours give? 　24) 150 (6 days, 6 hours. 　　　　5
　　　　　　　　　　　　 ―――― 　　　　　　　　 ――――
　　　　　　　　　　　　　6 　　　　　　　　　　　 150

And I find that they give **6** days and 6 hours for the Age of the Moon, as was required.

S. Why did you fay, if 24 give 30?

T. Becaufe that from Change to Change we reckon 30 days, (altho it wants fome hours and minutes of it) and every day is of 24 hours; and therefore to find how many days (Lunar) the 5 hours fignifie you muft Multiply 30 by 5, and divide the Product by 24, and the Quotient will fhow you the Age of the Moon, as in the precedent Example.

PROP. VII.

How to find the Cycle of the Sun, and Dominical Letter.

S. WHAT do you mean by the *Cycle* of the *Sun* ?

　　T. The Cycle Solar is a Revolution of 28 Years, which being ended begins again at the Unit, becaufe the Dominical Letter, is then in the fame order that it was at firft.

S. How do you find out the juft number of the Cycle of the Sun?

T. I find it by adding 9 to the propofed Year, and dividing it by 28, for the Remainder of the Divifion is the Cycle of the Sun; but if there remaineth nothing, the Cycle Solar will be 28.

I defire to know the Cycle of the Sun, for the Year 1684.

The Year propofed 1684
Add 9
　　　　　　The Sum 1693

　　　　　　　　　　　　　　　　　　　Which

Which being divided by 28 ,   28) 1693 (60
                                           13

There remaineth 13 for the Cycle of the Sun required.

S. What was this Cycle invented for?

T. This Cycle was invented, more to find by it the *Dominical* or *Sunday Letter*, than to show any *Change* of the Suns Motions.

S. What must I do to find the Dominical Letter?

T. To find the Dominical Letter; you must first be informed, that the Cycle of the Sun begins always (with us) at a Year *Bissext*, that is to say at two Letters, to wit, G F; and so on counting all the Letters backward, as you will better understand by the following Table, which at any time you may make your self, or at least as much of it, as you need, for knowing the Dominical Letter: As for Example, If you would know the *Sunday* (or Dominical) Letter for the Year 1684, whose Cycle Solar is 13.

You must begin your Table as in the Margin, and you will find that the Dominical Letter for the Year 1684, is F and E, since the Cycle of the Sun is 13, had it been more, you must have continued the Table in the same order, and the last Number of the Cycle will show you the Letter answerable to it.

S. Why do you begin to count the Cycle Solar from a *Bissext* Year, and the Letters *backward?*

T. It is only to imitate the Ancient *Romans* whose Custom it was.

S. Is there *no ways* to find out the Dominical Letter with *less trouble?*

T. Yes, if you can remember these Seven Latin Words: *Gratis, Filius, Eternus, Dei, Cœlum, Bonis, Addit:* Which must be counted upon the 4 Fingers of the Left hand, thus: For the *First* Year of the Cycle Solar, you must say upon the end of your fore Finger *Gratis Filius:* For the *Second* Year, upon the middle Finger, *Eternus:* For the *Third* upon the fourth Finger, *Dei:* For the *Fourth* Year upon the little Finger, *Cœlum:* Then beginning again at the fore Finger, say *Bonis Addit*, and so forth, untill the *Last* Number of the Cycle Solar, which then will shew you the Dominical Letter; provided you do not forget to count always two Letters upon the fore Finger; which will show you the Bissext Year, but the other Three (Fingers) only the common.

Cycle.	Sunday Let.
1	G F
2	E
3	D
4	C
5	B A
6	G
7	F
8	E
9	D C
10	B
11	A
12	G
13	F E

H R. QP.

PROP. VIII.

How you shall find what Day of the Week the Month begins.

S. **H**OW shall I find what Day of the Week the Month begins?

T. To find what Day of the Week the Month begins, you must learn by heart this Latin verse:

Astra Dabit Dominus Gratisque eabit Egenos,
Gratia Christicolæ, Feret Rurea Dona Sideli:

(or any 12 *English* words that begins with the same Letters that these do) which you must apply to the 12 Months of the Year, as *Astra*, for **January**; *Dabit*, for **February**; *Dominus*, for **March**; and so forth until **December**: The first Letter of every word, sheweth the Day of the Week on which the Month begins. For it is the Custom to assign the *Seven* first Letters of the Alphabet, to the *Days* of the Week; and to begin *January* with the first Letter of it A, and so to continue in Alphabetical Order 'till the last Day of the Year which ends with the Letter G, and therefore knowing with what Letter the Month begins, it is very easie to know on what Day of the Week it is, for you need but reckon from the Dominical Letter; as you will better understand by this

Example.

In the Year 1685*, the Dominical Letter is* D. *I demand what Day of the Week falls the first of* July?

Answer. Since *July* (by the precedent Rule) begins with the Letter G, and the Dominical Letter is D: I say that *July* begins on a *Wednesday*, for from D to G in Alphabetical Order, (as you must count it) there are Three Letters signifying the Third Day from the Dominical Letter, which can be no other than *Wednesday*: As I said.

S. Are not these *Seven* Letters the *same* that serve for the *Dominical Letter*?

T. Yes, all the difference I know, is that the Dominical Letter serveth for *Sunday*, and the six others for the *rest* of the Week in Alphabetical Order, *Contrary* to the Order of Dominical Letters which are counted *backward.*

S. Do they not differ also in Name?

T. Yes, for the Letter that serveth for *Sunday* is called the *Dominical Letter*; but those that serve for the rest of the Week, are called *Ferial*, because of the Holy Days that fall on them.

S. Can you tell by these Ferial Letters, on what Day of the Week any other Day of the Month falleth?

T. Yes, very easily, for the same Day of the Week that the Month begins, the 8*th.* the 15*th.* the 22*th.* and the 29*th.* begins also. This being understood, it must needs be very easie to know it: However I will give you an   *Example.*

In the Year 1685, I would know on what Day of the Week the 24th. of July falls.

Anſwer. Since the 22th. of *July* falls on a *Wedneſday* (as I ſaid) as well as the firſt Day of the Month. The 24th. muſt needs fall on a *Saturday.*

PROP. IX.

How to find the Moveable Feaſts.

S. WHAT do you Call *Moveable Feaſts?*
T. The Moveable Feaſts are thoſe that do not fall every Year on the ſame Day of the Month. As *Eaſter* Day, *Rogation* Sunday, *Aſcention* Day, *Whit* Sunday, and the like.
S. How ſhall I *find* theſe Moveable Feaſts?
T. The way to find what *Day* of the Month they fall on, is very eaſie; the main or chief thing being only to know the *Day* of the *New Moon* in *February*, for by it they are all known. *Shrove Sunday* being always the next *Sunday* to the Change of the Moon. *Shrove Tueſday* the next *Tueſday.* After *Shrove Sunday*, *Quadrageſſima* (or the *firſt Sunday* in *Lent*) is the next *Sunday* to *Shrove Tueſday.* *Eaſter Day* ſix Weeks next after *Quadrageſſima.* *Rogation Sunday* five Weeks or 35 Days, next after *Eaſter.* *Aſcention Day* four Days next after *Rogation Sunday.* *Whit Sunday* ten Days next after *Aſcention Day.* *Trinity Sunday* ſeven Days next after *Whit Sunday.* *Corpus Chriſti* four Days next after *Trinity Sunday.* The firſt *Sunday* in *Advent*, is the fourth *Sunday* before *Chriſtmas.* *Septuageſſima* is the third *Sunday* before *Quadrageſſima*, or firſt *Sunday* in *Lent.* *Quinquageſſima* is the next *Sunday* before *Quadrageſſima.* *Sexageſſima* is the next *Sunday* before *Quinquageſſima.*

PROP. X.

How to find the Time of High-water in any Harbour.

S. HOW ſhall I find the Time of *Full-ſea* in any Harbour?
T. To find the true Time of High-water (or Full-ſea) in any place, you muſt firſt find by the following Table of Tides, (if not by your Experience) what Moon maketh a Full-ſea in it, the Day of her Change or Full; for that being known, you may with

D eaſe

eafe find the Time of Full-fea at any other time: Only by Multiplying the Days of the Moons Age by 48, and dividing the Product by 60, and adding to the Quotient and reft of the Divifion the Hours and Minutes for the Tide.

Example.

If on the Day of Change or New Moon, it is Full-fea in the Downs at 1 Hour 30 Minutes, (as the Table of Tides fheweth) I demand what time it will be Full-fea there when the Moon is 8 Days Old?

Days of the Moons Age 8
Multiply by 48
The Product . . . 384

Which being divided by 60,   60) 384 (6 hours 24 minutes.
                                          24

gives 6 h. 24 m. To which I add, 1 h. 30 m. for the Tide on the Day of Change, and it makes 7 h. 54 m. for the Time of Full-fea required.

S. Why do you *Multiply* the Moons Age by 48, and *Divide* the Product by 60?

T. I Multiply it by 48 to reduce it into Minutes, becaufe the Moon and Tide abateth every Day 48 Minutes; and I divide it by 60, to reduce it into Hours, becaufe 60 Minutes make an Hour.

S. Is there no *fhorter* way to do it by?

T. Yes, for fhortnefs you may Multiply the Days of the Moons Age by 4, and divide the Product by 5, and adding for the Tide, or Day of Change, you fhall have the Time of Full-fea required, but if it exceeds 12, you muft Subtract the 12, and the Remainder fhall be the time of High-water.

Example.

If on the Day of Change it is Full-fea at London at Three of the Clock, I would know what time it will be Full-fea there, the Moon being 12 Days Old?

Days of the Moons Age 12
Multiply by 4
The Product 48

Which being divided by 5,   5) 48 (9 hours 36 minutes.
                                        3

giveth

9	36
3	00

giveth 9 hours 36 minutes: To which I add 3 hours for the Day of Change.

The Sum is . . . 12 h.   36 m.

From which I Subtract 12   00

Remaineth 36 m. for the Time of High-water required.   00   36 m.

S. Why do you *Multiply* by 4, and *Divide* by 5?

T. I Multiply by 4, for to reduce the Days of the Moons-Age into fifth parts of an hour; for 48 Minutes are $\frac{4}{5}$ or 4 fifths of an hour: And I divide the Product by 5, to reduce it into Hours; and the Remainder (of the Division) I Multiply by 12 to reduce it into Minutes, because what Remaineth are fifth parts of an-hour, or 12 Minutes each.

S. I observe that both your Examples are when the Moon is in her *increase*, but what must I do when it *decreaseth?*

T. You must (for shortness sake) take only the Days since the Full Moon, which being Multiplied and Divided as before, will show you the Age of the Moon, and when she comes to the South: To which if you add for the Tide or Change, the Sum will be the Time of Full-sea required.

But take Notice, that in Rivers this Rule faileth of some Minutes, chiefly about the latter end of the first and third quarter, and the beginning of the second and fourth, about which time it is High-water sooner than the Rule showeth (because of the weakness of the Tide at that Time, and the length of the River) and therefore when you will find the Time of High-water at *London*, or the like place, a good way from the Sea. Do not fail to make use of this Table, which shows what you must Subtract from the Time found by the precedent Rule; and the Remainder will be the Time of High-water required.

Example.

The Moons Age.					H.	M.
1	15	16	29		00	00
2	13	17	28		00	05
3	12	18	27		00	10
4	11	19	26	Subtract.	00	20
5	10	20	25		00	30
6	9	21	24		00	45
7	8	22	23		01	00

The Moon being 6 Days Old, I find by the Rule that it is High-water at *London*, at 7 of the Clock, 48 Minutes past: But my Table shows that I must Subtract 45 Minutes from it; the Remainder 7 Hours 3 Minutes, is the true Time of High-water at *London*.

S. What is the meaning of the Moons making *Full-sea* in an Harbour?

T. The meaning is, that the Moon is then come to that Rumb (or Point of the Compass) which agrees with the hour of the Day, or true Time of Full-sea (in that place) when she is either in her Change or Full.

S. Do you mean the *Rumb* that the Horizontal or ordinary Sea Compass showeth?

T. No, I mean a Compass Equinoxial or Parallel to the Equator, and therefore when we say that a *North-east*, and *South-west* Moon, maketh a Full-sea at 3 of the Clock in an *Harbour*, we mean that when the Moon is come where the Sun is at 3 of the Clock, it will then be Full-sea there; for it is the hours Circle which determineth the Time of Full-sea, and not the Azimuth as a great many think.

S. What do you mean by an *Equinoxial* Compass or *Parallel* to the Equator?

T. I mean a Compass by which one may know exactly the Hours, whose Pin, and Point of the socket, points to the Pole, and moves as it were the Axletree of the World: Therefore the South part of it must be raised as high as the Equinoxial or Equator is (to us, or) above our Horizon, as you will better understand by this Figure.

THE EQUINOCTIAL COMPASS.

S. Why

S. Why do moſt Pilots then ſet the Sun and Moon by the *Horizontal* or *ordinary* Compaſs, ſince there is an errour?

T. 'Tis becauſe it is a Cuſtom amongſt them, not only to expreſs by it how things bear from them, but alſo the hours of the day, by allowing three quarters of an hour or 45 Minutes to every Rumb or Point of the Compaſs, (which are in all 32) and therefore ſince the Moon paſſeth through them all in 24 hours, you muſt alſo learn what every Rumb or Point yields, ſince it muſt be added to the Days of the Moons Age for to find the Time of High-water.

S. Is it true, that the North and South yield 12 hours?

T. Yes, and they are the *only Rumbs* that ſhow the *true hours* of the Day by the Horizontal Compaſs: However I deſire you to learn well by Art the following Table. Since you cannot well put a Ship into an Harbour without it, as the General Tables of Tides are made.

	H.	M.
North by Eaſt and South by Weſt, or	0	45
North North Eaſt and South South.Weſt., or	1	30
North Eaſt by North and South Weſt by South, or . . .	2	25
North Eaſt and South Weſt, or	3	00
North Eaſt by Eaſt and South Weſt by Weſt, or . .	3	45
Eaſt North Eaſt and Weſt South Weſt, or	4	30
Eaſt by North and Weſt by South, or	5	25
Eaſt and Weſt, or	6	00
Eaſt by South and Weſt by North	6	45
Eaſt South Eaſt and Weſt North Weſt	7	30
South Eaſt by Eaſt and North Weſt by Weſt	8	25
South Eaſt and North Weſt	9	00
South Eaſt by South and North Weſt by North	9	45
South South Eaſt and North North Weſt	10	30
South by Eaſt and North by Weſt	11	25
North and South	12	00

S. I underſtand now pretty well this Table, but how muſt I *make uſe* of it?

T. You muſt make uſe of it thus: Suppoſe you know by the Table of Tides or Experience that a *North North Eaſt* and *South South Weſt* Moon makes Full-ſea in the Downs, a place where you would know the true Time of Full-ſea when the Moon is 10 Days Old. You muſt firſt reduce the 10 Days of the Moon into Hours as you have been taught, and you will find it to be 8 Hours, to which adding 1 Hour 30 Minutes that the *North North Eaſt* and *South South Weſt* yields (as the Table ſhows you) there comes 9 Hours 30 Minutes for the Time of Full-ſea required.

Another

Another Example.

If a North Eaſt and South Weſt Moon makes a Full-ſea at London-Bridge*, What Time will it be Full-ſea there when the Moon is* 22 *Days Old?*

Becauſe the Moon decreaſes I count only the Days from her Full, which I find to be 7, for 15 and 7 makes 22; theſe 7 Days I reduce into Time, or Hours and Minutes, by Multiplying them by 48, and dividing by 60, or elſe Multiplying by 4 and dividing by 5, and the Quotient and reſt of the Diviſion will ſhow that theſe 7 Days (changed into Time) give 5 Hours 36 Minutes, to which I add 3 Hours for a *North Eaſt* and *South Weſt* Moon; (as the Table ſhoweth;) and the Time of Full-ſea at *London* will be by the Rule at 8 Hours 36 Minutes when the Moon is 22 Days Old: But becauſe it is a good way from the Sea, I muſt Subtract an Hour from it, as before taught; and the Remaining 7 Hours 36 Minutes will be the true Time of High-water.

```
          7 Days
      By  4              5 ) 28 ( 5 Hours 36 Minutes.
  Product  28                ——    3 Hours for North Eaſt, &c.
                             3
                                   8 Hours 36 Minutes.
                  Subtract   1
                            ——
                             7 Hours 36 Minutes.
```

Another Example.

If a South South Eaſt and North North Weſt Moon, makes Full-ſea in North Yarmouth Roa.*, What time will it be Full-ſea there when the Moon is in her Change or Full ?*

Since the *South South Eaſt* and *North North Weſt* Rumb yields 10 Hours 30 Minutes, I ſay that at 10 Hours 30 Minutes it will be Full-ſea in North *Yarmouth* Road as was required.

S. Doth not the Sea flow more, by one Point of the Compaſs, in the *Spring Tides* than in the *Neap Tides ?*

T. Yes, in Harbours in Rivers that have any indraught, and are of ſome diſtance from the Sea, as *Graveſend, London,* or the like.

PROB.

PROP. XI.

The Hour of Full-sea, and Age of the Moon being known, how to find out at any time what Moon makes a Full-sea in any Harbour.

S. HOW shall I find by the *Age* of the Moon and *Hour* of Full-sea, what *Moon* makes a *Full-sea* in an Harbour?

T. To find what Moon makes a Full-sea in an Harbour, you must first reduce the Days of the Moon into Hours (as you have been taught) which Hours must be Subtracted from the Hour of Full-sea, and the Remainder will show you what Moon makes a Full-sea there, (when she is in her Change or Full) as you will better understand by this Example.

Suppose I be in an Harbour where it is Full-sea at 10 of the Clock, and the Moon is 5 Days Old, How shall I know what Moon makes a Full-sea in that place?

To know it, I Multiply the 5 Days of the Moon by 4, and it makes 20, which divided by 5 (to reduce it into Time) gives 4 Hours, that Subtracted from 10 the Hours of Full-sea there remains 6 Hours, and therefore I know that an East and West Moon makes a Full-sea there, they being the Rhumbs that yield 6 Hours, as the Table showeth.

S. How must I do when the Hours of *Full-sea* are *less* than the Hours of the *Moon?*

T. When that happens you must add 12 (Hours) to the Hour of Full-sea, and Subtract from it the Days of the Moon converted into Hours, as before.

Example.

Suppose I am in a Road where it is Full-sea at half an Hour past Two of the Clock, and the Moon is 5 Days Old, and I would know what Moon makes a Full-sea in that place?

To do it, I must first convert the 5 Days of the Moon into Hours, therefore I Multiply 5 by 4 and it makes 20, which divided by 5 gives 4: This Subtracted from 14 Hours 30 Minutes (for 2 Hours 30 Minutes the Time of Full-sea being less than 4, I add 12 Hours to it) there remains 10 Hours 30 Minutes; then I call to mind what Rhumb yields 10 Hours 30 Minutes, and I find it to be a *South South East* and *North North West:* Therefore I conclude that a North North East and South South West Moon makes a Full-sea in that Road, as was required.

PROP.

PROP. XII.

Knowing what Moon makes a Full-sea in an Harbour, and the Hour of High-water, how to find out the Age of the Moon. (Provided you know whether it be before or after the Full.)

YOU muſt Subtract the Hours of Full-sea when the Moon is in her Change or Full, from the Hour of High-water, and the Remainder will be the Hours of the Moon, which being converted into Days will ſhow you the Moons Age in her increaſe, and you are to remember to add 12 to the Hour of High-water or Full-ſea, (as you did before) when you cannot Subtract without it.

Example.

Suppoſe I am in an Harbour South Eaſt and North Weſt, and that it is Full-ſea at 5 of the Clock, (the Moon in her increaſe) I would know the Moons Age?

First I call to mind that a South Eaſt and North Weſt Moon yields 9 Hours, which I muſt Subtract from 5 Hours the Time of Full-ſea, but becauſe I cannot, I add 12 (Hours) more to it which makes 17, from which I Subtract 9, and there remains 8 Hours, which I Multiply by 5, (to reduce it into fifth parts of an Hour) comes 40, which I divide by 4, (to reduce it into Days of the Moon) comes 10, which ſhoweth that the Moon is 10 Days Old, as was required.

S. What muſt I do when the Moon decreaſeth?

T. You muſt do as before, only you muſt add 15 Days more to the Days of the Moon, and therefore inſtead of 10 Days in the foregoing Example ſhe ſhould have been 25 Days Old, if ſhe had been decreaſing.

A

A TABLE *of Tides, showing what Moon makes a Full-sea on the Coasts and Harbours of* England, Scotland, Ireland, France, Spain, Portugal, Holland, Flanders, *and other Places.*

North and South, or 12 Hours.

Days.		H.	M.		Setting of the
0	15	12	0	At the North Foreland; on Beachy Shore; at	*Tides upon*
1	16	0	48	Orfordneſs; at Dover Peer; at the Shooe, Light,	*the ſame*
2	17	1	36	and Kentiſh Knock; Spits, and a long the Swin;	*Point.*
3	18	2	24	half Tide at Newport; half Tide at Portſmouth,	
4	19	3	12	and the Iſle of Wight; in the Sleeve between	From the
5	20	4	00	Uſhant and Silly; in the Road of Gibralter; on	Neſs to Bo-
6	21	4	48	the Coaſt of Flanders; at the Jutland Iſlands,	leign.
7	22	5	36	before the hever, Eider, and Elve on the Coaſt of	
8	23	6	24	Holland; in the Condado; before Enchuyſen,	
9	24	7	12	Horn, and Urck; Dunkirk; at Bolem and Grave-	
10	25	8	00	ling; before Gherbrough, and the Race of Blan-	
11	26	8	48	quet; at Bosford Laplandie; deſired Port in	
12	27	9	36	Amerique Auſtralis; and from Cape Quentin to	
13	28	10	24	Bojador in Barbary.	
14	29	11	12		
15	30	12	00		

North by Eaſt and South by Weſt, or 45 Minutes.

Days.		H.	M.		Setting of the
0	15	12	45	At Rocheſter and Maldon; at Garnſey, thwart	*Tides upon*
1	16	1	33	of Beachy in the Offing; at Winchelſey, within	*the ſame*
2	17	2	21	the Maes; within Terveer; in the Chamber of	*Point.*
3	18	3	09	Rie; Weſt-end of the Nower; at Fluſhing;	
4	19	3	57	North Caen.	From Bo-
5	20	4	45		leign to the
6	21	5	33		Some; from
7	22	6	21		Staples to
8	23	7	09		Boleign.
9	24	7	57		
10	25	8	45		
11	26	9	33		
12	27	10	21		
13	28	11	09		
14	29	11	57		
15	30	12	45		

s   E   *North*

North North East and South South West, or 1 Hour 30 Minutes.

Days.	H. M		Setting of the Tides upon the same Point.
0 15	1 30	Before the River of Thames; thwart of Don-	
1 16	2 18	ginefs; from the Weſt end of Wight; without	
2 17	3 06	Calice and Blackneſs; in Bluet; before South	
3 18	3 54	Yarmouth, in the Road of the Downs; before the	From Calice to Boleign.
4 19	4 42	Fen in the Channel; at Berwick; before the Maes	
5 20	5 30	and Goree; before Terveer; the Weilings, on the	
6 21	6 18	Coaſt of Zealand; at Horn, Edam, and before	
7 22	7 06	Camfer, at Army, Ramkins, and Camfere; at	
8 23	7 54	Bell Iſle, under holy Iſland, Tinmouth, Graveſ-	
9 24	8 42	end; and at Corpus Chriſti Point; on the Coaſt	
10 25	9 30	of Finmarchie; Motzoren and Iſland Cadenox;	
11 26	10 18	and from the Straits to Cape Quintin.	
12 27	11 06		
13 28	11 54		
14 29	12 42		
15 30	01 30		

North Eaſt by North and South Weſt by South, or 2 Hours 15 Min.

Days.	H. M		Setting of the Tides upon the same Point.
0 15	2 15	Before the Maes; before the Weilings;	
1 16	3 03	St. Andrews; Denby; without Funtnay; and	
2 17	3 51	without Bluet: Cape Caribbe in America Au-	
3 18	4 39	ſtralis. (Or South Amerique.)	
4 19	5 27		Between Ca-
5 20	6 15		lice and Do-
6 21	7 03		ver: From
7 22	7 51		Dunkirk to
8 23	8 39		Graveling:
9 24	9 27		From Sta-
10 25	10 15		ples to Fe-
11 26	11 03		ram; From
12 27	11 51		Dartmouth
13 28	12 39		to Exmouth
14 29	1 27		
15 30	2 15		

North

North East and South West, or 3 Hours 00 Minutes.

Days.	H.	M		Setting of the Tides upon the same Point.
0 15	3	00	At London, Amsterdam, Roterdam, Dort, Zerickzee; before New Castle; without the Banks of Flanders between Calice and Dover; in Robin Hoods Bay; before the Tees and Hartlepool; before Conquet, the Pens, Groy, Armentiers, Use, Killiars, Porthus; the River Bourdeaux; the South Coast of Britain, Gascoin, and Poitou; the Coast of Biscai, Galicia, Portugal, and Spain; before the River of Nants, and before the Bay of Tinmouth; North Cape, from the Race to the Pole head; Quarter Tide at Flambrough head; on the West Coast of Ireland, at Boekness and Orkness, in Shotland and Fair Isle; the Island Cogen and the Rivers Mouth of Pecora; Cape parter, Cape Cruel, Cape Cantin, and Cape Matas, and Roxo, Black Cape, and from the Equator to the Cape of good Hope in Afrique; at Tenerif in the Canary Islands; and the Coast of Chily in America Australis.	Setting of the Tides upon the same Point.
1 16	3	48		
2 17	4	36		From Cape de Hague to Alderney, through the Race of Alderney; from Garnsey to the Caskets; from Milford to Ramsey.
3 18	5	24		
4 19	6	12		
5 20	7	00		
6 21	7	48		
7 22	8	36		
8 23	9	24		
9 24	10	12		
10 25	11	00		
11 26	11	48		
12 27	12	36		
13 28	1	24		
14 29	2	12		
15 30	3	00		

North East by East and South West by West, or 3 Hours 45 Minutes.

Days.	H.	M		Setting of the Tides upon the same Point.
0 15	3	45	Between Dover and Calice; at the Maes; at Roven, Silly; before St. Matthews Point; at Brest; in the Sound; between Ushant and the Main; before the Bass, at St. Martin, before Rochell, before Brouage, the River of Bourdeaux within the Haven; on the Coast of Spain, Portugal, Galicia; the South side of Britain, Gascoin, and the West Coast of Ireland; at Huntcliffoot; half Tide at Flambrough head; quarter Tide between it and Bridlington Bay.	Setting of the Tides upon the same Point.
1 16	4	33		
2 17	5	21		
3 18	6	09		From Struysaert to Deep; from the Lizard to the Start; from Cape Cleer unto Londey.
4 19	6	57		
5 20	7	45		
6 21	8	33		
7 22	9	21		
8 23	10	09		
9 24	10	5		
10 25	11	45		
11 26	12	33		
12 27	1	21		
13 28	2	09		
14 29	2	5		
15 30	3	45		

East

Eaſt North Eaſt and Weſt South Weſt, or 4 Hours 30 Minutes.

Days.	H. M		Setting of the
0	15	4 30	*Tides upon*
1	16	5 18	*the ſame*
2	17	6 06	*Point.*
3	18	6 54	
4	19	7 42	From Oſt-
5	20	8 30	end to S. Ca-
6	21	9 18	teline; from
7	22	10 6	Berchſleur
8	23	10 54	to Struy-
9	24	11 42	ſart; the
10	25	12 30	Breſound,
11	26	1 18	
12	27	2 06	
13	28	2 54	
14	29	3 42	
15	30	4 30	

In all the South Coaſt of Ireland, as Cape Cleer, Baltimore, King Sale, Corke, Youghhall, Waterford, Dungarvan, within Mounts Bay; in the Sea of Wales and Severn; in Falmouth, in Mouſe hole, Sept Iſles; without the Haven in the Broad Sound; without the Fourn; at the Bay within Uſhant; in the Bree Sound and Vourd; the Gleſts of Texel, Bloy, and S. Matthews, and at Calice in the Creek; before Humber, Flambrough, Scarbrough, and Abberwark.

out and in; from Cape Cleer to the Iſle of Salteas; between Londy and the Holms unto Briſtol, from Silly to the Lands end of England; from the Start to Portland.

Eaſt by North and Weſt by South, or 5 Hours 15 Minutes.

Days.	H. M		Setting of the
0	15	5 15	*Tides upon*
1	16	6 03	*the ſame*
2	17	6 51	*Point.*
3	18	7 39	
4	19	8 27	From the
5	20	9 15	Iſle of Baſs
6	21	10 03	to the
7	22	10 51	Fourn; from
8	23	11 39	the Dorſes
9	24	12 27	to Cape Cleer;
10	25	1 15	
11	26	2 03	
12	27	2 51	
13	28	3 39	
14	29	4 27	
15	30	5 15	

In all the Havens on the South Coaſt of Ireland; and in the Bay of Carnarvan; in Milford, Ramſey, Falmouth, Foy and Torbay; Plymouth, Dartmouth; between Silly and the Lizard; in Wales; thwart of Londy, before Lin; at the Mouth of Severn; at the Spurn, New Caſtle, and Humber; at Moonles and Caldy; from Cape Roxo to the Equator 15 Minutes leſs.

to Cape Cleer; from Silly to the Lizard; from Portland to Wight; from Wight to Beachy.

Eaſt and Weſt, or 6 Hours 00 Minutes.

Days.	H. M		Setting of the
0 15	6 00	At Hull, Wells, Weymouth, Londy, and the	*Tides upon*
1 16	6 48	Holms; at Briſtol, Waterford, and Abermorick;	*the ſame*
2 17	7 36	before Bremen, Teſſel, and Hamburg; at Saint	*Point.*
3 18	8 24	Mallows, St. Powls in the Haven; at Blackney,	
4 19	9 12	and Concallo; before Bourdeaux; without Uſhant,	From the
5 20	10 00	and Silly in the Channel; at Lin half Tide; at	Caskets to
6 21	10 48	Archangel and Entry of Divina; at Kebeck in	Berchfleur;
7 22	11 36	Canada, and at the River-mouth of the Ama-	from the
8 23	12 24	zones in Amerique Auſtralis. (If you add 15	Landſend of
9 24	1 12	Minutes more.)	England to
10 25	2 00		the Lizard.
11 26	2 48		
12 27	3 36		
13 28	4 24		
14 29	5 12		
15 30	6 00		

Eaſt by South and Weſt by North, or 6 Hours 45 Minutes.

Days.	H. M		Setting of the
0 15	6 45	At Weymouth Key, and Briſtol Key; between	*Tides upon*
1 16	7 33	Foy and Falmouth in the Channel; before St. Ni-	*the ſame*
2 17	8 21	cholas and Podeſemske, in Ruſſia; Foulneſs; at	*Point.*
3 18	9 09	Garnſey half Tide.	
4 19	9 57		From the
5 20	10 45		Iſle de Baſs
6 21	11 33		to Marwa-
7 22	12 21		nen along
8 23	1 09		the ſhore.
9 24	1 57		
10 25	2 45		
11 26	3 33		
12 27	4 21		
13 28	5 09		
14 29	5 57		
15 30	6 45		

East South East and West North West, or 7 Hours 30 Minutes.

Days	H.M		Setting of the Tides upon the same Point.
0	15	7 30	
1	16	8 18	Thwart of Plymouth, and of the Start, in the Channel; at the Lizard by the Land; between
2	17	9 06	Mouſe hole and Falmouth; in the Offing; in the
3	18	9 54	midſt of the Channel; at the Neſs by Wierin
4	19	10 42	ghen; in the Road of the Texel; at the entrance
5	20	11 30	of the Emes, or the River of Emden; before
6	21	12 18	the Coaſt of Frizland and the Fly; Milford
7	22	1 06	Haven, at Cape Cleer; Florida, in Amerique;
8	23	1 54	and at the Iſle Kilden.
9	24	2 42	
10	25	3 30	
11	26	4 18	
12	27	5 06	
13	28	5 54	
14	29	6 42	
15	30	7 30	

Setting of the Tides upon the ſame Point. ———— From the Iſland Bryack to S. Maloes; from Berchfleur to Seynhead.

South East by East and North West by West, 8 Hours 15 Minutes.

Days	H.M		Setting of the Tides upon the ſame Point.
0	15	8 15	
1	16	9 03	Thwart of the Iſland Wight in the Channel,
2	17	9 51	without the Caskets in the Channel; between
3	18	10 39	the Wight and Beachy by the Shore; without the
4	19	11 27	Fly; to the Weſtward of the Foreland, in Saint
5	20	12 15	Magnes Sound; Yarmouth; at St. Hellens;
6	21	1 03	Machnels Caſtle; Dublin and Lambey; and Cape
7	22	1 51	St. Mary; Cape Sera Lione in Afrique 15
8	23	2 39	Minutes more.
9	24	3 27	
10	25	4 15	
11	26	5 03	
12	27	5 51	
13	28	6 39	
14	29	7 27	
15	30	8 15	

Setting of the Tides upon the ſame Point. ———— Behind Garnſey in the Fair way; without the Seven Iſlands.

South

South East and North West, or 6 Hours 00 Minutes.

Days.	H. M		Setting of the
0 15	9 00	At the Race of Portland; at the Eaſt end of	Tides upon
1 16	9 48	Wight; between Garnſey and the Caskets; with-	the ſame
2 17	10 36	in the Seyn; before Cromer, Winterton, and Yar-	Point.
3 18	11 24	mouth; Friez and Wieringer flat; on the Coaſt	
4 19	12 12	of Friezland; before the Eaſtern and Weſtern	Between
5 20	1 00	Emes; before the Fly and Scholhalgh; at Eg-	Morlaix
6 21	1 48	mont and Harlem of Baſs; before the Caskets and	and the
7 22	2 36	Garnſey; at Orkney, Dumbar, and Kildnie, at	Treacle
8 23	3 24	Seven Clifts and fair Iſles; at Home-head, and	Pots; in
9 24	4 12	thwart of Plymouth and Dartmouth; Iſle of	the Bay of
10 25	5 00	Man and Cateneſs; and three Rivers in Canada,	Bennyt.
11 26	5 48	(or North Amerique.)	
12 27	6 36		
13 28	7 24		
14 29	8 12		
15 30	9 00		

South Eaſt by South and North Weſt by North, or 9 Hours 45 Minutes.

Days.	H. M		Setting of the
0 15	9 45	Thwart of Leyſtaff without the Banks; the	Tides upon
1 16	10 33	Needles at the Iſle of Wight, in the Channel	the ſame
2 17	11 21	thwart of Wight, the Caskets, thwart of Garn-	Point.
3 18	12 09	ſey in the Channel; at Leyſtaff; Chamberneſs;	
4 19	12 57	Dunnoſe, Tergou, Orfordneſs, and Albrough;	
5 20	1 45	and at Cape Blanc in Afrique.	Before Con-
6 21	2 33		calo and Iſle
7 22	3 21		of St. Mi-
8 23	4 09		chael.
9 24	4 57		
10 25	5 45		
11 26	6 33		
12 27	7 21		
13 28	8 09		
14 29	8 57		
15 30	9 45		

South

South South East and North North West, or 10 *Hours* 30 *Minutes.*

Days.	H.	M		Setting of the Tides upon the same Point.
0 15	10	30	In North Yarmouth and Leyftaff Road; at St. Hellens and the Cows; at Orfordnefs, and Harwich without the Banks; before the River of Thames; between the Ifle of Wight and the Main; at Bulleyn, Deep, and Seynhead, in the Fofle of Caen; at Struyfart, and all the Coaft of Normandy and Picardy, Calice Road, in the Frith; at Leyftaff, quarter Tide; Harwich, Dover and the South Foreland; in the Downs, and Chambernefs Road, between Orford and Orwel Waves; at Senegal.	
1 16	11	18		From Berchfleur to Alhonga, from Cape Dorfey to the Ifland Dardan.
2 17	12	06		
3 18	12	54		
4 19	1	42		
5 20	2	30		
6 21	3	18		
7 22	4	06		
8 23	4	54		
9 24	5	42		
10 25	6	30		
11 26	7	18		
12 27	8	06		
13 28	8	54		
14 29	9	42		
15 30	10	30		

South by Eaft and North by Weft, or 11 *Hours* 15 *Minutes.*

Days.	H.	M		Setting of the Tides upon the same Point.
0 15	11	15	At Hampton, Portfmouth, and Dunnofe; before the Haven of Caen, at Cows and Orfordnefs within the Sands; Fair Ifle Roads, Harwich within; between Cripple-Sand and the Cryel; between the Naze and Warehead of Lower; in the Chamber and Gore end; before Margate, and in the Frith.	
1 16	12	03		From Saint Matthews's Point to the Bake Ovens; from Fontnay to Saint Matthews's Point.
2 17	12	51		
3 18	1	39		
4 19	2	27		
5 20	3	15		
6 21	4	03		
7 22		51		
8 23	5	39		
9 24	6	27		
10 25	7	15		
11 26	8	03		
12 27	8	51		
13 28	9	39		
14 29	10	27		
15 30	11	15		

The

The Description and Use of the Tide-Table.

The first and second Columns contain the Age of the Moon, and the third and fourth Columns show the Hours and Minutes of Full-sea; in the fifth and great Column, you have the Names of places, and the Title at the Top of it shows when the Moon comes to each Point of the Compass at Full and Change; the sixth and last, showeth the setting of the Tides upon the same Rhumbs or Points of the Compass, as you will better understand by this

Example.

I would know what Moon makes a Full-sea in the Downs, and at what Hour it cometh to such Point of the Compass ?

I look for the Downs, and find it under the Title of *North North East*, and *South South West*; by which I know that at the Full and Change the Moon cometh to that Point of the Compass at 1 Hour 30 Minutes, and that it is then Full-sea in that Road, as was required.

Another Example.

I would know what Time it will be Full-sea in North Yarmouth Road, when the Moon is 9 Days Old?

I look for North Yarmouth Road, and find it under the Title of *South South East*, and *North North West*, and because the Moon is 9 Days Old, I look in the first Column of that page for 9, and right against it in the third and fourth Columns, I find 5 Hours 42 Minutes, for the Time of Full-sea in North Yarmouth Road when the Moon is 9 Days Old; as was required.

Another Example.

I would know what Time or Hour it will be Full-sea in Plymouth, when the Moon is 20 Days Old?

I look for Plymouth, and find it under the Title of *East by North*, and *West by South*, and because the Moon is 20 Days Old; I look in the second Column of the same page for 20, and right against it in the third and fourth Columns I find 9 Hours 15 Minutes for the Time of Full-sea in Plymouth, when the Moon is 20 Days Old, as was required.

Note, *That the Names of Places in the* Tide-Table *are expressed as in the* Great Wagoner, *as being most familiar to Sea-men.*

F

Geometrical Problems:

SERVING

For the Construction of the Figures contain'd in the following Books.

PROBLEM I.

A Right-Line being given, how to draw another Right-Line Parallel to it from a Point given.

THE Line given is B C, unto which it is required to draw another Parallel to it at A; first open your Compass, and placing one Foot of it on the Point given at A, with the other describe an Arch just touching the given Line B C, then keeping your Compass so open, place one Foot of it upon some part of the given Line B C, for Example at D, and with the other Foot describe an Arch of a Circle on the same side the Point is of; then draw the Line A E; so that it takes in the Point and only touches the Arch, and it will be Parallel to the Line B C at the distance required.

This way is expedite, but not truly Geometrical, and therefore I have subjoyned also the following way prescribed in Geometry.

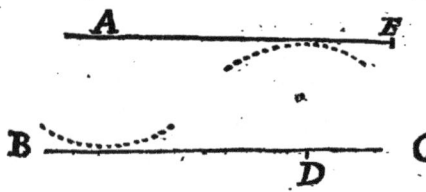

The Line given is A B, and it is required to draw another Parallel to it, that takes in the Point C, first take with your Compass the distance from A to C, and placing one Foot in B, with the other describe the Arch D E, then take with your Compass the given Line A B, and placing one Foot in the given Point C, with the other Foot describe the Arch F G, which

which cuts the firſt Arch in the Point H, then placing your Rule upon the Point C and H, draw a Line, and it ſhall be Parallel to the Line A B, as was required.

P R O B L. II.

How to Erect a Perpendicular on the end of a Right-Line given.

THE Line given is B C, and it is required to Erect the Perpen-dicular C F, firſt open your Compaſs to any diſtance, and placing one Foot in C, and the other in a Point above the given Line, as for Example, in D, with the other Foot (which was upon C) de-ſcribe a Circle, which cuts the given Line, as for Example, in E, and beſides touches the Extremity of it, as in C; place your Rule on D E, and obſerve the Point F, where it will alſo cut the Circle; for a Line drawn from C to F, will be the Perpendicular required: Which may alſo be Erected thus, firſt open your Com-paſs as before unto any diſtance, ſuppoſe unto the diſtance C E, then ſet one Foot of your Compaſs in the Point C, and with the other draw the Arch E D F, then ſet one Foot of your Compaſs in the Point E, and with the other draw the Arch D, then placing one Foot of your Compaſs in D, with the other draw the Arch G H, place your Rule on E D, and obſerve the Point I, where it will alſo cut the Arch G H, for a Line drawn from I to C will be a Perpendicular erected on the end of the Right-line given B C, as was required.

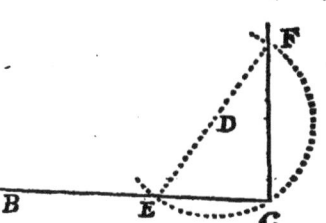

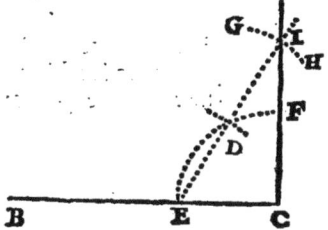

P R O B L.

PROBL. III.

How to raise a Perpendicular from the middle of a Line given.

THE Right-line given is **C D**, upon which from the Point **A**, you are to raise a Perpendicular **A B**; firſt from the Point given **A**, take with your Compaſs equal parts on the given Line, to wit, **A F, A G**; then opening your Compaſs wider from the Point F and G; draw two Arches, and from the Point B (where they cut one another) draw a Line to the Point **A**, the Line **B A**, which will be Perpendicular, as was required.

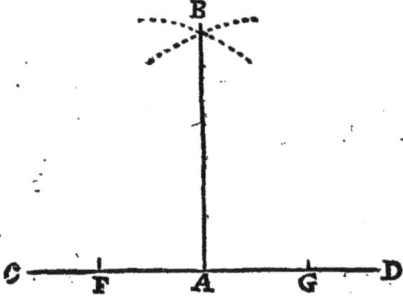

PROBL. IV.

How to let fall a Perpendicular from any Point aſſigned, upon a Right-line given.

THE Line given is **A B**, and the Point aſſigned is **C**, from whence you are to let fall a Perpendicular: Firſt, from the Point **C**, to the Line **A B**, draw any Line **C D**, and divide it into two equal parts in the Point **E**, then open your Compaſs to the Extent **E C**: Set one Foot of it at the Point **E**, and with the other Foot deſcribe the half Circle **C F D**, cutting the given Line in the Point **F**, then draw the Line **C F**, and it will be a Perpendicular to the (given) Line **A B**, as it was required.

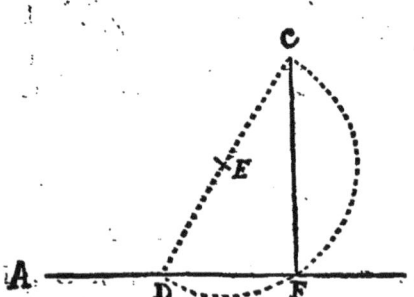

Another

Another way.

Set one Foot of your Compass in C, and with the other Foot defcribe an Arch, which cuts the Line given in two Points, as for Example, in D, E; then open your Compafs more then half the Extent DE, and fetting one Foot of your Compafs on the Point D, defcribe a little Arch under the firft, as directly under the Point given C as you can; do as much from the Point E; and thefe two Arches will cut or crofs one another in F, then placing your Rule on the Points C F, draw the Line C G, and it will be a Perpendicular, as was required.

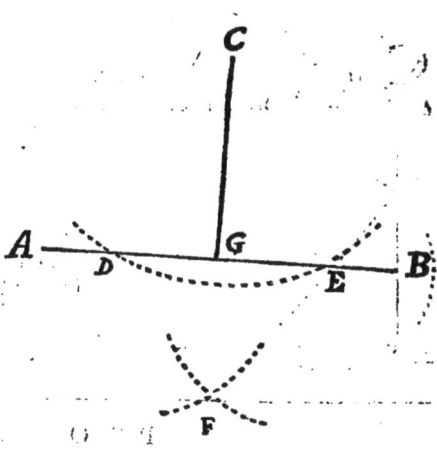

PROBL. V.

How to divide a Right-line given into two equal Parts, and a Circle into four.

THE Right-line given is F G, which is to be divided into two equal Parts: Open your Compafs wider than the half of the Line given, then fet one Foot of it on the Point G, and with the other Foot defcribe an Arch on both fides of the Line; do as much from the Point at F, and they will cut one another in H and L, from which Points draw a Line and it will cut the Line given exactly in the middle, and fo it will be divided into two equal Parts, as was required.

And in like manner you may divide a Circle into Four equal Parts, as the Figure fhoweth.

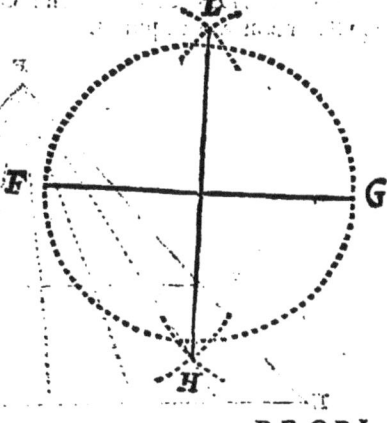

PROBL.

PROBL. VI.

How to find if a Line be Perpendicular upon another.

Uppose you would know if the Line AB is Perpendicular upon the Line AC, draw first any Line BC, then from the Point D, (in the middle of it) and at the distance or extent DC or DB, describe a Segment of a Circle BAC, and if that Circle pass through the Point A, the Line BA is Perpendicular upon the Line AC: For the Angle is a Right-angle; but if it pass beyond it, the Angle is Obtuse; and if it cuts the Lines AB, AC, short of the Angle A, the Angle is Acute, and so it cannot be a Perpendicular.

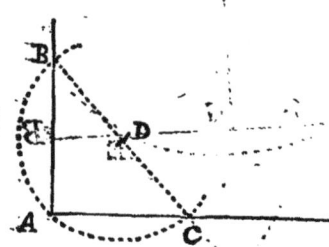

PROBL. VII.

How to divide a Line given into as many equal Parts as you will.

THE Line given is BC, which you are to divide into Seven equal Parts: First draw under BC another Line Parallel to it, as ED, and in this second Line which is to be longer than the first, take with your Compass as many equal Parts as you will divide the Line BC into, then from the first and last Points of those Divisions, to wit, from D and from E, draw Lines which just touch the Extremities or ends of the Line which is to be divided, and they will cut or cross one another in the Point F, to which if you draw Lines from all the Divisions of the Line ED the Line given BC, will be divided by them into as many equal Parts as was required.

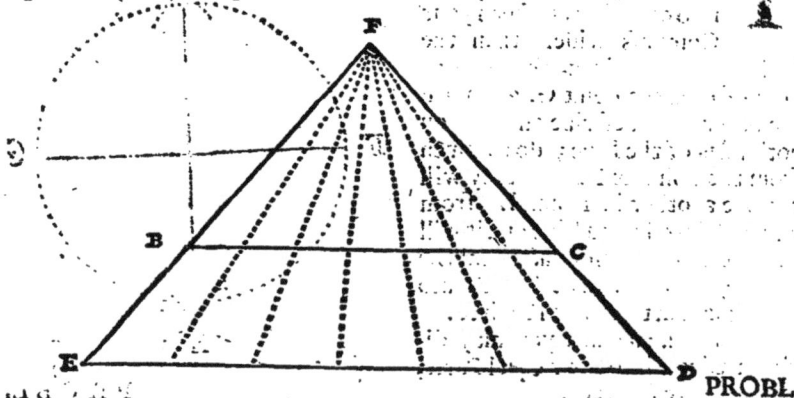

PROBL.

PROBL. VIII.

How to make an Angle (from a Point in a Line given) equal to an Angle given.

THE Line given is AB, and the Point is A, from which you are
desired to describe an Angle equal to the Angle given FDG.
(Take notice, that the middle Letter D, sheweth the Angle.)
First open your Compass at a
convenient distance, and setting one Foot
on the Point D, with the other Foot describe an Arch which cuts the two sides
of the Angle given in F and G, then
without altering your Compass (which
must still be open at the same distance)
remove the Foot from D, to the Point
given A, and with the other Foot describe the Arch BC, then take with
your Compass the distance FG, and set
the same from B to C, then draw the
Line AC, and the Angle CAB, will
be equal to the Angle FDG, as was
required.

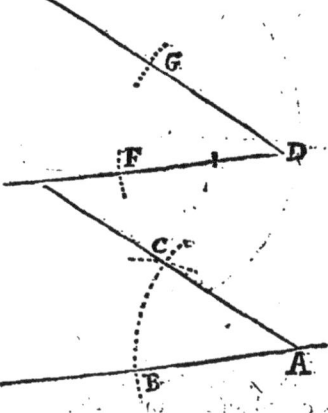

PROBL. IX.

How to divide an Angle given into two equal Parts.

THE Angle given is LHI, and it is required to divide it into
two equal Parts: First open your Compass at a convenient
distance, and from the Point H, describe the Arch IL, then
from the Point I
and L, describe the two
Arches which cross one another in M, draw the Line
HM, and the Angle given
will be divided as was required; for the Angle LHM
and MHI, will be equal.

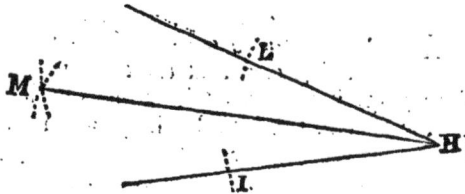

Of

Of a Circle.

THE *Center* of a Circle is the *Point* in the very midst of it, as for Example, the Point A: The *Diameter* is a Right-line drawn through the Center, which divides the Circle into halfs or two equal Parts, as B C:

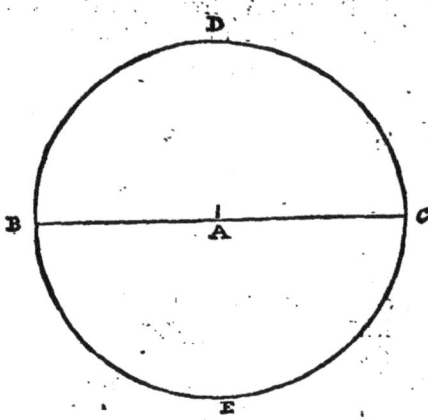

The *Semidiameter* is half of the Diameter, as A B: The *Circumference* is the Circular or round Line B D C E: Every *Circle* of a Sphere is divided into 360 equal Parts called Degrees, and the reason why Astronomers have chosen that Number, is because it can be Subdivided into most Parts without any Fraction or Remainder; as for Example, the half contains 180 Degrees, the Quarter 90 Degrees, the Third 120, the Fifth 72, the Sixth 60, the Eighth 45, the Ninth 40, and so forth.

Note, That what is neither Circle, half Circle, nor quarter Circle; is called an *Ark* or *Arch*, which is reckoned or estimated by Degrees and Minutes as well as the Circle, so that when we come to Treat of an Arch or Ark, you are to understand by it a certain Number of Degrees and Minutes of a Circle.

PROBL. X.

How to find the Center of a Circle.

THE Circle given is ABCD, whose Center is to be found out: First draw a Line (where you will) within the Circle as F G, and divide it into halfs or two equal Parts, as in or by Probl. VI. Taking care that the Perpendicular Line which divides it, be drawn until it touch the Circle on both sides of it, as in the Points I and K, then divide the Line I K into halfs in the Point L, and that Point will be the center required, for if from it you draw Lines to the Circumference, you will find them equal.

PROBL

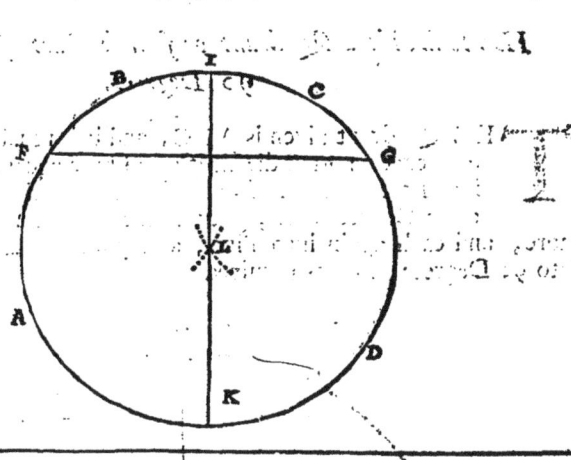

PROBL. XI.

How to finish a Circle begun, or else describe a Circle, which shall pass through Three Points given. (Which are not situated in a Right-line.)

THE three Points given are R, S, T, and it is required to describe a Circle, whose Circumference shall pass through those Three Points: Draw two Right-lines that conjoyn the Three Points given, as R S and S T; then divide each Line in halfs, or two equal Parts (as in Problem V.) and draw the Perpendiculars, (which divide the said Lines in halfs) so that they crofs or cut one another, as here in A, and that Point is the Center, upon which if you set one Foot of your Compafs, and extend the other to any of the given Points R, S or T, you may defcribe a Circle whofe Circumference fhall pafs through the Three Points given, as was required.

But if an Arch had been given, and it was required to finifh the Circle, all you have to do is only to mark Three Points in the faid Arch, and to work as before.

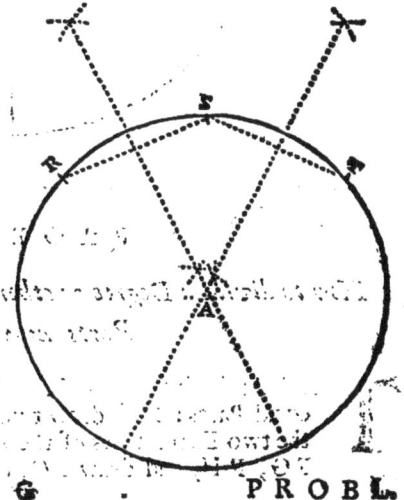

PROBL.

PROBL. XII.

How to divide a Quadrant or fourth Part of a Circle into
90 Degrees.

THE Quadrant given is A B C, and it is required to divide it into
90 Degrees: First divide it into Three equal Parts, and each Part
into Three more, and so it will be divided into Nine equal Parts
of 10 Degrees each, then divide each of those Parts into Two
more, and each again into Five, and your Quadrant will be divided
into 90 Degrees, as was required.

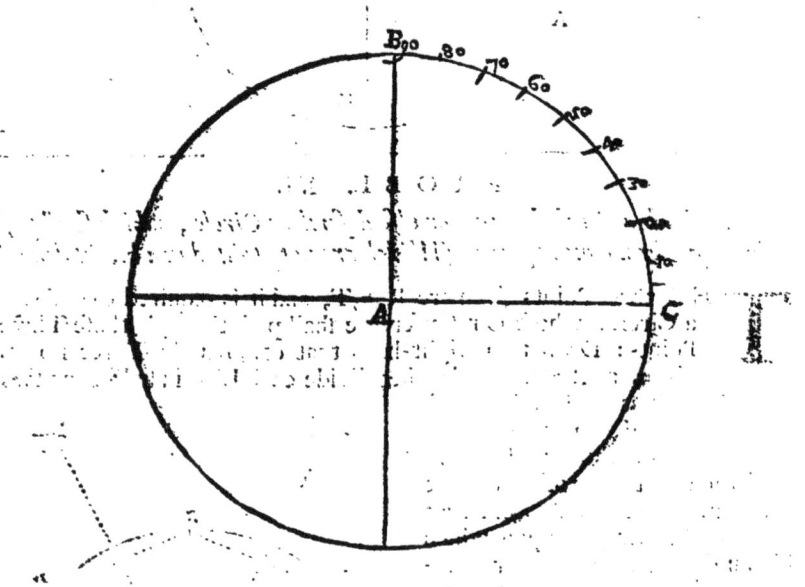

PROBL. XIII.

How to divide a Degree or other Part of a Circle into as many
Parts as you please.

THE Arch given is G H, and it is required to divide it into Six
equal Parts: First draw two Right-lines from the Center F, to
the two Extremities of the Arch which is to be divided, to wit,
F G, F H, and draw F G, to what length you think convenient,
making,

making or marking a Point in it at L; then by Problem XI. or rather with an Inftrument called a Bow, draw an Arch that takes in the Three Points F H L; and divide that Part of it which is between L H, into Six equal Parts, to wit, in the Points N, O, P, R, S; from which Points draw Lines to the Center F; and the Arch given G H, will be divided into Six equal Parts, as was required. *Note*, This is not Mathematical, but may ferve for Practice.

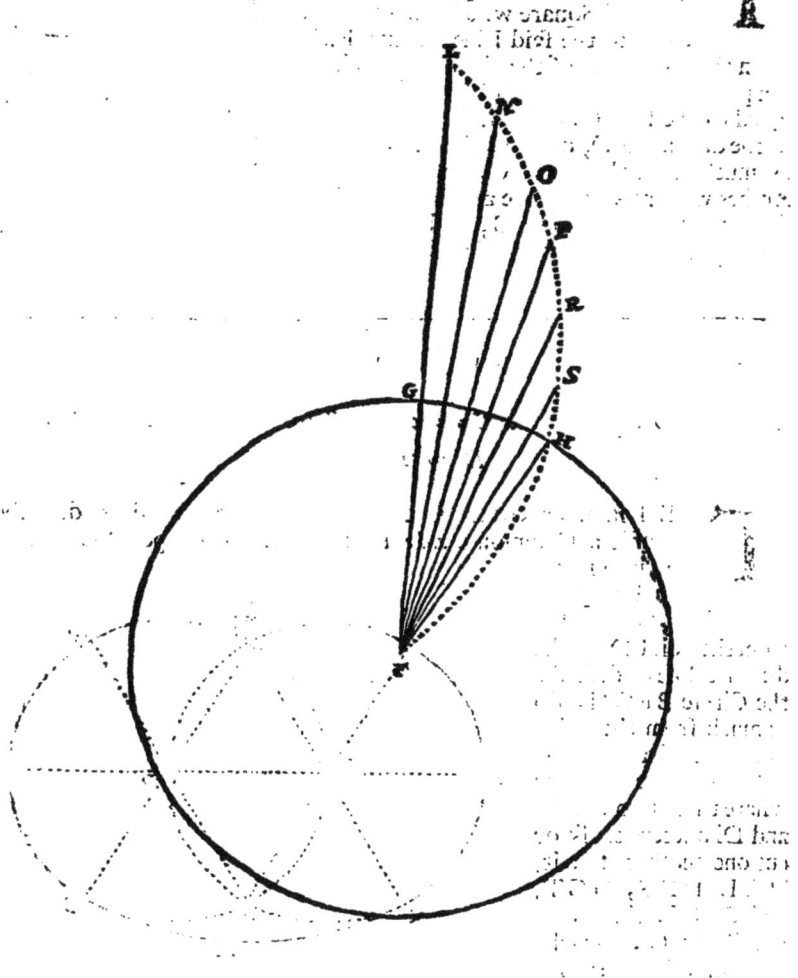

PROBL.

PROBL. XIV.

How to make a Geometrical Square upon a Right-line given.

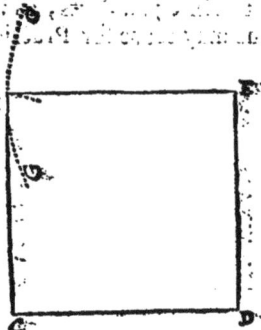

THE Line given is C D, and it is required to make upon it a Geometrical Square whose sides shall be equal to the said Line given: First upon the Extremity of the given Line, raise a Perpendicular (as by Prob. II.) that shall be equal to the Line C D, and from the Point F, at the distance F D, describe the Arch G; do as much from the Point C, and those two Arches will cross or cut one another in I, then draw the Lines I F and I C, and the Square will be C D F I, as was required.

PROBL. XV.

How to describe the Common or Cylindrical Oval about a Diameter given.

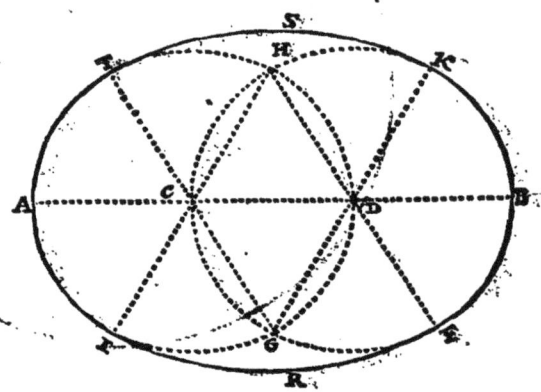

THE Diameter Given is A B, and it is required to describe about it a Common *Oval*: First divide the proposed Diameter A B, into Three equal Parts in the Points, C and D; from the Point D, at the distance D B, describe the Circle B F G H, do as much from the Point G, and draw Lines through the Points where those two Circles and Diameters cross or cut one another, to wit, H C I, H D F, G C T, and G D K, then from the Point G, at the distance G T, describe the Arch T S K, and from the Point H, the Arch I R F, and the Oval will be made, as was required.

PROBL. XVI.

How to defcribe a long Oval upon a Diameter given.

THE Diameter given is CD, and it is required to defcribe about it a long Oval: Firft, divide the propofed Diameter into Four equal Parts in the Points F G H, from F, at the diftance F D, defcribe the Circle D E G, do as much from G and H, then draw the Perpendicular L G M, and from the Extremities of it, to wit, from L and from M, draw (through the Points of Interfection) the Lines L F K, L H O, M F R, M H N, then from the Point M, at the diftance M N defcribe the Arch N S R, and from the Point L, at the diftance L F K, draw the Arch O S K, and the Oval will be made, as was required.

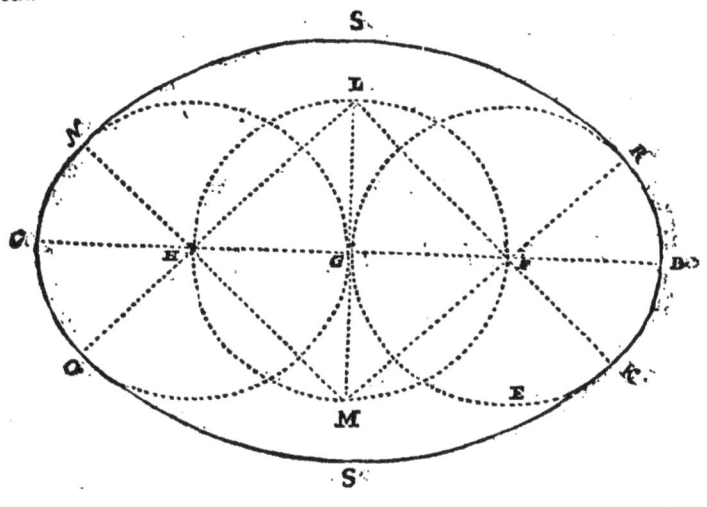

PROBL: XVII.

How to defcribe a true and perfect Oval.

THE Diameters given are A B and C D, and it is required to defcribe about them a perfect Oval: Firft crofs the two Diameters Perpendicularly in the very midft E; then take with your Compafs the half of the greater Diameter, to wit, E B, and placing

one

one Foot in D, defcribe the fecret Arch
G H L, fet upon G and L and D,
Three Pins, or Nails, and put a Thread
about the faid Pins, then pulling off
the Pin at D, put in the room of it a
Stile, and turning the Thread about as
the Figure fhows your Stile will defcribe
a true Oval.

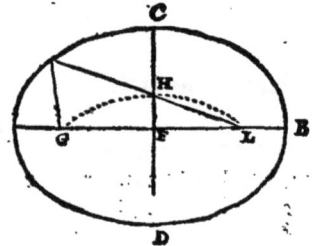

The End of the Firft Book.

THE

THE
Compleat ART
OF
NAVIGATION.

THE SECOND BOOK.

The Principles of the Sphere and of Astronomy,
necessary to Navigation.

Definition of the Sphere.

S. **I**S it necessary for a Sea-man to understand the *Sphere?*

T. Yes, if he designs to be an Artist in *Navigation*, for without some knowledge in *Astronomy*, he will never pass for an able Pilot; for if the *Sun* and *Stars* are to be his Guide, it is very necessary he should understand their *motions*; to prevent those *Errors* to which the *ignorant* are subject.

S. If it be so, I will follow your advice, and therefore pray tell me what a *Sphere is?*

T. A Sphere is a *Solid* and *Round* Body; covered with one only Superficies, (or upper Face) in the middle of which, is a Point called a *Center*, from which all Right-lines being drawn to any Part of the Superficies (or upper Part thereof) are of equal length.

S. What do you mean by a *Solid* Body?

T. I mean a Body, that hath *Length*, *Breadth*, and *Depth*; according to the signification of the Word.

S. Why,

S. Why do you say, *Covered with one only Superficies?*

T. It is becaufe it differs from all *Polygons*, as *Pyramids*, *Cubes*, and others, which have feveral Superficies or upper Faces; as for Example, a Cube or Dye.

S. Why do you say that *all Right-lines drawn from the Center to any Part of the Superficies, are equal?*

T. It is to fhew that it is a perfect Round Body; for if it was not exactly Round, the Lines drawn from the Center to any Part of the Superficies, would not be *equal*.

S. What is it, that this Sphere reprefents?

T. It reprefents the *frame* of the *whole World*.

S. Upon what is it turned about?

T. Upon the *two Extremities* of its *Axletree*, where we imagin to be two very firm, and immoveable Pins, otherwife called *Poles*, from the *Greek Word Poli*, and *Polo*, from which it is derived.

S. What is the *Axletree* of the World?

T. It is a *Diameter*, or Right-line, drawn (through the *Center* of the Sphere) from one Pole to the other, about which the World continually turneth.

PROPOSITION I.

Into how many Parts the World is divided, and what each Part containeth.

S. INTO *how many* Parts is the World divided?

     T. The World is divided into *Two Parts*, to wit, *Celeftial* and *Elemental*.

     S. What is it that the *Celeftial* Part containeth?

T. According to *Ptolomy*, the Celeftial Part containeth *Eleven* Heavens.

S. Pray *Name* them to me in Order, beginning with the *neareft* to us.

T. The loweft and neareft to us, is the Heaven of the *Moon*, whofe Character is . . . .   ☽

Next above it, is that of *Mercury* . . . .   ☿

Above *Mercury*, is that of *Venus* . . . .   ♀

Above and next to *Venus*, is that of the *Sun* . . . . .   ☉

Next above the *Sun*, is that of *Mars* . . . . . .   ♂

And above *Mars*, that of *Jupiter* . . . .   ♃

And above *Jupiter* that of *Saturn*, which is the laft and higheft Planet . . . .   ♄

Next

Next above *Saturn* is the *Firmament* or Eighth Heaven, which contains
the *fixed Stars.*   Above the *Firmament* is the *Cryftalline* Heaven, and above
the *Cryftalline* Heaven, is the firft mover (or *Primum Mobile*) which is the
Tenth Heaven.   The Eleventh and higheft, is the *Imperial Heaven*, where
it 'is thought that God refide.

S.   What doth the Elemental Part contain?

T. It containeth the Four Elements: The *Earth*, the *Water*, the *Air*,
and the *Fire*; which is the higheft Element and neareft to the Moon,
as this Figure fheweth.

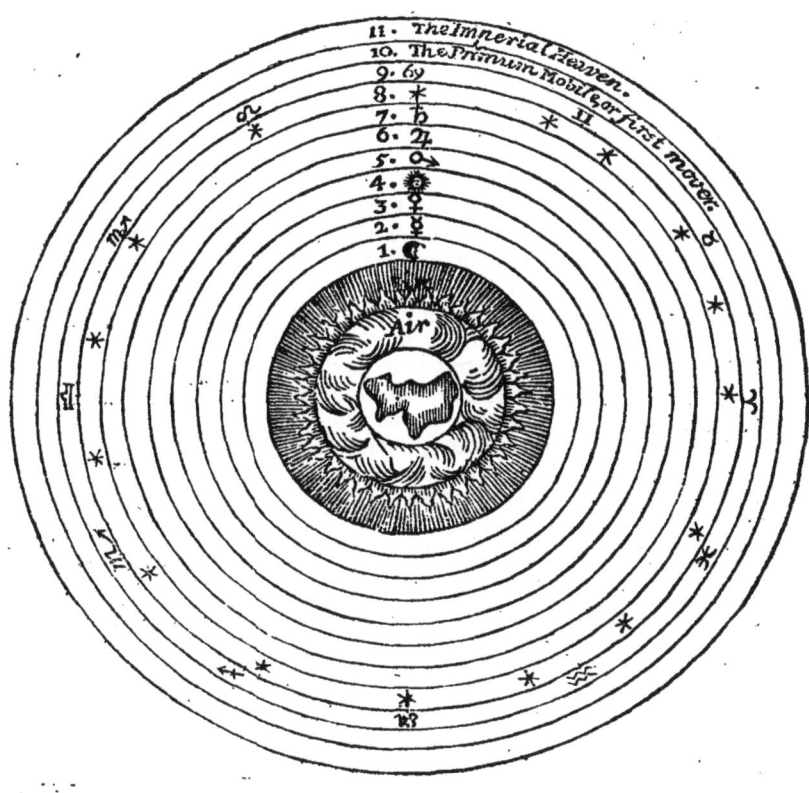

PROP. II.

The Motions of the Heavens; their Revolutions; the Cause of their Names; why the Zodiack is said to be in the Primum Mobile; and if the Sun and Stars are in any Sign or not.

S. **D**O all the Heavens move?

    T. No, for the General Opinion is, that the *Imperial Heaven* doth not move, but all the rest do.

    S. What Motion hath the *Primum Mobile?* (or first Mover?)

    T. It hath but one from East to West, but so swift and violent, that altho it be Prodigiously Great, it makes nevertheless its Revolution (or Turn) in 24 Hours: Turning also in the same Time, and in the same Manner, the Nine other Heavens that are under it; altho contrary to their own proper Motion, which is from West to East; and that is the Cause that they cannot make their own proper Revolution, but in a much longer Time; and so much the longer, the nearer they are to it, as it shall be made plain to you in its due place.

    S. How do you call that motion of the *Primum Mobile?*

    T. It is called *Diurnal*, or daily moving, because that Heaven finisheth its Revolution in 24 Hours, which is a (*natural*) Day.

    S. You will oblige me very much, to make me understand how the first mover (or *Primum Mobile*) can turn the other Heavens contrary to their own proper movings.

    T. It is very easie to make you understand it, if you consider that it is caused by the great swiftness and violence of the Diurnal moving, that forces them to turn, as if you took a small Insect called a *Beetle*, and placing it on a (turning) Wheel to move towards you from West to East, and the while you turn the Wheel swiftly from East to West: For by so doing you will turn the Beetle about many times contrary to its own course, before it can get about the said Wheel.

    S. Pray give me some other Example, to make me understand well these two movings.

    T. I think that this is intelligible enough, however I will satisfie you; admit then you are in a Ship, Sailing from East to West with a good Wind and fresh Gale; and that in the same time, you and the rest of the Company should walk or move from the Prow or Fore-part of the Ship, to the Stern, it is certain, that besides the common moving, which carries you altogether Westward, you have another particular to your selves and contrary to the first, since you move from West to East, and there-
fore,

fore it being the fame with the Stars, I hope that you now underftand well how the *Primum Mobile* may Turn the other Heavens contrary to their own proper movings.

S. Is there any Star in the *Primum Mobile* or firft mover?

T. No, there is none, this Heaven being of a Subftance moft Pure and Clear as well as the Imperial Heaven.

S. Is the Ninth Heaven alfo without any Stars, and of the fame Subftance that the *Primum Mobile* is of?

T. Yes, but it differs from the *Primum Mobile*, in that this hath *two* movings; (and the other hath but *one*,) one upon the Poles of the World according to the Diurnal or Daily moving, and the other from Weft to Eaft, upon its own Poles; turning fo flowly about, that many are of opinion that it makes but a Degree in a Hundred Years.

S. In what Time then do you think that it makes its Revolution?

T. The middle opinion is, that it is 36000 Years in finifhing its Revolution: Thus *Ptolomy.* But *Alphonfus* fays 49000 Years, and *Copernicus* only 25000.

S. Why will they have the *Zodiack* to be in the *Primum Mobile*, feeing there are no Stars in it?

T. It is to the end that it may be above all the Planets and fixed Stars, that by it one may the better know under what Sign they are all. (However there are fome that fancie it only on the Superficies of the Firmament.)

S. What! are they not in the *Signs* themfelves, and do not we ufe to fay, *that the Sun is in fuch a Sign?*

T. It is true, but for all that, it is not meant fo, by the Aftronomers; but only that the Sun is *under* fuch a Sign.

S. Do you know the Caufe why the Ninth Heaven is fo long a making its Revolution?

T. It is becaufe this Heaven is next to the *Primum Mobile*, which carrieth it about contrary to its own Courfe, with fo much fwiftnefs and violence that it cannot make its own proper Revolution fo foon as the Heavens that are further from it, and are lefs.

S. Why is it called the *Cryftalline* Heaven?

T. Becaufe of the clearnefs thereof.

S. How do you call the Eighth Heaven?

T. It is called the *Firmament*, and it contains all the fixed Stars.

S. Why is it called *Firmament?*

T. Either becaufe it is the Foundation or Ground of the fixed Stars: Or elfe becaufe the Stars which it contains, are fixed and firm in it.

S. Why do you call them *fixed Stars?*

T. Becaufe thofe Stars are firm and faftened in the Firmament, as a knot is in a Board or Plank.

S. How doth this Heaven move?

T. It

T. It moves Two ways, as I have said: For firft, it Turneth about every Day from Eaft to Weft, upon the Poles of the World, according to the moving of the *Primum Mobile:* (called Diurnal moving:) Secondly, it moves from Weft to Eaft, upon the Poles of the Ecliptick, and that, as flowly as the Ninth Heaven, which makes but a Degree in 100 Years; and its Revolution in 36000 Years.

S. How do you call the Seventh Heaven?

T. It is called *Saturn.*

S. Why *Saturn?*

T. Becaufe it contains the Planet *Saturn* of whom it takes its Name.

S. How long time is it a making its own proper Revolution (that is to fay from Weft to Eaft upon the Poles of the Ecliptick?)

T. The General opinion is, that it is near 30 Years, and becaufe there wants but fome few Days of it, moft Authors Write that *Saturn* is 30 Years making his Revolution.

S. How long is *Jupiter* making its own Revolution?

T. Twelve Years, tho to fpeak more exactly he wants fome Days of it.

S. And *Mars,* how long is he?

T. *Mars* is 2 Years: But the *Sun, Venus,* and *Mercury,* make all Three, their own proper Revolution, in 365 Days and almoft fix Hours. As for the *Moon* which is the loweft Planet, fhe is but 27 Days and 5 Hours, making her own Revolution.

PROP. III.

The thicknefs and Diftance of the Heavens; how the Stars appear through them, and the Magnitude of the Planets and fixed Stars.

S. OUT of Curiofity, I fhould be glad to know the opinion of Aftronomers, concerning the *Thicknefs* of the Heavens and of their *Diftance* from us; for it vexes me when I hear people talk of thofe things, and cannot Difcourfe (with them) about them.

T. You do well to fay out of Curiofity, for elfe I fhould be loth to lofe time about it, knowing how little ufeful it is to *Navigation:* But fince you are fo defirous of it; and that fome others may be of the fame mind: I will tell you the general opinion, which is that the Heaven of the Moon containeth.

In

In Thickneſs	105222 Miles.
The Heaven of *Mercury*	253372
The Heaven of *Venus*	3274494
The Heaven of the *Sun*	343996
The Heaven of *Mars*	2630800
The Heaven of *Jupiter*	1859654
The Heaven of *Saturn*	19604454 Miles.

S. If the Heavens are ſo Thick, how is it poſſible, that we can ſee the Stars as we do; Principally ſince they are yet higher than the *Seven* Heavens, and how comes it, that they appear to us as if there was no Heaven before them.

T. It is becauſe the Heavens are ſo Clear and Tranſparent, that the Stars appear with eaſe through them, becauſe of their great Light; which appears to us little only, becauſe of their great Diſtance from us: For the Heavens that are before them, are no more a hindrance to their Light; than a clear Cryſtal Glaſs would be to many great Flambeauxs (or Torches) lighted in a dark Night; if at a Diſtance Proportioned to their Light, the ſaid Cryſtals or Glaſſes, were placed before them; for I do not believe that it would hinder you from ſeeing their Light, or diſtinguiſhing them.

S. Was it of neceſſity, that the Heavens ſhould be ſo Thick?

T. Yes, for elſe they could not have contained each one his Planet or Star, for the opinion of the Philoſophers is, that there is no (fixed) Star, but which is far greater in Compaſs about than the Earth.

S. What do they think of the Planets?

T. It is thought that they are alſo greater about than the Earth (or both Sea and Land together) except *Venus*, *Mercury*, and the *Moon*.

S. That ſurpriſes me very much, and I ſhould be glad to know each one's Magnitude.

T. The uſual opinion has been that *Saturn* is 79 Times greater than the Earth: *Jupiter* 81 Times: *Mars* 2 Times: And the *Sun* 167 Times: But as for *Venus*, it is thought 28 Times leſs than the Earth; *Mercury* 3143 Times: And the *Moon* 39 Times.

S. Since you compare the Magnitude of the Planets, to that of the Earth; pray let me know its Magnitude.

T. It is very fit you ſhould know it; therefore Multiply 360 Degrees (her Compaſs about) by 60 Miles (which each Degree contains) and the Product will be 21600 Miles, for the Compaſs of the whole Earth in Miles.

S. Can you tell me the opinion of the moſt Learned, concerning the Diſtance between us, and the Firmament, and other Heavens; as thoſe of the Sun and Moon?

To It

T. It is not hard to Anfwer your requeft, for the opinion of the famous Aftronomer *Ticho-Brahe* (a *Dane*) is, that from the Center of the Earth to the Moon, there is 56 Semidiameters of the Earth. And from the fame Center to the Sun 1142 Semidiameters, or 3923912 Miles; and to the Firmament 14000 Semidiameters, or 48104000 Miles.

S. How many Miles doth the Semidiameter of the Earth contain?

T. It contains 3436 Miles, for the Diameter containeth 6872 (Miles.)

S. Do all Aftronomers agree that the Firmament is 48104000 Miles Diftant from the Center of the Earth?

T. No; for *Copernicus*, and others, are of opinion that the Firmament, is fo far from us, that no Meafure can reach to it; grounding their Reafons upon *Aftronomy*, which teacheth, that proportionably to the Diftance of the Heavens or Planets from us, they finifh or end their proper Revolutions (upon the Poles of the Zodiack) in lefs or longer Time: As for Example, the *Moon* which is neareft to us, ends her Revolution in 27 Days and almoft 8 Hours; which the Sun cannot do but in 365 Days and almoft 6 Hours, becaufe he is further from us; the fame is to be underftood of any other Planet proportionably to its Diftance from the Earth; and therefore if *Saturn* (which of all the Planets is the furtheft from it,) cannot end his proper Revolution but in 30 Years; and the fixed Stars (according to the moderate opinion) in lefs than 36000 Years, we may very well believe, that they are in a Heaven at an exceeding great Diftance from us. For, if (according to the opinion of the famous *Ticho-Brahe*) Saturns greateft Diftance from the Earth be 14000 Semidiameters of it, or 48104000 Miles; which *Maurolyc* (in his *Appendix of Cofmography*) and feveral Aftronomers have augmented Four Times more: I fay to keep the fame Proportion with the Diameters, (or Semidiameters) which are in the Periods or Circumferences of their Revolutions, (according to what I have faid) the Stars fhould be Diftant from the Earth 57724800000 Miles; that is to fay, 1200 times further from the Earth than *Saturn*, according to the Reafon or Proportion of the 30 Years of *Saturn*, to the 36000 Years of the Stars; which yet according to *Maurolyc* would amount to more: And therefore I leave it to you to judge, if in fo prodigious a fpace one can be affured of any thing, either concerning the Diftance or Magnitude of the Stars, when *Ticho-Brahe* himfelf (who in other Things pretends to be exact) durft not determine any thing but by Conjecture.

S. What is a *Planet*?

T. A Planet is nothing elfe but a *wandring* Star which moves in a Heaven by it felf, for it is derived of a *Greek* Word, which fignifies to *wander* or *go a ftray*, becaufe they are fometimes at a wide Diftance, and at other Times near to each other.

P R O P.

P R O P. IV.

Of the Circle, into how many Parts it is divided; what those Parts are called; how many sorts of Circles there are in a Sphere; and the Reason of it?

S. WHAT is a Circle?

T. A Circle is a flat Figure, Termined by one only Line called the Circumference, (or Periphery,) in the midst of which, there is a Point called a Center; from which all Right-lines drawn to the Circumference of it are equal.

S. Into how many Parts is a Circle divided?

T. It is divided into 360 Parts, called *Degrees*, and each Degree is divided into 60 Parts more called *Minutes*, and some times those Minutes are divided into 60 Parts more called *Seconds*; those Seconds into 60 other Parts called *Thirds*; and so on until *Tenths*; and to distinguish them one from another, we use to place a small Cypher (°) upon or after the Degrees, and upon or after the Minutes an Accent (') and so on until Tenths, as in this Example: 16°, 40', 2", 20''', 15'''', 28''''', &c. Which you must Read thus, 16 Degrees, 40 Minutes, 2 Seconds, 20 Thirds, 15 Fourths, and 28 Fifths.

S. How many *sorts* of Circles are there in the Sphere?

T. There are *two sorts*, greater and lesser.

S. What is the *greater* Circle?

T. It is a Circle which divides the Sphere into Two equal Parts, and hath no other Center than that of the Sphere.

S. How many Miles doth a Degree of those greatest Circles on the Earth contain?

T. It contains 60 Miles, (*Viz.* as many Miles as Minutes.)

S. Why do you call them Miles?

T. Because each Part containeth a Thousand Geometrick Paces, for a Mile signifies only a Thousand such paces.

S. What is a Geometrick pace?

T. A Geometrick pace is 5 Feet, and therefore a Mile contains 5000 Feet.

S. What is the *lesser* Circle?

T. It is a Circle which divides the Sphere into Two unequal Parts; and hath another Center than that of the Sphere: All the great Circles are equal, and cut one another into two equal Parts; but the lesser Circles are unequal, except those which are equally Distant from the Center of the Sphere.

S. How many great Circles are there in the Sphere:

T. According to some Authors there are Six; but others take particular notice but of four. Those that count Six, reckon thus, the *Zodiack*, the

the *Equator*, (or Equinoxial,) the *Meridian*, the *Horizon*, and the *two Colures*; but those that allow but four do not reckon the *two Colures*; because they are not necessary, but count them as *Meridians*; however they shall be defined in their due places.

S. Why do they imagine *so many* Circles on the Sphere?

T. It is to come to the most perfect and exact knowledge of it; for when a thing is too great, we use to divide it into several Parts, to the end that knowing *every Part* by it self, we may with more ease come to the knowledge of the *Whole*.

PROP. V.

Of the Equator. (*Or Equinoxial.*)

S. **W**HAT is the *Equator*?

T. The *Equator* is a great Circle (commonly called by Pilots, the *Equinoxial Line*,) which is in every Part, equally distant from the two Poles of the World; and therefore it divideth the Sphere into two equal Parts, as this Figure shows.

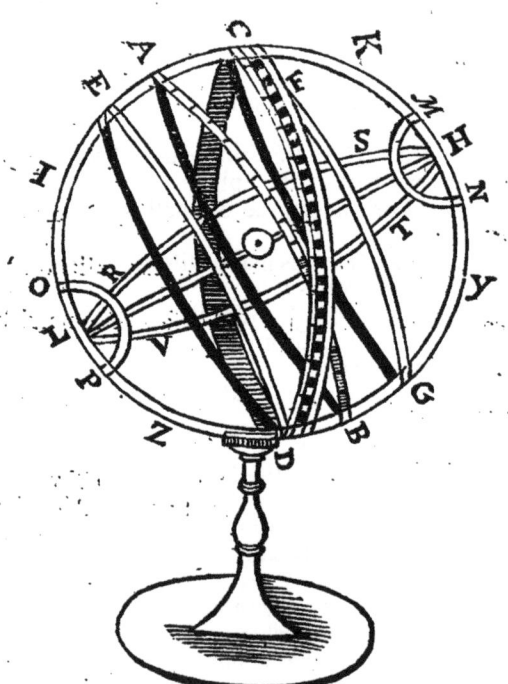

The

The Circle A B is the Equator.
H the Pole Arctick. (or North Pole.)
L the Pole Antarctick. (or South Pole.)
C D the Zodiack.
F G the Tropick of *Cancer.*
E D the Tropick of *Capricorn.*
M N the Circle Arctick.
O P the Circle Antarctick.
R S T V the Colure of the Equinox.
I K Y Z the Meridian and Colure of the Solstice.

S. Why is it called *Equator?*

T. Because the Sun being in that Circle (which happens twice every Year) about the 10th. of *March,* and 12th. of *September,* the Days and Nights are of an equal length through all the World; for the Horizon and Equator are two great Circles which cut one another in the very midst, from whence it followeth, that the Sun (which that Day moves round that Circle) takes as much Time to run through that part which is under the Horizon, as it doth the part above it; and that is the cause that it is called *Equator,* from the Latin Words, *Equator dici & noctis.* Which signifies the *equal Proportioner* of the Day and Night: For the Day is nothing else but the *Time* that the Sun is *above* the Horizon, and the Night the *Time* that he is *under* it.

S. To what use serveth this Circle?

T. It hath many necessary uses; for it sheweth the *Declination* and *Right-Ascention* of the Sun, and fixed Stars; and the Diurnal (or Daily) motion of the *Primum Mobile,* which finisheth its Revolution in 24 hours, which hours are to be Measured by the Degrees of the Equator; allowing 15 Degrees thereof to an Hour: It is also of Excellent use in Geography, for it marks the Distance of Places both in Latitude and Longitude, since it is from it, that we begin to reckon the Degrees of Latitude going towards either of the Poles; and the Degrees of Longitude are counted upon that Circle.

S. How do you call the Poles of the Equator?

T. One is called *Pole Arctick,* (or North Pole,) and the other *Pole Antarctick,* (or South Pole.)

S. Why do you call the North Pole *Arctick?*

T. It is called *Arctick* from the nearest Constellation to it, which in *Greek* is called Ἄρκτος, and in *Latin Artos,* which signifies the *Bear;* and therefore when we call it *Pole Arctick,* it is the same as if we should call it the *Pole of the Bear.*

S. What signifies the Word *Antarctick?* (as you call the South Pole?)

I   *T.* It

T. It fignifies the contrary, or oppofed to the Arctick; and there-fore when we call it *Pole Antarctick*, it is the fame as if we fhould call it the *Pole oppofite to the Arctick.*

PROP. VI.

Of the Zodiack; Ecliptick Line, Signs, and other things that depend thereon.

S. WHAT is the *Zodiack?*

T. The *Zodiack* is a great and broad Circle, which di-videth the Sphere into two equal Parts, by cutting the E-quator at Oblique-angles (or flopingly) in two Points; one in the beginning of *Aries*, and the other in the beginning of *Libra*; fo that one half of this Circle declineth towards the North, as far as the Tropick of *Cancer*; and the other half towards the South as far as the Tropick of *Capricorn*, as in the precedent Sphere or Figure. *Prop.* V.

S. Why do you call it the *Zodiack?*

T. Becaufe that Circle is divided into 12 Signs, which have every one the Name of fome Animal, (except *Libra*;) or elfe becaufe the Sun (which moveth always in the midft of this Circle) gives Life to the World; for it is derived either of the *Greek* Word *Zodion*, which fignifies a little Beaft, or elfe of *Zoe*, which fignifies Life.

S. Who firft divided the *Zodiack* into 12 Parts?

T. The *Egyptians.*

S. How did they go to work to divide it?

T. They made ufe of a Water Hour-glafs, like thofe made now with Sand: At firft they let the Water run (without hindrance) 'till the Heavens had gone once about; then they divided it equally into 12 Parts, and then by running each Part fucceffively, they did alfo divide the Fir-mament into 12 equal Parts.

S. How did they do, to know under what Sign each Star is?

T. They did obferve the time that a fixed Star did appear at the Horizon, letting one of thofe Parts of Water run in the fame Time, untill the laft drop; then they took notice of fome other Star, that begun to appear, letting run for the fecond Time the like quantity of Water; and then feeing fome other Star arifing, they let the Third part of Water run as before; and fo on, untill they had divided the Fir-mament into 12 equal Parts; and for fear of difordering thofe Parts, and to know them another Time with more eafe; they did diftinguifh them by fome Marks, and gave them different Names; and that is the reafon that thofe 12 Parts of the Zodiack, are called the 12 *Signs* of the Zodiack.

S. How

S. How are thefe Signs feverally named, and with what Characters are they Marked?

T. The firft is called, *Aries* (or the Ram) and is thus marked   ♈

The fecond, *Taurus* (or the Bull)   ♉

The third, *Gemini* (or the Twins)   ♊

The fourth, *Cancer* (or the Crab)   ♋

The fifth, *Leo* (or the Lion)   ♌

The fixth, *Virgo* (or a Virgin)   ♍

The feventh, *Libra* (or the Ballance)   ♎

The eighth, *Scorpio* (or the Scorpion)   ♏

The ninth, *Sagittarius* (or the Archer)   ♐

The tenth, *Capricornus* (or the He Goat) ♑

The eleventh, *Aquarius* (or the Pourer of Water) ♒

The twelfth, *Pisces* (or the Fishes)   ♓

S. Which do you call the Northern Signs?

T. The Six firft that I have named, to wit, *Aries*, *Taurus*, *Gemini*, *Cancer*, *Leo*, and *Virgo*.

S. Why are they called Northern Signs?

T. Becaufe they are in that part of the Zodiack which declines Northward, or is on the North fide of the Equator, the other Six are alfo called Southern Signs, becaufe they are in the other half of the Zodiack which declineth towards the South.

S. How many Degrees doth each Sign contain?

T. Every Sign contains 30 Degrees; fo that the Sun is 30 Days (or very near it) in each Sign, moving or getting only one Degree in a Day.

S. What Time doth the Sun enter into thefe Signs?

T. The Sun enters into *Aries*, the 10th. of *March*.
Into *Taurus*, the 10th, of *April*.
Into *Gemini*, the 11th. of *May*.
Into *Cancer*, the 11th. of *June*.
Into *Leo*, the 13th. of *July*.
Into *Virgo*, the 13th. of *Auguft*.
Into *Libra*, the 13th. of *September*.
Into *Scorpio*, the 13th. of *October*.
Into *Sagittarius*, the 12th. of *November*.
Into *Capricornus*, the 11th. of *December*.
Into *Aquarius*, the 10th. of *January*.
Into *Pisces*, the 8th. of *February*.

I 2   *S.* How

S. How broad is the Zodiack?

T. The breadth of the Zodiack is of 16 Degrees, that is to say, 8 Degrees on each side of the Ecliptick.

S. Why do they imagine (and represent) the Zodiack so broad?

T. To show us that part of the Firmament, under which the Planets move; and to the end, that it may contain the 12 Signs before named.

S. What is the *Ecliptick*?

T. It is a Circular Line in the very midst of the Zodiack; which showeth the Sun's course; being the only place, in which he moveth, and keepeth constantly.

S. Why is it called *Ecliptick*?

T. Because that all the Eclipses of the Sun and Moon happen under this Line; for as the Sun never goeth out of it, it is impossible there should be any Eclipse, but when the Moon is under the same.

S. Is not the Ecliptick Line divided into several Parts?

T. Yes, it is divided into 360 Degrees, (or 12 times 30, which is the same.)

S. Where doth this Division begin?

T. It begins at the Equinox of *March*, or first Point of *Aries*; from which Point the Longitude of the Stars are also counted, according to the Succession of the Signs; that is to say, from *Aries* to *Taurus*, and so on.

S. How much doth the Ecliptick decline from the Equator?

T. Its Declination or distance from the Equator is at present 23 Degrees and near 31 Minutes; (for it wants but 10 Seconds;) so that the Equator and Ecliptick (by cutting one another obliquely) make oblique Angles on both sides the Equator of 23 Degrees 31 Minutes.

S. To what use serveth the Ecliptick Line?

T. It serveth, to distinguish and to mark the time of Years, Months, Seasons; and the Degrees at which the Stars rise and set: It showeth also the time of the Eclipses; and by its obliquity (with the Horizon) we know the cause of the inequality of Days and Nights.

PROP. VII.

Of the Horizon.

S. **W**HAT is the *Horizon*?

T. The Horizon is a Circle which determineth, and boundeth our sight (and divideth that Part of the Heaven which we see, from that which we cannot see) according to the signification of the *Greek* Word, *Horizon*.

S. How many kinds of Horizons are there?

T. Two kinds, one called *Rational* and the other *Sensible*.

H 2

S. What

S. What is the *Rational* Horizon?

T. It is a great Circle, every part of which, is diftant from our *Zenith*, and *Nadir*, 90 Degrees; and therefore we may fay, that it divideth the Heavens and Earth into Two equal parts; fo that one half of the Heaven is always above this Circle, and the other half under it, fo that it cannot be feen; from whence we may conclude, that the Zenith, and Nadir, are the true Poles of the Horizon.

S. Why is it called *Rational?*

T. Becaufe we do not fee it, as we do the Senfible Horizon, but only conceive it by Reafon, it being impoffible that our Eyes fhould difcover all that extent, or diftance, that is between it and us, becaufe of the roundnefs of the Sea, and Land.

S. To what ufe ferveth this Horizon?

T. It hath feveral ufes; for, it fhews us, the Stars that rife and fet, and informs us what Stars appear to us, and what not; it is alfo by this Circle, that we know the Artificial Days, from the Artificial Nights; being the meafure thereof. It is likewife of excellent ufe for finding the the height of the Pole, Sun and Stars; by which we come to the knowledge of the Latitudes. (That moft effential Part of Navigation.)

S. What is the *fenfible* Horizon?

T. It is a Circle which Limits, or bounds our fight, when being at Sea (or in a Plain) we look round about us as far as we can; for then it forms it felf, or appears, where it feems to us that the Sea (or Land) touches the Heavens; fo that we fee our felves environned or inclofed with it, and in the very midft or Center thereof.

S. Is there any *Difference* between thefe two Horizons as to the Sun and Stars? And is it not hard to know (when they rife) on which of the Horizons they appear to us?

T. No, as to that, there is no *fenfible* Difference, for they are both taken for the fame, becaufe the Semidiameter of the Earth is not at all confiderable, being compared to that of the Heaven.

S. When we go any Voyage, (or Travel) do we not change the Horizon?

T. Yes, from Time to Time you change the Horizon, for you difcover fome part of the Heavens which before you could not fee, becaufe of the Roundnefs of the Sea and Land.

S. Are there then more Horizons?

T. No, there are no other kinds, tho many Horizons of each kind; for altho one change his Horizon, he doth not change it in its kind; for in what Country or Place, foever we be, we can fee no other Horizon than the

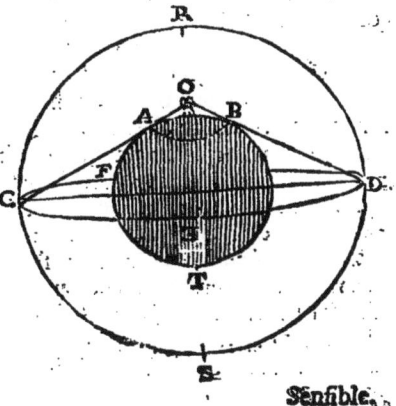

Senfible,

Senfible, (of which you read laft) becaufe we fee always half of the Heavens: The foregoing Figure will make you underftand it better, where the Globe A B T reprefents the Earth, the Circuit C R D S the Heavens, R the Zenith, S the Nadir, C E D F the Rational Horizon, O the place of the Spectators Eye, T the Antipodes, and O A B the fenfible Horizon, whofe Semidiameter Q A, or Q B, is not above Three Miles when his Eye is but fix Foot above the Surface of the Earth, but increafeth proportionally to the height of the Eye, or thing feen, for if his Eye was 10 Foot high, he might difcover 4 Miles, (fuppofing that the thing feen lies on the ground, as for Example, a Coal of Fire, or Lanthorn) if 20 Foot 5 Miles; but if the thing that you difcover is high above the Superficies of the Earth or Sea, you may fee it a great deal further, proportionably to its height.

PROP. VIII.

Of the Meridian.

S. **W**HAT is the *Meridian?*

 T. The Meridian is a great Circle paffing through the Poles of the World, and the Zenith, or Point Vertical to any body, or place.

 S. Why is it called *Meridian?*

 T. Becaufe it divideth the Day into two equal Parts, being as far from the Horizon or Sun Rifing to Noon or Mid-Day, as it is from Mid-Day to Sun Setting; for it is derived from the *Latin* Word *Meridies*, which fignifies Noon-Day. (Or Twelve of the Clock.)

 S. How many Meridians are there?

 T. There are as many as you can conceive or imagin Points in the Equator, (or Equinoxial,) neverthelefs the Geographers and Hydrographers (which are thofe that make Cards for Sea and Land) draw them but at 10 Degrees diftance one from another; and fome Time at 15 Degrees, to avoid the great number of Lines, which otherwife would fill their Cards, and hinder the chief ufe of them.

 S. If the Meridians pafs through the Zenith of ever Body, are there not as many Meridians as Zeniths?

 T. No, becaufe the Circles which pafs through the Poles of the World, and your Zenith (or place) pafs alfo through the Zeniths (or places) of many more befides your felf; for all places fituated North and South one from another, have the fame Meridian; and therefore if you fhould Travel never fo far (exactly) from North to South, or from South to North; you would not change your Meridian, for you would always be under the fame Circle: But it is not fo going from Eaft to Weft, or from

Weft

Weſt to Eaſt ; for then you would change your Meridian: for all places ſituated Eaſt and Weſt, one from another, differ in Meridians.

S. Doth not the Meridian that paſſes through the *Zenith*, paſs alſo through the *Nadir* ?

T. Yes, for if it paſs through the Poles and *Zenith*, it muſt needs paſs alſo through the *Nadir*.

S. Is it not Twelve of the Clock, as well when the Sun toucheth that Circle under the Horizon, as when it touches it above ?

T. Yes, every time that the Sun toucheth that Circle (which happens twice in 24 Hours,) it is Twelve of the Clock, but with this Difference, that when it touches it in our Hemiſphere, it is Noon-day, but when it touches it in the Hemiſphere of our Antipodes it is Mid-night, ſo that our Mid-day is their Mid-night, and our Mid-night is their Mid-day.

S. Is it Noon-day at the ſame Time, to all thoſe that are under the ſame Meridian?

T. Yes, and Mid-night too, and all other Aſtronomical Hours are the ſame to them all, but differ to thoſe who dwell Eaſt and Weſt one from another, becauſe they are under different Meridians; from whence it follows, that it will be ſooner Noon-day to thoſe that are moſt Eaſt ; for as the Sun Riſes Eaſterly, it is certain that he will come ſooner to their Meridian, than to thoſe that are moſt Weſt; which Difference ſhall be ſo much the greater, the more they are diſtant one from another.

S. If it be ſo, the Sun then always makes Noon-day to ſome people or other, ſince every Minute of the Day, it is on ſome Meridian.

T. It is true, and we may add that it is alſo always Riſing, always Setting, and always at the Hour of Mid-night, becauſe he is always over the Antipodes of ſome people or other, as well as in the Meridian, and Horizon of others.

S. How many Degrees muſt the Meridians be diſtant (Eaſt and Weſt) one from another, to differ an Hour by the Sun-Dial? (I mean to be Noon-day (or Twelve of the Clock) in one place, when it is but Eleven in the other place, ſituated more Weſterly.)

T. Their difference in Longitude muſt be 15 Degrees, or the two places muſt be 900 Miles Diſtant one from another ; for every Degree (or 60 Miles difference in Longitude) cauſes 4 Minutes difference upon the Sun-Dial: By which rules you will find that

Deg.

Deg.	Miles.		Min.		Deg.	Miles.		Min.	
2	120		8		11	660		44	
3	180		12		12	720		48	
4	240		16		13	780		52	
5	300	gives	20		14	840	gives	56	
6 or 360		24		15 or	900		60	
7	420		28		30	1800		2	hours.
8	480		32		45	2700		3	
9	540		36		60	3600		4	
10	600		40		75	4500		5	

And so forth.

S. Is it neceffary for a Pilot to know the Difference of Time between two Meridians?

T. Yes, for by it he may come to know his Longitude.

S. Amongft the Meridians is there not one more *remarkable* than the reft?

T. Yes, there is one that hath the name of *firft Meridian*, becaufe that from it (Eaftward) the Longitude is Counted upon the Equinoxial (or Equator.)

S. Whereabout in the Cards (or Globes) do they place this firft Meridian?

T. It is commonly placed with us at the Ifland *Gratiofa*, (one of the Iflands of the *Azores*,) but formerly it was placed at the *Canary* Iflands.

S. Why do they place it at the Iflands *Azores*, rather than any other?

T. Becaufe that thwart of it there is no *variation*; the Needle pointing there Right North and South to the Poles of the World.

S. Why was it firft placed at the *Canary* Iflands?

T. Becaufe thofe Iflands were the furtheft part of the World that was then difcovered.

S. Why are the Degrees of Longitude counted from the firft Meridian Eaftward, rather than Weftward?

T. becaufe *Ptolomy* and others did it before us, and we are willing to imitate them, altho they did it only, becaufe thofe Iflands were the moft Weftern-parts of the World known to them; and fo they chofe rather to reckon their Longitude Eaftward, from Country to Country that was known; than to begin upon that which they thought, or fuppofed to be nothing but Sea, and did not know at all.

S. What ufe hath the Meridian?

T. It hath feveral good ufes, for it is by it we know the greateft Altitude (or height) of the Sun and Stars, and their neareft diftance from our Zenith, their Declinations, and Right-Afcentions; it marks the Longitude, and it is by it we reckon our Latitude, and that we know

the

the Eaſt part of the World from the Weſt part of it; and the very midſt of the Artificial Days and Nights, from which we begin to reckon, or count the Hours.

S. What do you mean by Artificial Days and Nights?

T. The Artificial Day, is that Time that the Sun is above our Horizon, or from Sun Riſing to Sun Setting: And the Artificial Night, is that Time that the Sun is under our Horizon, or from Sun Setting to Sun Riſing.

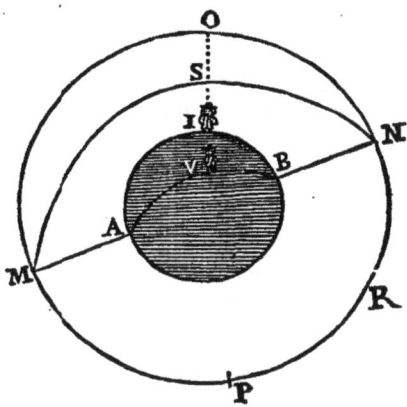

M N repreſents the Poles of the World.

O the Zenith, and P the Nadir of the Man at I.

The Circle N O M P R is the Meridian of the Man at I.

N S M is the Meridian of the Man at V.

You are to imagin the ſame Circles upon the Earth, that you do in Heaven: Therefore N O M and N S M being Celeſtial Meridians, A I B and A V B muſt be the Meridians upon Earth.

P R O P. IX.

Of the Colures.

WHAT are the *Colures* ?

T. The Colures are two great Circles, which cut or croſs one another at Right-angles in the Poles of the World, one of which is called *Colure* of the *Equinox*, and the other *Colure* of the *Solſtice*, as you may ſee in the foregoing Figure of the Sphere, in *Prop.* V.

K   *S.* Why

S. Why are they called *Colures* ?

T. Becaufe. where the Sphere is oblique, they never appear whole above the Horizon, and fo imagining or fuppofing that they are cut off, they are called Colures from *Colos* and *Oura*, which fignifies imperfect or maimed.

S. Why is one called Colure of the *Equinox* ?

T. Becaufe it cuts the Ecliptick in the beginning of *Aries* and of *Libra*, the two Points in which the Sun makes the Days and Nights equal through all the World, thence called the two Equinoxial Points.

S. What do you fay of the other Colure ?

T. I fay that it is called the Colure of the *Solftice*, becaufe it cuts the Ecliptick in the beginnings of *Cancer* and *Capricornus*, which are the two Solfticial Points.

S. Why is it called *Solfticial* ?

T. Becaufe that when the Sun is in any of thofe Points, (the 11th. of *June*, and 11th. of *December*) he feems to ftand, for he can go no further becaufe of the Tropicks which are his Bounds; and therefore we may well fay, that *Solftice* is derived of two Words *Sol* and *Statio*, which fignifie the Sun (and) ftanding.

S. Are not the Solftices diftinguifhed as well as the Equinoxes ?

T. Yes, for one is called the *Summer Solftice*, (when the Sun is at the Tropick of *Cancer* the Time of our longeft Days,) the other the *Winter Solftice*, (when the Sun is at the Tropick of *Capricorn* the time of our fhorteft Days.)

S. For what ufe are thofe four Cardinal Points; the two *Equinoxial* and the two *Solfticial* ?

T. Their ufe is to fhow us the *beginning* of the four Seafons of the Year, for the *Spring* begins at the *Vernal Equinox*, or beginning of *Aries*; (the 10 of *March*;) The *Summer*, at the *Summer Solftice*, or beginning of *Cancer*; (the 11th. of *June*;) *Autumn* at the *Autumnal Equinox*, or beginning of *Libra*; (the 12th. of *September*;) The *Winter* at the *Winter Solftice* or beginning of *Capricorn*. (The 11th. of *December*.)

PROP. X.

Of the Leffer Circles, the two Tropicks, and the two Polar Circles.

S. HOW many lefs Circles are there in the Sphere ?

T. There are Four, the two Tropicks, and the two Polar Circles.

S. Why are they called *leffer Circles* ?

T. Becaufe they are lefs than thofe which divide the Sphere into two equal parts.

Since

Since we are to imagin the same Circles upon the Earth as in the Heavens; I shall describe them so here, because it may be more easie for you to conceive them so, than in the precedent Sphere, *Prop.* V. of this Book.

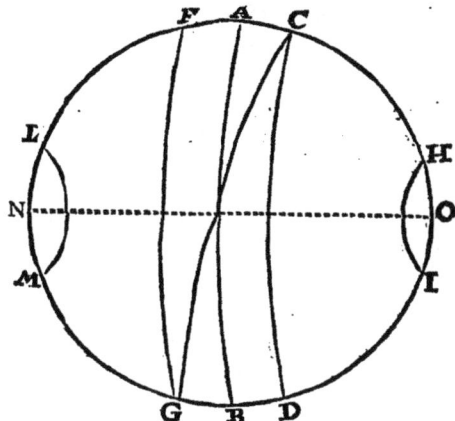

Admit you are in a Right-Sphere, (where the Poles are at the Horizon) A B is the Equinoxial.

C D the Tropick of *Cancer*, and F G the Tropick of *Capricorn*, both distant from the Equinoxial 23 Degrees 31 Minutes.

H I is the Circle Arctick, and L M the Circle Antarctick.

C G the Ecliptick.

The Points or occult Circle, N O is the Horizon.

The Interval or Distance between the two Tropicks C D and F G is the Torrid Zone.

The Interval or Distance between the Tropicks and the Polar Circles, to wit, between C H, I D, and L F, M G, are the two Temperate Zones.

And the Intervals between the Polar Circles and the Poles, are the Frigid or Cold Zones.

S. What are the *Tropicks?*

T. They are as I said two lesser Circles Parallel to the Equator (or Equinoxial) from which they are distant 23 Degrees 31 Minutes.

S. Why are they called Tropicks?

T. Because that when the Sun ariveth at either of these two Circles he turneth back again, for the *Greek* Word, *Tropos*, from whence the Word is derived, signifies *Turning*; and because one of the Tropicks passeth through the first Point or beginning of *Cancer*, it is called the *Tropick of Cancer*, and the other the *Tropick of Capricorn*, because it passeth also through the first Degree of that Sign.

S. For

S. For what ufe are the Tropicks?

T. Their ufe is to Limit or Bound the Torrid Zone, as well as the Suns Courfe, and alfo to mark the longeft and fhorteft Days and Nights.

Of the Polar Circles, otherwife called Circles Arctick and Antarctick.

S. WHAT are the *Polar Circles?*

T. They are alfo two leffer Circles diftant each from the Poles of the World 23 Degrees 30 Minutes.

S. Why are they called *Polar?*

T. Becaufe they are imagined to be in the Heaven exactly where the Poles of the Ecliptick Turn about the Poles of the World.

S. Why is one called the Circle *Arctick*, and the other the Circle *Antarctick?*

T. One is called Circle *Arctick* becaufe it paffeth through the Conftellation of *Urfa Major* or the great *Bear*; and the other is called *Antarctick*, becaufe it is *oppofite* to the Pole *Arctick*.

S. For what ufe are the Circles Arctick and Antarctick?

T. Their ufe is to Limit or Bound the Temperate Zones on the North and South fide of them.

PROP. XL.

Of the Zones.

S. WHAT is a *Zone?*

T. It is a Circular part of the Heaven and Earth, of Confiderable Breadth, comprehended between two Circles: See Figure, *Prop. preced.*

S. How many Zones are there?

T. Five, one called *Torrid*, two *Temperate*, and two *Frigid*, or extream Cold.

S. What is the *Torrid Zone?*

T. It is that part of the Earth comprehended between the two Tropicks.

S. Why is it called *Torrid?*

T. Becaufe of the Sun, which being always on fome part of it caufes there an exceeding great Heat, which made the Ancients fancy that all in it was Burnt, and therefore they called it the *Torrid Zone*, which is as much as to fay, the *Burnt Zone*.

S. What are the *Temperate Zones?*

T. They are Two parts of the Earth, one of which is comprehended between the Tropick of *Cancer* and the Circle *Arctick*, and the other between the Tropick of *Capricorn* and Circle *Antarctick*.

S. Why

S. Why are they called *Temperate?*

T. Becaufe they are not subject to the great Cold of the *Frigid* Zones, nor to the great Heat of the *Torrid.*

S. What are the *Frigid Zones?*

T. They are Two parts of the Earth Comprehended between the Circle Arctick and the North Pole; and between the Circle Antarctick and the South Pole.

S. Why are they called *Frigid?*

T. Becaufe of the great Cold they are fubject to, for the Sun ftrikes his Beams fo obliquely upon them, that they receive but little or no Heat by it, and therefore they are called *Frigid;* which fignifies Cold.

S. As I underftand then, they are the four leffer Circles which divide the Earth into five *Zones.*

T. It is true, for it is by the two Tropicks and two Polar Circles, that both the Heaven and Earth is divided into five Parts, (called *Zones.*)

PROP. XII.

Of the Parallels.

S. HOW do you call the *Circular* or *Right-lines* drawn Eaft and Weft upon the Globes and Charts.

T. They are called Paralleis.

S. Why are they called Parallels?

T. Becaufe that Word in *Greek* fignifies nothing elfe but a thing *equally diftant* in every part from another thing, and therefore that Name is very proper to all Circles or Lines drawn Eaft and Weft, becaufe they are in all their parts equally diftant from the Equinoxial Line.

S. Since Parallel fignifies *equally diftant* from fome other thing, is not that Name common to all the leffer Circles in the Sphere.

T. Yes, neverthelefs it is only attributed to thefe, becaufe the others (of which you have Read before) have Names proper for them.

S. How many kinds of Parallels are there?

T. Two, the firft of which are called the *Suns Parallels*, becaufe we imagin 180 Circles (or thereabouts) between the two Tropicks, one of which the Sun goes through every Day, becaufe of one Degree which he advances every Day in the Ecliptick. The fecond kind are called *Parallels of Latitude*, becaufe they ferve for knowing the Latitude of a place, and of this kind one may imagin as many as Parts or Points in the Meridian, except 21 on the North fide of the Equinoxial, and as many on the South fide of it which have a particular Name, being called *Parallel of the longeft Day*, becaufe that in places that are in the fame Latitude, or as far diftant from the Equinoxial as they are, the Days are longer by a quarter of an hour than at the next Parallel of that kind

towards

towards the Equator , for the further the Parallel is from the Equinoxial the longer the Days are, as this Table fhoweth you, their true diftance or proportion from it being fet down, as followeth :

	Latitude.		Latitude.
The firft is at . .	4° — 15′	The twelfth . .	41° — 20′
The fecond . . .	8 — 30	The thirteenth .	43 — 15
The third . . .	12 — 45	The fourteenth .	45 — 24
The fourth . . .	16 — 35	The fifteenth . .	48 — 40
The fifth . . .	20 — 30	The fixteenth . .	51 — 50
The fixth . . .	24 — 15	The feventeenth .	54 — 30
The feventh . .	27 — 30	The eighteenth .	56 — 30
The eighth . . .	30 — 45	The nineteenth .	58 — 20
The ninth . . .	33 — 40	The twentieth . .	61 — 10
The tenth . . .	36 — 24	The one and twentieth	63 — 16
The eleventh . .	39 — 00		

PROP. XIII.

Of the Azimuths.

WHAT are the *Azimuths?*

T. They are great Circles which pafs through the Zenith and Nadir, and every Part or Point of the Horizon, however they are not vulgarly reckoned or accounted amongft the great Circles of the Sphere.

S. Why are they called *Azimuths ?*

T. Becaufe we will imitate the Ancient Aftronomers who ufed to call them fo, I fuppofe more to fhow their Origine than any thing elfe, for the Word is *Arabick*, and fignifies only *Vertical Circles.*

S. For what ufe are the Azimuths ?

T. Their ufe is to fhow the Altitude or Height of the Sun and Stars, for their Altitude is nothing elfe but the Degree of the Azimuth comprehended between the Horizon and the Sun or Star : It ferveth alfo to fhow at what part of Heaven, or Point of the Compafs, the Sun and Stars are at any time; (being above our Horizon;) alfo to know the variation of the Compafs; and how two places bear one from another, fince the 32 Parts in which the Compafs is divided, are the fame as fo many haifs of Azimuths or Vertical Circles falling down flat as they appear on the Compafs, as you will better underftand by the following Figure.

S. Where do the Azimuths take their beginning, or from what part of the Horizon are they counted ?

T. From the Point at which the Sun Rifes when he is at the Equator or Equinoxial (which is the true Eaft and Weft) from whence they are accounted Southward, and you may imagin as many Azimuths as Points or at leaft Degrees in the Horizon, that is to fay 360.

S. What

S. What is the Azimuth of the Sun and Star in Computation?

T. It is nothing elfe but the Number of Degrees comprehended on the Horizon between the Azimuth which paffes through the Center of the Sun or Star to the Horizon, and the true Eaft and Weft Point mentioned in the preceding Anfwer.

S. By what means may I know the Azimuth of the Sun and Stars?

T. By the Latitude of your place, the Declination of the Sun or Stars, and their Altitude above your Horizon; for thefe three things being known you may thence eafily gather the Azimuth.

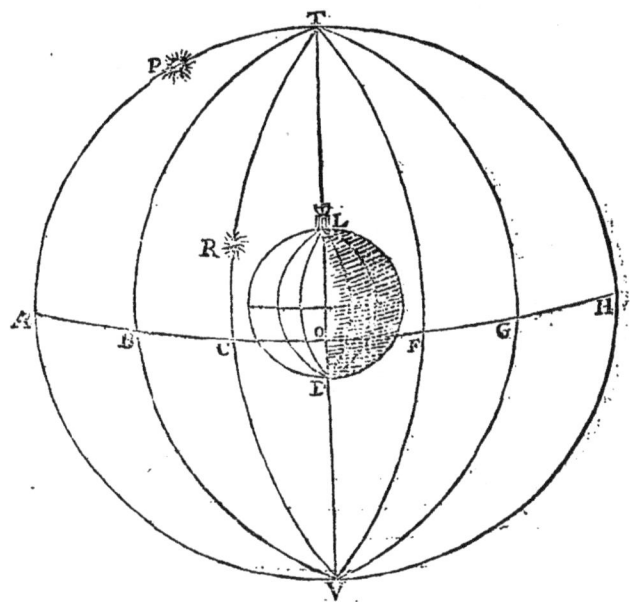

In this Figure, T being the Zenith of a Man at L, and ACPH the Horizon; the Azimuths will be the Circles A T, B T, C T, D T, F T, G T; and all others that pafs through the Zenith T and Nadir V, of which you may imagin as many as there are Points in the Horizon: Now fuppofing T A to be the Meridian, and O the Eaft Point where the Sun Rifeth when he is at the Equinox, and through which the firft Azimuth T D paffeth, the Cuftom is to count the Azimuths from O towards A which is the South; fo that the Meridian is the 9cth. Azimuth or Vertical Circle; the Rhumbs or Lines drawn upon the Fly of your Compafs, and Center of it, reprefent thefe Circles crouched down 'till they be flat: So that knowing what Degree of the Compafs the Sun is at, you know the Azimuth of the Sun, and it is the fame of the Stars, whofe Altitude is nothing elfe but the Number of Degrees of the Azimuth which are betwixt

betwixt the Horizon and the Sun: (or obferved Star:) As for Example, if the Sun were in P, his Altitude or Height would be A P, and the Complement of his Height would be the Arch P T, which is the Diftance of the Sun from the Zenith T; likewife if the obferved Star were in R, its Altitude would be C R, and the Complement of its Height, that is, its Diftance from the Zenith, would be the Arch R T. The Rifing Amplitude is an Arch of the Horizon comprehended between the Point the Sun or Star Rifes at, and that Point called the true Eaft: As for Example, if the Sun (or Star) fhould rife at the Point C, O being the true Eaft of the World, the Arch O C would be the Rifing Amplitude; the fame is to be underftood of the Setting Amplitude either Southerly, or Northerly, according to the Suns Declination; but of this more in the following propofitions, when we come to treat of the Amplitude.

PROP. XIV.

Of the Almicanterahs, or Almicanters.

S. **W**HAT are the *Almicanterahs*, or *Almicanters?*
    T. They are little Circles, Parallels to the Horizon, and drawn one above another from the Horizon unto the Zenith, as you may eafily underftand by this Figure, where G H being the Horizon, E I, R S, O P, M N, and the other Circles Parallel to the Horizon, are called Almicanters.

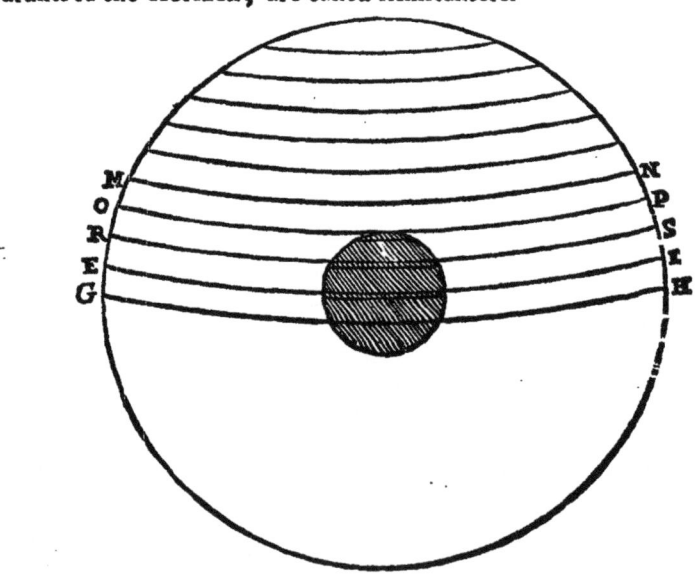

S. Why

S. Why are they called *Almicanterahs* ?

T. The Name was given by the *Arabian* Astronomers, when the Mathematical Sciences flourished in their Empire, and in their Language signifies only Circles of *Altitude*. Afterwards *Astronomy* spreading from them to us, brought along with it many such *Arabian* Words.

S. How many Almicanterahs are there ?

T. There are as many as one can imagin Points or Parts in the Azimuth, for they are *infinite* as well as the Meridians, nevertheless we seldom reckon *more* than there are Degrees from the Horizon to the Zenith, that is 90.

S. For What use are these Circles ?

T. Their use is but small, for they serve only to mark the Altitude of the Sun or Stars, which the Azimuth can do without them.

S. How are these Circles of Altitude counted ?

T. They are counted upon the Azimuth, or Vertical Circles, from the Horizon to the Zenith.

P R O P. XV.

Showing which of the Circles of the Sphere are most useful in Navigation; and of the Latitude, and Longitude.

S. OF all the *Circles* you have Treated of, pray tell me the most considerable for *Navigation*, that I may take particular notice of them.

T. The most considerable are the *Meridian*, the *Horizon*, and the *Equator*, (or Equinoxial.)

S. Why those Three Circles more than others ?

T. Because their use is for determining the *Latitudes* and *Longitudes*, which is the principal part of *Navigation*, as you will find when you come to Practice.

Of Latitude.

S. WHAT is the *Latitude* of a place ?

T. The *Latitude* is nothing else but the *Distance* of a place or Zenith thereof from the *Equinoxial*, and therefore when you will know where you are arrived, you must find out (by your Observation) how many Degrees and Minutes your Zenith is Distant from the Equator or Equinoxial, and that shall be your Latitude, which is always equal to the Height of the Pole. To define it more Astronomically, I say, that the Latitude is an Arch of the Meridian Comprehended between my Zenith and the Equator, and that the Etimology of the Word signifies only

L   *Breadth.*

Breadth. (Remember that the Latitude is always counted from the Equinoxial to the Poles of the World, so that there are 90 Degrees of Latitude North, and 90 more of Latitude South.)

Essay,

To put an instance, suppose now your Ship was in North Latitude 48 Degrees 30 Minutes: *I desire you would tell me what that means?*

S. I conceive it means that my Ship is Distant from the Equinoxial Northward 48 Degrees 30 Minutes.

T. You have it.

Of Longitude.

S. WHAT is the *Longitude* of a place?

T. The *Longitude* of a place, is nothing else but the *Distance* from the *first Meridian*, to the Meridian that passes through the Zenith or place whose Longitude you desire to know; but to define it as the Astronomers do; it is an Arch of the Equator (or Equinoxial) Comprehended between the first Meridian, and the Meridian that passes through the Zenith or Point Vertical of the place.

S. What signifies the Word *Longitude?*

T. It signifies the *Length*, (of the Earth.) And you are to remember that the Longitude is always counted from the first Meridian, some counting altogether Eastward, and other boths Eastward and Westward.

Essay, (of the Longitude.)

Suppose now that your Ship is arrived in the Longitude of 6 Degrees: *I demand what that means?*

S. That means that my Ship is Distant from the first Meridian Eastward 6 Degrees. But why is the Latitude of *several Denominations?* Or why do you make a Difference in the Latitudes, calling one Latitude *North*, and the other Latitude *South?*

T. It is only to know on which side of the Equinoxial we are, whether we be Northward or Southward of it.

PROP.

P R O P. XVI.

Of the Meridian Altitude of the Sun and Stars, and of their Declination.

S. WHAT is the *Meridian Altitude?*

T. It is the *greatest Height* of the Sun, or Star, above the Horizon, which happens every time that the Sun or Star comes to our Meridian, (from whence it is properly called *Meridian Altitude.*)

S. What fignifies the Word *Altitude?*

T. It fignifies only *Height*, and therefore when we call it *Meridian Altitude*, it is the fame as if we fhould call it *Meridian Height.*

S. Why do you take more notice of the *Meridian* Altitude than of any other?

T. Becaufe then it is the only Time to obferve the Latitude, or the Diftance of the Sun or Star from our Zenith.

S. What is the *Height* or *Altitude* of the Sun or Star?

T. It is the Degrees and Minutes comprehended between the Horizon and the Center of the Sun or Star: Or thus, It is the Degrees and Minutes of the Sun or Stars Height above the Horizon.

Of the Sun and Stars Declination.

S. WHAT is the *Declination?*

T. The *Declination* is nothing elfe, but the *Diftance* of the Sun or Stars from the *Equator*; but to define it as Aftronomers do, it is an Arch of the Meridian comprehended between the Equator, (or Equinoxial) and the Sun or Star.

S. If it be only the Diftance of the Sun or Star from the Equinoxial, it is then the fame as the Latitude of a place according to what you have already faid.

T. Altho it feems the *fame* there is a great *Difference*; for the Latitude of a place never Changes, but the Declination Changes every Minute.

S. Is it not the *Primum Mobile* (or ninth Heaven') which is the Caufe that the Declination of the Sun Changes fo often as it doth?

T. No, for if the Sun and Stars had no other moving than that of the *Primum Mobile* (from Eaft to Weft) their Declinations would be always the fame, for they would be always at the fame Diftance from the Equator; and it is their own proper moving upon the Poles of the Zodiack which is the Caufe of it; for as that Circle cuts or croffes the Equator flopingly, and the Sun and Stars move upon the fame Poles and to the

fame

same manner, they muſt of neceſſity be ſome time nearer or further from the Equator, becauſe of their oblique moving under it, and ſo the Declination muſt needs increaſe or diminiſh.

S. How much is the Suns greateſt Declination?

T. It is not (at preſent) above 23 Degrees 30 Minutes, for as ſoon as the Sun is come to the Tropicks, (which are Diſtant from the Equator but 23 Degrees 30 Minutes,) he returns towards the Equinoxial, and ſo by little and little his Declination diminiſheth.

S. Why do you make a Difference in the Declination, calling one Declination *North* and the other Declination *South*?

T. It is becauſe the Sun (and Stars) decline ſometimes on one ſide the Equator and ſometimes on the other, and therefore to know on what ſide he is, we give different Names to his Declination, for by it we know if we muſt Add it to, or Subtract it from our Obſervation to have our Diſtance from the Equinoxial, which as I ſaid before is our Latitude.

S. Can the Sun be without any Declination?

T. Yes, for every time that he is at the *Equinox* or firſt Points of *Aries* and *Libra*, (that is about the 10th of *March*, and 11th of *September*,) he hath no Declination, for he is then at the Equinoxial.

S. How much is the greateſt Declination of the Stars?

T. The Declination of the Stars differs according to their Latitude and Longitude, of which we ſhall ſpeak in its due place.

S. Is there nothing elſe to be known concerning the Suns Declination?

T. Yes, there is much more that a good Pilot or an Artiſt muſt know, but I ſhall reſerve it untill the Third Book, where it will find a more proper place than this, which is only deſigned for Principles; however I deſire you would take a particular notice of theſe Propoſitions, they being very neceſſary for Navigation.

PROP. XVII.

The Complement of the Altitude of the Sun, Stars, and Pole above our Horizon; alſo the Complement of the Latitude, of the Declination, and of the Sun or Stars Diſtance from our Zenith.

S. WHAT is the *Complement* of the *Altitude*?

T. It is the Degrees that the Sun or Star is Diſtant from our Zenith, which being added to the Degrees of Altitude makes up the 90 Degrees that our Zenith is Diſtant from the Horizon, for Complement ſignifies only that which makes up, completes or finiſhes, ſo that the Complement of the Height of the Pole

is

is also the Degrees Comprehended between the Pole and the Zenith, as you may better understand by this

Example.

Suppose that the Height of the Pole or Altitude of the Sun (or Star) were 50 Degrees; I say that its Complement is 40 Degrees, because there is no Number but 40, that can make up the 90 Degrees comprehended betwixt the Horizon and our Zenith.

S. What is the *Complement* of the *Latitude?*

T. The Complement of the Latitude is the Height of the Equinoxial above the Horizon; for you know that the Latitude is the Distance of our Zenith from the Equinoxial, and therefore if we are but 52 Degrees from it, the Height of the Equinoxial above the Horizon must needs be 38 Degrees, for else there would not be 90 Degrees from our Zenith to the Horizon as there are, and it is those 38 Degrees which in this Example I call the Complement of the Latitude, because (as hath been said) they complete or make up the 90 Degrees that our Zenith is Distant from the Horizon.

S. What is the Sun or Stars *Distance* from our Zenith?

T. It is the Degrees and Minutes comprehended between the Zenith and the Center of the Sun, or Star, so that their Altitude is always the Complement of their Distance from our Zenith, (and then the Degrees are accounted from our Zenith to the Horizon) and their Distance from our Zenith is the Complement of their Altitude, (and then the Degrees are accounted from the Horizon to the Zenith, I mean that the Degrees begin at the Horizon.)

S. What is the *Complement* of the *Declination?*

T. It is an Arch or Number of Degrees comprehended between the Sun or Star, and the nearest Pole of the World.

PROP. XVIII.

Of the Difference in Latitude, and in Longitude.

S. WHAT is the *Difference in Latitude?*

T. The Difference in Latitude is the Degrees and Minutes comprehended between two Latitudes.

S. How is the Difference of Latitude to be known?

T. By Subtracting the lesser Latitude from the greater, when they are both on the same side the Equator; that is, both North or both South; and by adding it when they are on different sides, the one being North Latitude and the other South.

Example.

Example.

If I depart from the Latitude North of 50 Degrees 30 Minutes, and find by my Observation that I am arrived in the Latitude North of 47 Degrees; the Difference of my Latitudes will be 3 Degrees 30 Minutes, for if I Subtract 47 from 50 Degrees 30 Minutes, the Remainder will be 3 Degrees 30 Minutes, for the Difference of my Latitudes.

S. Must I Subtract always the lesser Latitude from the greater, to know or have my Difference in Latitude?

T. Yes, when they are both on the same side of the Equinoxial, that is, either both North Latitude, or both South Latitude; but when they are not, or that one is North Latitude, and the other South, then they must not be Subtracted one from another, but added together, as this Example showeth.

If I depart from the Latitude North of 4 Degrees 15 Minutes, and I find by my Observation that I am arrived in the Latitude South of 3 Degrees 30 Minutes, my Difference in Latitude shall be 7 Degrees 45 Minutes; for 4 Degrees 15 Minutes, and 3 Degrees 30 Minutes; being added together will come to 7 Degrees 45 Minutes.

S. Suppose that from 4 Degrees 15 Minutes of Latitude North, you had Sailed just to the Equinoxial, what would you do then, for you could neither Add nor Subtract.

T. In that case the 4 Degrees 15 Minutes would be the Difference in Latitude.

Of the Difference in Longitude.

S. WHAT is the *Difference* in Longitude?

T. The Difference in Longitude is the Degrees and Minutes comprehended between two Meridians or between the Longitude of *your departure*, and the Longitude of the place you are *arrived* at.

S. How is the Difference in Longitude to be known?

T. Only by Subtracting the lesser Longitude from the greater, and the Remainder will be the Difference in Longitude; but you are to remember that when you cross the first Meridian, or one Longitude be on the East side of it, and the other on the West, or one on the West and the other on the East, you must always add 360 Degrees to the least of the Longitudes to have their Difference.

Example.

Example.

If I depart from the Longitude of 6 Degrees 30 Minutes, and Sail Westerly untill I am arrived in the Longitude of 358 Degrees: I say that my Difference in Longitude is 8 Degreees 30 Minutes, for if you add 360, to 6 Degrees 30 Minutes, there will come 366 Degrees 30 Minutes, from which if you Subtract 358 Degrees, (the Longitude you are arrived at) the Remainder will be 8 Degrees 30 Minutes, for the Difference in Longitude; (since I have Sailed the neareſt way from the Longitude of 6 Degrees, to the Longitude of 358 Degrees.)

Another Example.

If I depart from the Longitude of 357 Degrees, and Sail Eaſterly until I am arrived in the Longitude of 4 Degrees 15 Minutes; I say that my Difference in Longitude is 7 Degrees 15 Minutes, for if you add 360 Degrees to 4 Degrees 15 Minutes, there will come 364 (Degrees) 15 Minutes, from which if you Subtract 357 (Degrees) the Remainder will be 7 Degrees 15 Minutes for the Difference in Longitude, (since I have Sailed the neareſt way from the Longitude of 357 Degrees, to the Longitude of 4 Degrees 15 Minutes,) but you are to take notice that this is to be obſerved only by thoſe that reckon or count the whole Longitude of the World (360 Degrees) without dividing it: But for thoſe that will have the Longitude of ſeveral Denominations, calling that part of the World which is on the Eaſt ſide of the firſt Meridian Eaſt Longitude; and what is on the Weſt ſide of it Weſt Longitude, (becauſe it is more eaſie to them;) I ſay in that caſe they muſt obſerve the ſame things as for the Latitude; I mean adding the two Longitudes together, to have their Difference when they are of Different Denominations, ſince they count their Degrees from the firſt Meridian Weſtward in the ſame manner as they are counted on the Eaſt ſide of it, I mean by 1, 2, 3, and ſo forth, until 180.

PROP. XIX.

Of the Amplitude Ortive and Occaſive.

S. WHAT is the *Amplitude Ortive?*

T. It is the Degrees and Minutes that the Sun or any Star riſes Diſtant from the true Eaſt Point : Or elſe thus, the Amplitude is an Arch of the Horizon Comprehended

between

between the true Eaſt, and the Point that the Sun or Star Riſes at (either Northward or Southward) as was Demonſtrated to you in the Figure of the 13th. Propoſition.

S. Why do you ſay either Northward or Southward?

T. Becauſe the Amplitude differs according to the Declination, for when the Sun hath North Declination, his Amplitude is alſo North, but when his Declination is South, his Amplitude is alſo South: So that from the 10th. of *March*, to the 11th. of *September*, the Sun Riſes and Sets on the North ſide of the Equinoxial, or true Eaſt and Weſt Point, but from the 11th. of *September*, to the 10th. of *March*, he Riſes and Sets on the South ſide of it, and therefore there is Amplitude Ortive North, and Amplitude Ortive South; Amplitude Occaſive North, and Amplitude Occaſive South.

S. What ſignifies *Ortive?*

T. It ſignifies only Riſing or to Riſe, for it is derived from the *Latin* Word *Ortus*, and therefore when we call it Amplitude Ortive, it is the ſame as if we ſhould ſay the Riſing Amplitude.

S. What is the *Amplitude Occaſive?*

T. It is the Number of Degrees and Minutes that the Sun Sets Diſtant from the true Weſt Point: Or elſe thus, it is an Arch of the Horizon, comprehended between the true Weſt, and the Point that the Sun Sets at, (either Northward, or Southward.)

S. What ſignifies *Occaſive?*

T. Occaſive is derived from the *Latin* Word *Occaſus*, which ſignifies to Set, and therefore when we call it Amplitude Occaſive, it is the ſame as if we ſhould ſay, the Setting Amplitude.

S. For what uſe is the *Amplitude?*

T. Its chief uſe is for finding the *Variation* or *Declination* of the Compaſs, from the true North Point.

PROP. XX.

Of the Parallax, and Refraction.

S. **W**HAT is the *Parallax?*

    T. The Parallax is nothing, but the Difference betwixt the Altitude of the Sun in reſpect of the *ſenſible* Horizon, and the ſame Altitude of the Sun in reſpect of the *Rational* Horizon, as you may better underſtand by this Figure.

Admit

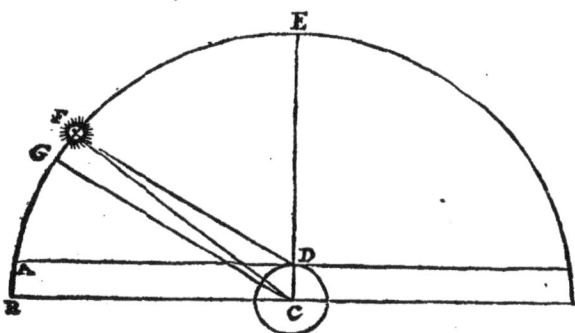

Admit the Center of the Earth to be C, and the Point on its Superficies (where you ſtand) D, from whence you obſerve the Altitude of
the Sun F, above the ſenſible Horizon A D, (Perpendicular to the Line
D C;) that Height will be the Angle F D A; but if you draw the
Rational Horizon B C, (Parallel to the ſenſible,) the Altitude of the
Sun taken in reſpect of the Rational Horizon, will be the Angle F C B;
and the Difference between theſe two Heights is what we call Parallax, and to determin this Difference, draw the Line G C Parallel
to F D.

Demonſtration.

It is plain (by the 28th. of the firſt of *Euclide*) that the Angle G C B
is equal to the Angle F D A, then the Angle F C G is the Difference
between the Angle of the true Height F C B, and the Angle F D A of
the obſerved Height. And becauſe the Angles Alternate F C G and
G F D are alſo equal; the Angle C F B will be equal to that Difference.

S. What can I learn by this Figure and Demonſtration?
T. You may learn to underſtrnd the Reaſon, why the Aſtronomers
would have the Pilots to add ſome Minutes to the Suns Altitude, proving by this Figure that the Suns Height appears leſs than it is, and
to Correct that error they have made Tables of its Parallax, by which
they ſhow what muſt be added to their Obſervations according to the
Height of the Sun above the Horizon.

'M   *S.* What

S. What is the *Refraction?*

T. The Refraction is the Difference (in Minutes) between the place where the Sun (or Star) is, and that where it appeareth to be, which Difference is caused by the continual vapors, and exhalations, which the Sun by his heat draws up into the Air, (from the Sea and Land) which makes him appear higher than he is at his Rising, or Setting, or when he is near the Horizon.

S. Are there Tables for the Refraction, as well as for the Parallax?

T. Yes, there are some, which show what Minutes Astronomers would have us to Subtract from the Sun or Stars Altitude, (or else add to its Distance from the Zenith;) but I cannot recommend them to you as very necessary in Voyages Southward, knowing that they are of no use (where or when the Sun is not far from the Zenith) because we observe only the Meridian (or highest) Altitude of the Sun, which in those parts is void of all sensible Refraction; however I shall insert them here for your Voyages to the Northward.

S. Why are they not as necessary in our Voyages to the Southward, as in those to the Northward?

T. Because that in Sailing to the South we come so much under the Sun, that its Meridian Altitude (which all Pilots observe to find their latitude) is most commonly exempt or void, of any sensible Parallax, and Refraction. But it is not so in Sailing Northerly, or Northward, to *Ice-land*, *Green-land*, or other places near the Poles; because often Times, the Meridian Altitude of the Sun is so low, or little, that there must needs be some sensible Refraction, and therefore they are here presented to you, to make use of them at such a Time.

A

A Table *of the Parallax of the Sun, according to the Observations* of Philip Lansberg , (*a Dutch-Man.*)

Altitude or Degrees from the Horizon.	Parallax in Minutes and Seconds. M. S.	Distance from the Zenith.		Altitude or Degrees from the Horizon.	Parallax in Minutes and Seconds. M. S.	Distance from the Zenith.		Altitude or Degrees from the Horizon.	Parallax in Minutes and Seconds. M. S.	Distance from the Zenith.
0	2.18	90		30	2.00	60		60	1. 9	30
1	2.18	89		31	1.58	59		61	1. 6	29
2	2.18	88		32	1.57	58		62	1. 4	28
3	2.18	87		33	1.56	57		63	1. 2	27
4	2.18	86		34	1.54	56		64	1. 0	26
5	2.18	85		35	1.53	55		65	0.58	25
6	2.17	84		36	1.52	54		66	0.56	24
7	2.17	83		37	1.50	53		67	0.54	23
8	2.17	82		38	1.49	52		68	0.52	22
9	2.17	81		39	1.47	51		69	0.49	21
10	2.16	80		40	1.46	50		70	0.47	20
11	2.16	79		41	1.44	49		71	0.45	19
12	2.15	78		42	1.42	48		72	0.43	18
13	2.14	77		43	1.41	47		73	0.40	17
14	2.14	76		44	1.39	46		74	0.38	16
15	2.13	75		45	1.38	45		75	0.36	15
16	2.12	74		46	1.36	44		76	0.33	14
17	2.12	73		47	1.34	43		77	0.31	13
18	2.11	72		48	1.32	42		78	0.29	12
19	2.10	71		49	1.31	41		79	0.26	11
20	2.10	70		50	1.29	40		80	0.24	10
21	2. 9	69		51	1.27	39		81	0.22	9
22	2. 8	68		52	1.25	38		82	0.19	8
23	2. 7	67		53	1.23	37		83	0.17	7
24	2. 6	66		54	1.21	36		84	0.15	6
25	2. 5	65		55	1.19	35		85	0.12	5
26	2. 4	64		56	1.17	34		86	0. 9	4
27	2. 3	63		57	1.15	33		87	0. 7	3
28	2. 2	62		58	1.13	32		88	0. 5	2
29	2. 1	61		59	1.11	31		89	0. 2	1
30	2. 0	60		60	1.09	30		90	0. 0	0

A Table of the Parallax of the Sun, according to the Observations
of the Famous Ticho-Brahe.

Altitude or Degrees from the Horizon.	Parallax in Minutes and Seconds. M. S.	Distance from the Zenith.	Altitude or Degrees from the Horizon.	Parallax in Minutes and Seconds. M. S.	Distance from the Zenith.	Altitude or Degrees from the Horizon.	Parallax in Minutes and Seconds. M. S.	Distance from the Zenith.
0	3.00	90	30	2.36	60	60	1.30	30
1	3.00	89	31	2.34	59	61	1.28	29
2	3.00	88	32	2.32	58	62	1.25	28
3	3.00	87	33	2.30	57	63	1.22	27
4	2.59	86	34	2.29	56	64	1.19	26
5	2.59	85	35	2.27	55	65	1.16	25
6	2.59	84	36	2.25	54	66	1.13	24
7	2.58	83	37	2.23	53	67	1.10	23
8	2.58	82	38	2.21	52	68	1. 8	22
9	2.57	81	39	2.19	51	69	1. 5	21
10	2.57	80	40	2.18	50	70	1. 2	20
11	2.56	79	41	2.16	49	71	0.59	19
12	2.56	78	42	2.14	48	72	0.56	18
13	2.55	77	43	2.12	47	73	0.53	17
14	2.54	76	44	2.09	46	74	0.49	16
15	2.54	75	45	2.07	45	75	0.46	15
16	2.53	74	46	2. 5	44	76	0.43	14
17	2.52	73	47	2. 3	43	77	0.40	13
18	2.51	72	48	2. 0	42	78	0.37	12
19	2.50	71	49	1.58	41	79	0.34	11
20	2.50	70	50	1.56	40	80	0.31	10
21	2.49	69	51	1.54	39	81	0.28	9
22	2.48	68	52	1.51	38	82	0.25	8
23	2.46	67	53	1.48	37	83	0.21	7
24	2.45	66	54	1.46	36	84	0.18	6
25	2.44	65	55	1.43	35	85	0.15	5
26	2.43	64	56	1.41	34	86	0.12	4
27	2.41	63	57	1.39	33	87	0. 9	3
28	2.39	62	58	1.36	32	88	0. 6	2
29	2.37	61	59	1.33	31	89	0. 3	1
30	2.36	60	60	1.30	30	90	0. 0	0

S. OI

S. Of thefe Two *Tables* (of the Parallax) which do you counfel me to make ufe of?

T. Truly it is no great matter which; for the *Hollanders* generally ufe that of *Lansberg*, as well for the Parallax, as for the Suns Refraction in their Practice at Sea, contrary to the Aftronomers, who prefer thofe of *Ticho-Brahe.*

The Ufe of the Tables of the Parallax.

T. **I**F you have obferved the Altitude of the Sun, you muft look for the Degrees of it in the firft Column on the Left-hand, (of the *Table* you make choice of) and over againft it in the Column of the *Parallax*, you will find the Minutes and Seconds of the Parallax, which muft be added to the Degrees and Minutes of the Suns Altitude to have its true Height: But if you have obferved the Suns Diftance from your Zenith, you muft look for the Degrees of it in the third Column, (over which is written the *Diftance from the Zenith*,) and over againft it on the Left-hand Column of the Parallax, you will find the Minutes and Seconds, which muft be Subtracted from the Degrees and Minutes obferved, and the Remainder will be the Suns true Diftance from your Zenith.

Example.

The Suns Meridian Altitude, by Obfervation being 20 Degrees, I require the true Altitude.

	D.	M.	S.
Altitude by Obfervation	20	00	00
Parallax (according to *Lansberg*) added	00	02	10
The true Meridian Altitude	20	02	10

Another Example.

The Suns Diftance from my Zenith by Obfervation being 70 Degrees, I require the Suns true Diftance.

	D.	M.	S.
The Suns Diftance by Obfervation	70	00	00
The Parallax (according to *Lansberg*) Subtracted .	00	02	10
Remains the true Diftance of the Sun from my Zenith	69	57	50

S. Why have you faid nothing of the Parallax of the fixed Stars?

T. It is becaufe they are fo far Diftant from us, that they have no fenfible Parallax.

A Table *of the Refractions of the Sun, according to the Observations of* Philip Lansberg, *and* Ticho-Brahe.

Lati-tude.	Lansberg. Refraction.		Ticho-Brahe. Refraction.		Diff. from the Zenith
	Min.	Sec.	Min.	Sec.	
0	34	00	34	00	90
1	26	00	26	00	89
2	21	00	20	00	88
3	18	00	17	00	87
4	15	45	15	30	86
5	14	00	14	30	85
6	12	30	13	30	84
7	11	15	12	45	83
8	10	05	11	15	82
9	09	05	10	30	81
10	08	15	10	00	80
11	07	15	09	30	79
12	07	05	09	00	78
13	06	40	08	30	77
14	06	19	08	00	76
15	06	00	07	30	75
16	05	42	07	00	74
17	05	24	06	30	73
18	05	07	05	45	72
19	04	50	05	00	71
20	04	35	04	30	70
21	04	16	04	00	69
22	04	00	03	30	68
23	03	44	03	10	67
24	03	28	02	50	66
25	03	12	02	30	65
26	02	56	02	15	64
27	02	40	02	00	63
28	02	24	01	45	62
29	02	09	01	35	61
30	01	54	01	25	60
31	01	39	01	15	59
32	01	24	01	05	58
33	01	09	00	55	57
34	00	55	00	45	56
35	00	44	00	35	55

A

A Table *of the Refractions of the Fixed Stars, according to the Observations of* Ticho-Brahe.

Altitude or Degrees from the Horizon.	Refract. in Minutes and Seconds.		Distance from the Zenith.		Altitude or Degrees from the Horizon.	Refract. in Minutes and Seconds.		Distance from the Zenith.
	M.	S.				M.	S.	
0	30	00	90		10	5	30	80
1	21	30	89		11	5	00	79
2	15	30	88		12	4	30	78
3	12	30	87		13	4	00	77
4	11	00	86		14	3	30	76
5	10	00	85		15	3	00	75
6	9	00	84		16	2	30	74
7	8	15	83		17	2	00	73
8	6	45	82		18	1	15	72
9	6	00	81		19	0	30	71
10	5	30	80		20	0	00	70

The Use of the Tables *for the Refraction.*

T. SInce the Refraction makes the Sun or Stars appear higher than they are, it must be Subtracted from the Sun or Stars Altitude; and therefore after you have observed their Height, you are to look for the Degrees of it, in the first Column of the Table (you make choise of) and over against it in the next Column, you will find the Minutes and Seconds of the Refraction, which must be Subtracted from the Degrees and Minutes of the Sun, or Stars Altitude, to have their true Height : But if you observe, or count only the Suns Distance from the Zenith, you must look for the Degrees of its Distance in the third Column (where you see written at the Top, the *Distance from the Zenith*) and in the next Column of the Refraction (on the Left-hand) you will find the Minutes and Seconds which must be added to the Degrees and Minutes observed ; and the whole will be the Suns true Distance from your Zenith.

Example.

Example.

The Suns Meridian Altitude by Obfervation being 10 *Degrees:*
I require the true Altitude.

	D.	M.	S.
The Suns Altitude by Obfervation	10	00	00.
The Refraction (according to *Lansberg*) Subtracted .	00	08	15
Remains the true Meridian Altitude	09	51	45

But if in the Obferved Altitude there is both Parallax and Refraction, you muft Subtract them one from another, and the Remainder of the Refraction being Subtracted as before from the Suns obferved Altitude, the Remainder will be the true Meridian Altitude; but if you will know the true Diftance of the Sun from your Zenith, you muft add the Remainder of the Refraction to the obferved Diftance of the Sun from the Zenith, and the Sum will be the Suns true Diftance, (from the Zenith.)

Example.

The Suns Altitude being by Obfervation 12 *Degrees: I require its*
true Height.

	M.	S.
The Refraction of the Sun, for the Altitude of 12 Degrees (according to *Lansberg*) is	7	5
The Parallax, Subtract	2	15
The true Refraction	4	50

	D.	M.	S.
Altitude by Obfervation	12	00	00
The true Refraction Subtract	00	04	50
The true Height of the Sun	11	55	10

Example 2.

The Suns Diftance from the Zenith being by Obfervation 78 *Degrees:*
I demand its true Diftance.

	M.	S.
The Refraction for the Diftance of 78 Degrees (or rather for its Complement) is according to *Lansberg* . . .	7	05
The Parallax is	2	15
The true Refraction	4	50

Diftance

	D.	M.	S.
Distance by Observation	78	00	00
The true Refraction to be added	00	04	50
The true Distance from the Zenith	78	04	50

Note, That in Navigation it is not necessary to account Seconds, except there be above 30, and then you are to add one Minute more than is marked in the Table, but else we take no notice of them: As in the precedent Example, having found the Refraction to be 7 Minutes 5 Seconds, and the Parallax 2 Minutes 15 Seconds; you should have Subtracted only the 2 Minutes of Parallax from the 7 Minutes of Refraction, without taking notice of the Seconds, because they are under 30, but if you had found the Refraction (for Example) 7 Minutes 45 Seconds, and the Parallax 2 Minutes 50 Seconds; then you should have Subtracted 3 Minutes of Parallax, from 8 Minutes of Refraction, because the Seconds are above 30, and that you must add one Minute more for them than your Table shows: For thus your Computations will be more easie and expedite, than if you should reckon by Seconds.

A Table *of the Refraction of the Sun, to be made use of without Parallax.*

The Altitude.	Min.	Distance from the Zenith.	The Altitude.	Min.	Distance from the Zenith.
0	31	90	11	7	79
1	23	89	12	6	78
2	17	88	13	6	77
3	14	87	14	5	76
4	13	86	15	5	75
5	12	85	16	4	74
6	11	84	17	4	73
7	10	83	18	3	72
8	8	82	19	2	71
9	8	81	20	2	70
10	7	80	21	1	69
11	7	79	22	1	68

The Use of this Table *is the same as that of* Ticho *and* Lansberg.

N   S. Can

S. Can you assure me by your own experience of the Refraction of the Sun and Stars?

T. I cannot assure you of it by my own experience (because my Voyages have been to the South) but the *Hollanders* can, having had a very sensible Example of it, in a Voyage they made to *Nova Zembla*, in the Year 1596. For being forced to Winter there because of the Ice (the Sea being Frozen) they saw the Sun 14 Days sooner than they should have done; the Refraction making the Sun appear above the Horizon, when they were sure (by his Declination, and the Height of the Pole) that he was yet under it.

S. If it be so, there is some reason to believe it, but I wish you would prove the Refraction by some Demonstration to take off all doubts, and convince me of that truth.

T. To satisfy you in that, put a Ring, Half a Crown or the like, into a Bason, and go backward until you can see it no more, then stand, and order somebody to fill the Bason with Water; which done, the Ring or Half-Crown will appear very plain to you, and that if you go yet farther backward; the reason is, because the Water is thicker than the Air, through which only you saw it at first; and therefore if Bodies under Water, seen through the several transparent substances of Water and Air, appear higher than they really are; you may thence imagin, that the Sun or Stars seen through the different transparent substances of the Heavens and Air may be refracted, so as to appear higher than they likewise really are.

S. Hath the Moon any Parallax?

T. Yes, and for my Part I should rather Counsel you to make use of the Table of the Parallax for her, than for the Sun; it being the opinion of several Modern Authors, that the Sun has no sensible Parallax (because of his great Distance from us) however I have given you the Tables of those Famous Men who are not of that mind, to try the truth of it by your own experience, when you come to places convenient for Observation, and whose Latitudes are well known.

A Table of the Parallax of the Moon according to the Observations of Ticho-Brahe.

Altitude.	Semidiameters of the Earth.										Distance from the Zenith.
	52	53	54	55	56	57	58	59	60	61	
	Distance of the Moon.										
	92028	93798	95568	97337	99107	100877	102647	104416	106186	107956	
	Parallax.										
	M.	M.	M.	M.	M.	M.	M.	M.	M.	M.	
0	66	65	64	63	61	60	59	58	57	56	90
3	66	65	64	62	61	60	59	58	57	56	87
6	66	65	63	62	61	60	59	58	57	56	84
9	65	64	63	62	61	60	59	58	57	56	81
12	65	64	62	61	60	59	58	57	56	55	78
15	64	63	62	61	60	59	58	57	56	55	75
18	63	62	61	60	59	58	57	56	55	54	72
21	62	61	60	59	58	57	56	55	54	53	69
24	61	60	59	58	56	55	54	54	53	52	66
27	60	58	57	56	55	54	53	52	51	51	63
30	58	57	56	55	54	53	52	51	50	49	60
33	56	55	54	53	52	51	50	49	48	48	57
36	54	53	52	51	50	49	48	48	47	46	54
39	52	51	50	49	48	47	47	46	45	44	51
42	50	49	48	47	46	45	45	44	43	42	48
45	47	46	46	45	44	43	42	42	41	40	45
48	45	44	43	42	42	41	40	39	39	38	42
51	42	41	41	40	39	38	38	37	37	36	39
54	40	39	38	37	37	36	35	35	34	34	36
57	37	36	35	35	34	33	33	32	32	31	33
60	34	33	32	32	31	31	30	30	29	29	30
63	31	30	29	29	28	28	27	27	26	26	27
66	27	27	26	26	25	25	25	24	24	23	24
69	24	24	23	23	22	22	22	21	21	20	21
72	21	20	20	20	19	19	19	18	18	18	18
75	17	17	17	16	16	16	16	15	15	15	15
78	14	14	13	13	13	13	13	12	12	12	12
81	11	10	10	10	10	10	9	9	9	9	9
84	7	7	7	7	7	6	6	6	6	6	6
87	4	3	3	3	3	3	3	3	3	3	3
90	0	0	0	0	0	0	0	0	0	0	0

A Table *of the Refraction of the Moon,* *according to the* Observations *of* Ticho-Brahe.

Altitude or Degrees from the Horizon.	Refraction. Min.	Distance from the Zenith.		Altitude or Degrees from the Horizon.	Refraction. Min.	Distance from the Zenith.
0	33	90		21	5	69
1	25	89		22	5	68
2	20	88		23	4	67
3	17	87		24	4	66
4	15	86		25	3	65
5	14	85		26	3	64
6	14	84		27	3	63
7	13	83		28	2	62
8	12	82		29	2	61
9	11	81		30	2	60
10	11	80		31	1	59
11	10	79		32	1	58
12	10	78		33	1	57
13	9	77		34	1	56
14	8	76		35	1	55
15	8	75		36	1	54
16	7	74		37	1	53
17	7	73		38	1	52
18	6	72		39	0	51
19	6	71		40	0	50
20	5	70		41	0	49

The Ufe of thefe Tables are the fame as thofe of the *Sun*, and therefore need no farther Explanation ; and for the reafon before-mentioned, I have omitted the Seconds.

P R O P.

P R O P. XXI.

Of Climes.

S. WHAT is a *Clime?*

T. It is a Space of the Earth Comprehended between two Parallels, so that on one side of it the longest Day in the Year surpasses by half an hour the longest Day in the Year on the other side.

S. How many *Climes* are there?

T. There are but 24 on each side of the Equinoxial; the furthermost of which ends at that part of the Earth where the longest Day in the Year is of 24 Hours.

S. In what Latitude is it that the longest Day in the Year is of 24 Hours.

T. It is in the Latitude of 66 Degrees 30 Minutes, or in the places situated right under the Polar Circles (otherwise called the Circles Arctick, and Antarctick) and no further; because the longest Day is not to be accounted any more by Hours, but rather by Days, Weeks, and Months, insomuch that they who dwell right-under the North-pole have six Months of Light or Day, whilst the Sun is in the Northern Signs or hath Declination North; and six Months Dark or Night, when the Sun is in the South Signs or hath Declination South; and contrary-wise they that dwell right-under the South-pole have six Months Day, the Sun being in the six Southern Signs, or having Declination South; and six Months Night, whilst the Sun remains in the six Northern Signs, or hath Declination North.

S. Are the six Months of Night as Dark as our Nights are commonly to us?

T. No, for as the Sun never goeth lower under the Horizon of those People than 23 Degrees (or thereabout) the Twilight or Dawning of the Day appears almost always, and there is one part of that time that it is considerably Light, when the Sun appears to them by Refraction, which happens when the Sun is near the Horizon.

S. How can you tell in what Climat a place is situated?

T. By Subtracting the length of the Equinoxial Day (12 Hours) from the longest Day of the proposed Place, and doubling the Remainder to reduce it into half hours, which will show in what Climat you are, or your place is situated: As for Example, Suppose that at *Cambridge* the longest Day in Summer be of 16 Hours 30 Minutes; if you Subtract 12 Hours from it, (the length of an Equinoxial Day) the Remainder will be 4 Hours and a half, (for 30 Minutes is half an hour) which being doubled makes 9 half hours, by which you know that *Cambridge* is in the
Ninth

Ninth Clime from the Equinoxial Northward, since *Cambridge* has North Latitude, and the Climes are accounted as the Latitudes, I mean from the Equinoxial towards the Poles.

S. How do you know the length of the longest Day of a place?

T. By the Latitude of the proposed place and the Suns greatest Declination, which being known, you may easily find out by the Globes and other ways, the length of the longest Day in the Year, which you will find to agree with the following Table.

A Table *showing the Longest Day in every Degree of Latitude.*

Latit. Degrees.	Longest-Day H.	M.	S.	Latit. Degrees.	Longest-Day H.	M.	S.
1	12	03	28	30	13	56	16
2	12	06	56	31	14	01	12
3	12	10	24	32	14	06	08
4	12	14	00	33	14	11	12
5	12	17	28	34	14	16	24
6	12	20	56	35	14	21	52
7	12	24	48	36	14	27	20
8	12	28	00	37	14	33	04
9	12	32	36	38	14	37	36
10	12	35	12	39	14	44	56
11	12	38	48	40	14	51	12
12	12	42	41	41	14	57	44
13	12	46	08	42	15	04	24
14	12	49	44	43	15	11	20
15	12	53	28	44	15	18	40
16	12	57	20	45	15	26	08
17	13	01	04	46	15	34	08
18	13	04	46	47	15	42	24
19	13	08	56	48	15	51	04
20	13	12	48	49	16	00	08
21	13	16	48	50	16	09	24
22	13	21	04	51	16	19	52
23	13	25	04	52	16	30	52
24	13	29	20	53	16	41	52
25	13	33	35	54	16	54	08
26	13	38	00	55	17	07	04
27	13	42	24	56	17	21	04
28	13	46	16	57	17	36	16
29	13	51	36	58	17	52	48

Latit.

Latit. Degrees	Longest-Day H.	M.	S.		Latit. Degrees	Longest-Day D.	H.	M.
59	18	10	48		74	96	17	00
60	18	30	36		75	104	01	04
61	18	53	20		76	110	07	27
62	19	18	24		77	116	14	22
63	19	48	40		78	122	17	06
64	20	24	24		79	127	09	55
65	21	10	32		80	134	04	58
66	22	20	40		81	139	31	36
					82	145	00	43
					83	151	02	06
					84	156	03	03
Degrees	Days	H.	M.		85	161	05	23
67	24	01	40		86	166	11	23
68	42	01	16		87	171	21	47
69	54	16	25		88	176	05	29
70	64	13	46		89	181	21	58
71	74	00	00		90	187	06	39
72	82	06	36					
73	89	04	58					

P R O P. XXII.

Of the Position of the Sphere.

S. IS the Sphere in the *same Position* to all People, in respect of the Horizon?

T. No, for to some People the Sphere is *direct*, to some others it is *oblique*, and to others it is *Parallel*.

S. When is it called a *Direct Sphere*?

T. It is when both the Poles of the World are at the Horizon, as in the following Figure, which sheweth how the Earth is posited to all those who dwell under the Equator, (or Equinoxial.)

Thus

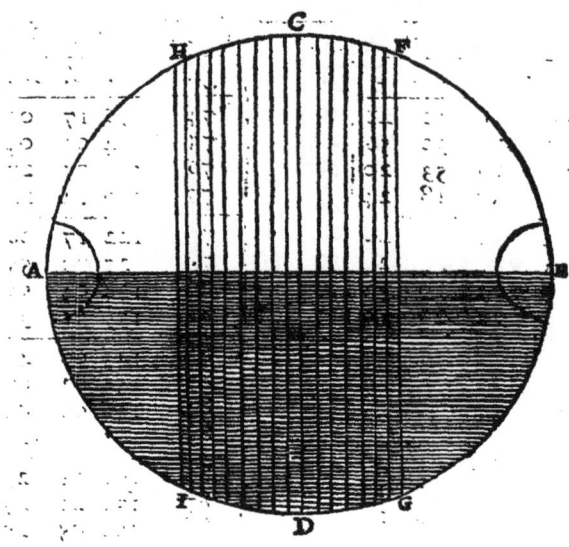

B is the North Pole.
A the South Pole.
A B the Horizon.
C D the Equinoxial.
F G the Tropick of *Cancer*.
H I the Tropick of *Capricorn*.   The Circles between H C, C F, and I D, D G, are the Parallels that the Sun deſcribes in making his Revolution; that is to ſay one every Day, at one Degree (of the Ecliptick) Diſtance one from another.   The Arches under the Horizon A B are the Night Arches; and thoſe above it, the Day Arches.

S. Why is it called a *Direct Sphere* ?

T. It is becauſe the Sun, Stars, and other Celeſtial Bodies, aſcend directly above the Horizon, and deſcend as directly under it.

S. What is there in a *Direct Sphere* to be taken notice of ?

T. You are to take notice of the equality of the Days and Nights, (that each of them is of 12 Hours) and the Sun, Stars, and other Celeſtial Bodies are as long above the Horizon, as they are under it.

S. When is it called an *Oblique Sphere* ?

T. It is when one of the Poles is elivated above the Horizon, and that the Horizon cuts the Equator (or Equinoxial) and all the Parallels Obliquely, ſlopingly, or at Oblique-angles, as in the following Figure, which ſhoweth how the Earth is poſited to thoſe Nations that dwell in an Oblique Sphere.

S. What

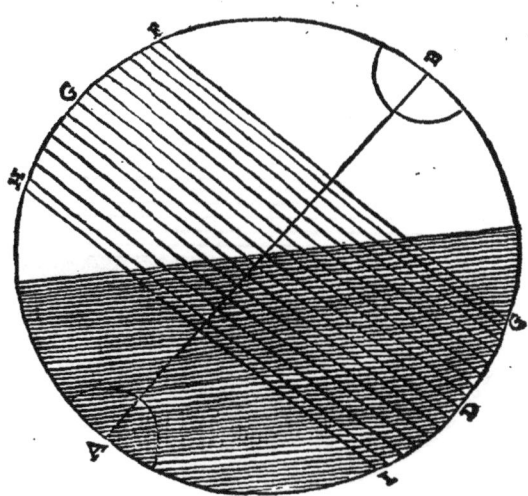

S. What muſt I take notice of in the *Oblique Sphere?*

T. You are to take notice that the Sun and Stars have (in reſpeƈt of the Horizon) Oblique and unequal Aſcentions, and Deſcentions, and that the Days and Nights are unequal (except when the Sun is at the Equinox.)

S. Why are not the Days and Nights unequal, as well when the Sun is at the Equinox, as in a Parallel?

T. It is becauſe the Equator, is not divided unequally by the Horizon as the Parallels are, but only into two equal Parts; from whence it followeth, that when the Sun is at the Equator, (or Equinoxial) he is as long under the Horizon, as above it; and therefore the Day and Night will be then of equal length.

S. Why is it called an *Oblique Sphere?*

T. It is becauſe the Sun, Moon, and Stars perform their Diurnal Motions in Circles Oblique to the Horizon.

S. When is it called a *Parallel Sphere?*

T. It is when one of the Poles is in the Zenith, and the other in the Nadir, and that the Equator (or Equinoxial) is Parallel to the Horizon; as in the Figure following, which ſhoweth how the Earth is poſited to thoſe whom we will ſuppoſe to dwell under the Pole.

O

S. What

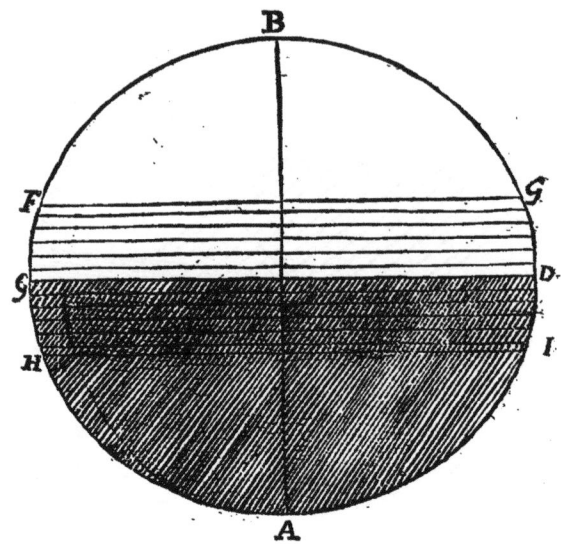

S. What do I learn by this *Parallel Sphere?*

T. You learn that, there is 6 Months of Day-light, and 6 Months of Night; for if it is Day all the time that the Sun is seen above the Horizon, you cannot doubt but from the 10th. of *March*, to the 12th. of *September*, it will be Day to those that dwell in a Parallel Sphere, under the Pole Arctick, (or North Pole) because the Sun is all that time on the North-side of the Equator, and is above their Horizon; and on the contrary, it will be Night to them from the 12th. of *September*, to the 10th. of *March*; because the Sun will be then on the South-side of the Equator, and under their Horizon, (except some few Days that the Sun appeareth by Refraction;) but to those who dwell under the South-pole (if any do) it is quite contrary, for they have 6 Months Day when the Sun hath Declination South, and 6 Months Night when he hath Declination North.

S. Why is it called a *Parallel Sphere?*

T. It is because the Sun and other Celestial Bodies (in the Diurnal Revolution of the Heavens) move Parallel to the Horizon.

P R O P.

P R O P.　XXIII.

Of Eclipses.

S. WHAT signifies the Word *Eclipse?*

T. It signifies to want Light, and to be darkened and hidden from our sight.

S. How many sorts of *Eclipses* are there?

T. Two sorts, one of the Sun, and the other of the Moon.

S. When is it that those Eclipses happen?

T. They happen when the Sun and Moon are at the same Time under the Ecliptick; or when they meet at the Head or Tail of the Dragon.

S. What do you mean by the *Head* and *Tail* of the *Dragon?*

T. I mean nothing else but those two Points on the Ecliptick under which the Moon passeth, in making her own proper Revolution; and therefore who understands well what the Equinoxes are, may very well understand what the Head and Tail of the Dragon are in the Ecliptick; for as the Equinoxes are only two Points in the Equinoxial which mark the intersection of two Circles; to wit, that of the Ecliptick and Equinoxial: So the Head and Tail of the Dragon are but two Points in the Ecliptick, which mark the Intersection of the Ecliptick and the Circle which the Moon describeth, called her *Deferent.*

S. What is an Eclipse of the *Moon?*

T. It is when the Sun and Moon are Diametrically opposed one to another, and that the Earth is exactly betwixt them both.

S. What is the cause of the Eclipse of the *Moon?*

T. It is the Earths coming between her and the Sun, for by that interposition the Moon is hindered from the Light of the Sun, from which she borrows her own Light, for she shines only by the Suns Light falling upon her; and therefore cannot shine when the Earth comes between and stops that Light. So that you are to reckon the Darkness of the Moon in an Eclipse to be the shadow of the Earth falling upon her.

S. When is it that the Sun and Moon are Diametrically opposed one to another?

T. It is when a Line drawn from the Center of the Sun to that of the Moon, passeth through the Center of the Earth.

S. Is not the Moon opposed to the Sun every time that she is in her Full?

T. Yes, but not Diametrically.

S. Do the Eclipses of the Moon happen always when she is in her Full?

T. Yes; and the Eclipses of the Sun only when the Moon is in Conjunction with him.

　　　　　　　　　S. When

S. When is the Eclipfe of the *Sun?*

T. It is when the Moon is between the Sun and the Earth.

S. Are the Eclipfes of any ufe in *Navigation?*

T. Yes, if by them the Longitude may be known, as fome would have it; but to tell you my opinion, I doubt very much of it, fince the moft Famous Aftronomers (as *Ticho-Brahe*, *Ptoleme*, *Longo-Montanus*, *Metius*, *Keplerus*, *Hortenfius*, *Regiomontanus*, *Lansbergius*, and many more,) do not only differ much one from another in their Obfervations, but are found in a confiderable error, by the known Diftance of feveral Places, as you may fee in the 12th. Book of *Fourniers Hydrographie*, where you will alfo find the caufe of that error, too long here to relate.

PROP. XXIV.

Of the Latitude of the fixed Stars.

S. WHAT is the *Latitude* of the *Stars?*

T. The Latitude of the Stars, is nothing elfe but the Diftance of any Star from the Ecliptick, (as the following Figure fhoweth,) and therefore thofe Stars which are Diftant from it Northerly, are faid to have Latitude North, but thofe which are Southerly, or betwixt the Ecliptick and South Pole of the Zodiack, have Latitude South.

S. Is it true that the Latitudes of the Stars never Change?

T. Altho the Latitude of the Stars be not alltogether unvariable or unchangable, it is fo little, that in 400 Years their change is infenfible, for as their greateft change (of Latitude) is but of fome Minutes, it becomes fenfible in its parts but in a very long time; which is the reafon that moft Authors write that the Latitudes of the Stars never change, (being not worth ones while to take notice of it.)

AB

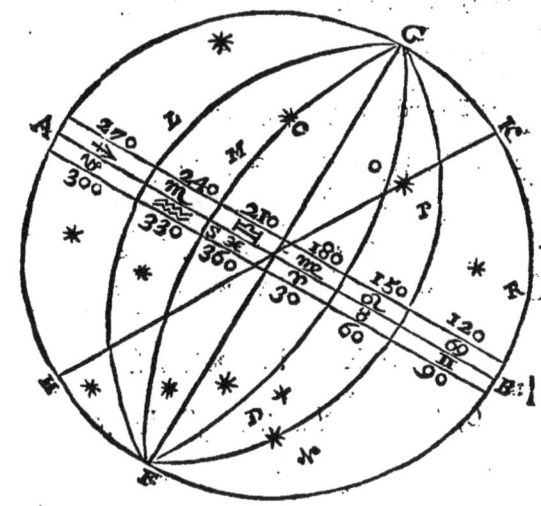

A B (in the middle or the Zodiack) is the Ecliptick.

K and H are the Poles of the World.

G F the Poles of the Ecliptick.

G A F, G L F, G M F, G O F, G P F, G R F, are the 6 Circles of position, which divide the Firmament into 12 equal Parts; by which we know under what Sign any Star is situated, (altho it be out of the Zodiack) for the Circles of Position pass through the beginning of every Sign, as this Figure showeth.

The Latitude of a Star is counted from the Ecliptick towards either of the Poles of the Zodiack, so the Arch C S, is the Latitude of the Star at C, the Longitude of a Star is to be counted in the Ecliptick; from the first Point of *Aries* to the Star, according to the succession of the Signs, that is to say, from *Aries* to *Taurus*; and so forward.

P R O P. XXV.

Of the Longitude of the Stars.

S. **W**HAT is the *Longitude* of the *Stars?*

F. The Longitude of the Stars is nothing else but the Distance of any Star from the Vernal Equinox or first Point (or beginning) of the Sign *Aries*; (according to the Order of the Signs;) or thus, it is an Arch of the Ecliptick Comprehended

hended between the firſt Point of *Aries*, and the Circle which paſſeth through the Body of the Star and Poles of the Zodiack, as you will better underſtand by the precedent Figure.

S. Do the fixed Stars Change their Longitude?

T. Yes, for it is the opinion of the beſt Aſtronomers that their Longitude increaſes 51 Seconds every Year; which is ſomething more than a Degree in 71 Years.

S. If it be ſo, the Table of the Stars Longitude ſhould be Corrected every 4th. Year if not oftener.

T. It is true, and I will ſhow you in its proper place how to Correct them, as well as the Tables of the Suns Declination.

PROP. XXVI.

Of the Declination of the Stars.

S. **W**HAT is the *Declination* of the *Stars?*

     T. The Declination of the Stars is the ſame as the Declination of the Sun, to wit, the Diſtance of any Star from the Equinoxial, either Northward or Southward.

S. Do the fixed Stars Change in Declination as well as in Longitude?

T. Yes, and that is the chief reaſon that the Tables of the Stars Declination muſt be Corrected from time to time; for the newer your Tables are, the more exact you will be in your Obſervations.

PROP. XXVII.

Of the Right Aſcention of the Sun and Stars; Oblique Aſcention; and Aſcentional Difference.

S. **W**HAT is the *Right Aſcention* of the *Sun*, (or *Stars?*)

     T. It is the Point (or Degree and Minute) of the Equinoxial which comes with the Sun or Star to the Horizon, and that Riſes (or goeth down) with it, in a Right Sphere.

S. What ſignifies *Aſcention?*

T. It ſignifies only to Aſcend or riſe up, for it is derived of the *Latin* Word *Aſcentio.*

S. Why

S. Why do you call it *Right Afcention?*

T. It is called Right; becaufe that in a Right Sphere (where the Horizon is Right) the Stars and other Celeftial Bodies, Afcend Right (or Perpendicularly) above the Horizon.

S. Doth the Right Afcention fignifie any thing elfe but the Point of the Equinoxial, which rifes with the Sun or Star in a *Right Sphere?*

T. No, for by the Right Afcention we mean nothing elfe but the Point or Degree of the Equinoxial, which paffeth by the Meridian with the Sun or Star, in what part of the World foever we be, as the following Figure fheweth.

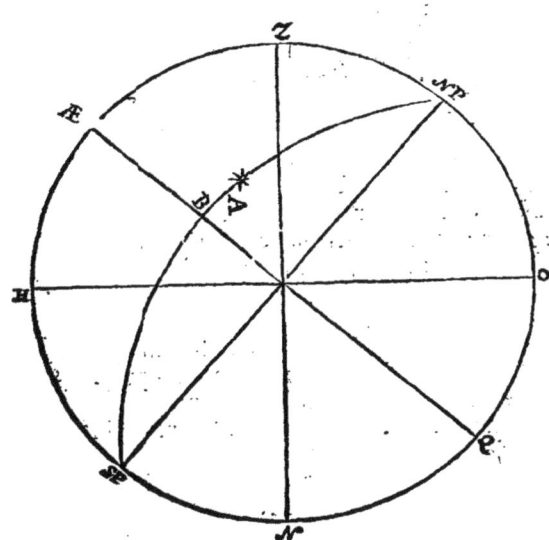

Wherein let the Sun or Star be conceived at A, then is the Point B (where a Meridian N A S paffing through the Point, cuts the Equator Æ B Q,) the Point of its Right Afcention : And the whole Arch upon the Equator intercepted between that Point and the beginning of *Aries,* is call'd the Right Afcention of the Sun or Star reprefented at A.

S. How can you call it Right Afcention where the Horizon is Oblique, and the Sun and Stars *Afcend Obliquely?*

T. Well enough, for in the Oblique Sphere, the Meridian does the Office of a Right Horizon, fince the two Poles are in the Meridian, and the Sun and Stars (carried by the *Primum Mobile*) cut or crofs the Meridian Perpendicularly, or at Right-angles, in the fame manner as when they Rife or Afcend above a Right Horizon; which is a very convenient
          thing

thing for Pilots and Aftronomers, becaufe they make their beft and chiefeft Obfervations when the Sun and Stars are upon the Meridian.

S. This would feem plain enough to fome People; but for my part I do not yet underftand it well, pray make it more intelligible if you can?

T. To make it plainer, and more eafie; you muft confider, that the Equinoxial and the Zodiack are two great Circles which cut one another into two equal parts, in the two Points called the Equinoxes, in the firft Points or beginnings of *Aries* and *Libra*; now it is from the beginning of (the Sign) *Aries*, that we begin the Right Afcention, according to the order of the Signs; that is, from Weft Eafterly: And therefore the Right Afcention of the Sun (or Stars;) is nothing elfe, but the Point or Degree and Minute of the Equinoxial, that comes to the Meridian with the Sun, (or Stars) which Degrees, are accounted from the beginning of *Aries*, to that Point; and as the Equinoxial reprefents to us the Time, and that by its Motion we reckon our Days, (fince a natural Day is but the Revolution of (all) the Equinoxial about the World) it followeth that all the parts of the Equinoxial pafs fucceffively one after another by our Horizon and Meridian; and therefore the Right Afcentions (which are but thofe Points of the Equinoxial, that pafs or come with the Sun or any Star upon our Meridian) come likewife to our Meridian one after another in the fame manner as all the Degrees of the Equinoxial do; that is to fay, 15 Minutes of a Degree in one Minute of an hour; one Degree in 4 Minutes of an hour; and 15 Degrees in an hour. So that for Example, a Star whofe Right Afcention differs from another of 15 Degrees, will come to our Meridian one hour fooner, or later than the other Star; and more or lefs proportionably, to the Degrees of Difference in their Right Afcentions.

S. As I underftand there is no great Difference between the Right Afcentions, and the Longitudes of the World.

T. If you underftand well what the Longitudes of the World are, you muft needs underftand what the Right Afcentions are: For I know no Difference at all; fince the Longitudes like the Right Afcentions are accounted upon the Equinoxial, and begin at the firft Meridian; which in the Right Afcention is half of the Colure of the Equinox which paffeth by the beginning of *Aries*; which Colure of the Equinox is alfo a Meridian, fince it paffeth through the Points of the Equinoxes, and the Poles of the World; from which the Right Afcentions go, always increafing, Eaftward untill they return to the firft Meridian or Colure of the Equinoxes, where we count 360 Degrees of Longitude or Right Afcention: And therefore if you can well diftinguifh the Longitude of a place; you muft needs likewife underftand the meaning of the Right Afcention of the Sun or of any Star.

S. What is the ufe of the Right Afcention of the Stars?

T. Its ufe is chiefly to find what time the Stars will come upon our Meridian, that by it we may obferve our Latitude; it ferveth alfo to find the hour of the Night, as fhall be fhown to you in the Fourth Book.

S. What

S. What is the *Oblique Afcention?*

T. The Oblique Afcention is the Point, or Degree of the Equinoxial, that rifes or goes down with the Center of the Sun, or Star, in an Oblique Sphere.

S. What is the *Afcentional Difference?*

T. The Difference Afcentional, is the Difference betwixt the Right and Oblique Afcention; that is to fay, the number of Degrees contained betwixt that Point of the Equinoxial that rifes with the Center of the Sun, or Star; and that Point of the Equinoxial that comes to the Meridian with the Center of the fame Star, (or Sun.)

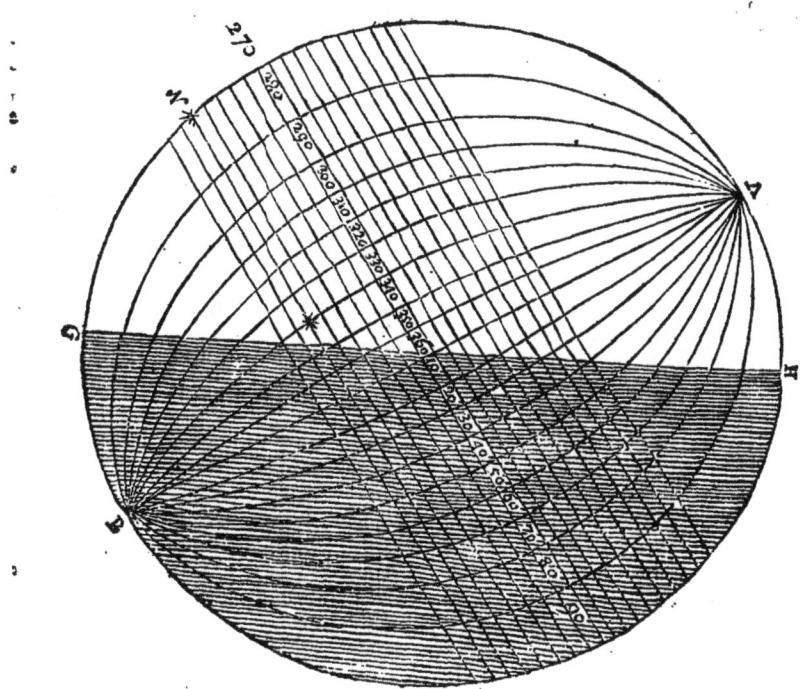

Admit that A N G B is the Meridian of the place you are at.

The Right Afcention of the Star at N, (on your Meridian) is 270 Deg. G H being the Horizon.

The Oblique Afcention of the fame Star at O, is 340 Degrees, and the Afcentional Difference of the fame Star is 70 Degrees, for if you Subtract 270 Degrees, its Right Afcention, from 340 its Oblique Afcention, there will remain 70 Degrees for the Afcentional Difference of the fame Star.

P R O P.

PROP. XXVIII.

Of the Stars, their Substances and Motions.

S. WHAT is the opinion of the Philofophers concerning the Subftance of the *Stars?*

T. Their opinion is that they are of the fame Subftance with the Heavens in which they are.

S. If it be fo, how come they to differ fo much from the reft of the Heaven in which they are placed?

T. It is becaufe the reft of that Heaven is not fo thick, and fo cannot receive and retain the Light of the Sun as the Stars can, by which means only they become vifible to us.

S. Why are not the Stars feen as well in the Day time, as in the Night?

T. It is becaufe they are darkened by the great Light of the Sun, (from whom they borrow their Light) for every body knows the greateft Light darkens the leaft.

S. Do not the fixed Stars move otherwife than the Heaven in which they are placed?

T. No.

S. If the Stars have no Motion of themfelves, how comes it then to pafs, that fome times they feem nearer to the Meridian, and at other times further off?

T. It is becaufe of the feveral Motions of the Firmament in which they are placed, and not that they change place; for as they are fixed in that Heaven, and always at the fame Diftance one from another, it is altogether impoffible.

S. What do you think of the *Sun, Moon,* and other *Planets,* do they change place or not?

T. Yes, they do; but not by their own Motions, for it is only through the Motion of the Heaven in which the Planet is fixed.

PROP. XXIX.

Of the chief Stars; their Magnitude; and into how many Conftellations they are divided.

S. HOW many Stars do Aftronomers take notice of?

T. Of 1377 Stars, of which 1241 appear in our Horizon, (according to the Catalogue left to us by *Tycho,* and other Ancient Aftronomers,) and 136 about the South Pole, which were

were obferved by *Hootman*, (an Inhabitant of *Sumatra* in the *Eaft-Indies*,) and not long fince by the Ingenious Mr. *Hally*, who corrected their Latitudes and Longitudes.

S. How did they come to the knowledg of thofe Stars?

T. By their Latitude and Longitude, which they did obferve with Inftruments made for that purpofe.

S. How did they do to know them another time?

T. They reduced them into Conftellations, and to fuch images, as to their fancy, fuch number of Stars together did beft reprefent, and then gave them Names to find them with more eafe another time.

S. Who were they that firft reduced the Stars into Conftellations and images?

T. The *Egyptians*, when they made the divifions of the Heavens as you have already read.

S. Can you prove the Antiquity of thofe Conftellations?

T. Yes, with eafe, fince in the 38 Chap. of *Job*, verf. 31 and 32, there is mention made of the *Pleiades*, *Orion*, and *Arcturus*.

S. Into how many Conftellations are thefe 1377 Stars reduced?

T. Into 62 Conftellations, of which Twelve are the Signs of the Zodiack, 23 more are in the North Hemifphere, and 27 in the South.

S. What are the Names of all thefe Conftellations?

T. Although I have already given you the Names of the Twelve Signs; yet in Anfwer to your Queftion, I fhall repeat them again, in the Method and Words of the Ingenious Sir *Jonas Moor* the Elder, in his Chapter of *Cofmography*.

The Conftellations on the North-fide of the Zodiack are 23, viz.

1. The *Leffer Bear* of 10 Stars, whereof 2 of the Second, 1 of the Third, 3 of the Fourth, 1 of the Fifth, and 3 of the Sixth Magnitude. This Conftellation is next to the North Pole, that Star in the Tip of the Tail will be but 2 Degrees and 14 Minutes from the Pole in the Year 1700, and will come nearer and nearer to the Pole for about 400 Years, when it will be within half a Degree of it, and then it will depart from it again. This is called the *Pole-ftar*, the Sea-ftar, becaufe obferved by Mariners, and is of the fecond Magnitude. You muft obferve, that both in the *Greater Bear*, and in this of the *Leffer*, there are in either of them a Wain, called by us *Charles's-Wain*, made of 7 Stars each, (which is the firft thing you muft learn) in both which the Wain is fancied by 4 Stars, (the two lowermoft whereof reprefent the Wheels) and the Horfes by 3; the Fore-horfe is reprefented for the Pole-ftar, and the Brighteft in the Wain, is called by Sea-men the Brighteft of the *Guards*, and thefe are to be perfectly known. To find the Place of the Pole by this Conftellation, you may fancy the Pole-ftar and the next Horfe to make an Equilateral Triangle with the Pole, on that Part towards the Bright-ftar called the *Guard*, and it will point near the Pole it felf.

P 2   2. The

2. The *Great Bear* of 35 Stars, whereof 7 of the Second, 3 of the Third, 12 of the Fourth, 8 of the Fifth, and 5 of the Sixth. Of thofe 4 which make up the Wain, that in the Bear's fhoulder is called *Dubhe*, the Thill-horfe is called *Alioth*, they are all of the fecond Magnitude. Obferve, that the two Stars called the *Guards* of the Greater Bear, and by imagining a Line to be extended by thofe two Stars, you will find the Pole-ftar; and alfo, that the Pole it felf lies between the Thill-horfe *Alioth* and the Pole-ftar.

3. *Draeo*, or the *Dragon*, a Conftellation of 35 Stars, that lies wreathing betwixt the two Bears; it has but 1 Star of the fecond Magnitude, which follows the laft but one in the Tail, it hath 10 Stars of the Third, and is notable, becaufe it hath Stars in every one of the Twelve Signs, and for that the Pole of the Ecliptick lies in the very middle of this Conftellation.

4. *Cepheus*, a King of *Ethiopia*, a Conftellation that has not any noted Star either of the Firft or Second Magnitude in it, it contains 21 Stars.

5. *Bootes*, the Keeper of the Bear, or *Arctophylax*, has in it 32 Stars, whereof 1 is of the Firft Magnitude betwixt his Legs, called *Arcturus* by the *Greeks*, and *Arimech* of the *Arabs*, a noted Star.

6. The *Northern Crown*, or *Ariadne's Crown*, has in it 21 Stars, whereof 1 is of the Second Magnitude, called the *Bright Star* in the Crown.

7. *Hercules*, with his *Club*, watching the Dragon, contains 62 Stars, whereof none of the Firft or Second Magnitude, there are 9 of the Third, whereof that in his Head called *Ras Algethi* is the moft noted.

8. The *Harp*, or *Vultur Cadens*, of 15 Stars, whereof 1 is of the Firft Magnitude, called *Lucida Lyra*, or the Bright Star in the Harp.

9. The *Swan*, of 40 Stars, whereof 1 is of the Second Magnitude, near the Tail.

10. *Caffiopea*, who was the Mother of *Andromeda*, and fits here in her Chair, fhe has in her Breaft a bright Star of the Third Magnitude, called *Scheder*; there are in this Conftellation only 28 Stars, according to *Baierus* and others.

11. *Perfeus*, the Son of *Danae*, cleared *Andromeda*, and brought away *Medufa*'s Head; it contains 42 Stars, whereof 2 are the Second Magnitude, one in his Left fide called *Algenib*, the other in *Medufa*'s Head called *Algol*, the reft are of the Fourth, Fifth, and Sixth Magnitudes.

12. *Auriga*, the Carter, of 40 Stars, whereof one at his Back of the Firft Magnitude, called the *Goat-ftar*, *Hircus*, and *Capella*.

13. *Serpentarius*, that holds the Serpent, contains 30 Stars, whereof one of the Second Magnitude in his Head.

14. The *Serpent* of 35 Stars, whereof one only of the Second Magnitude in its Neck.

15. *Sagitta*, or the *Dart*, of 8 Stars, but none of any confiderable bignefs.

16. The

16. The *Eagle*, or *Flying Vulture*, of 27 Stars, whereof only one is of the Second Magnitude in its Neck, called *Vultur volans*, or *Aquila*.

17. The *Dolphin*, of 10 small Stars, none of the First or Second Magnitudes.

18. The *Lesser Horse*, containing 4 Stars of the Fourth Magnitude.

19. *Pegasus*, or the *Flying Horse*, a fair Constellation of 23 Stars, whereof 4 are of the Second Magnitude, that in the Tip of the Wing is called *Markab*, these said 4 Stars make a Square.

20. *Andromeda*, or the Chained Woman, Freed and Married to *Perseus*, containing 27 Stars, whereof 3 are of the Second Magnitude, the first in the Head, the second in the Girdle, and the third in her Leg.

21. The *Triangle*, of 6 small Stars.

22. *Bereniee's Hair*, of 13 Stars, all of small Magnitudes.

23. *Cor Caroli*, a small Constellation, formerly *informes*, added by the Worthy and Loyal Knight Sir *Charles Scarbrough*, of 3 Stars, situate betwixt the *Great Bear* and the last Constellation *Coma Berenicis*, whereof that of the Second Magnitude is called *Cor Caroli* in Memory of King *C H A R L E S* the Martyr.

The Constellations in the Zodiack are 12 *, viz:*

1. *Aries*, the *Ram*, the Leader of the Flock, containing 19 Stars; that which is most noted, is that in his Ear of the Third Magnitude, from whence many Astronomical Tables were formerly Calculated, and from this *Copernicus* accompted the Procession of the Equinoctial.

2. *Taurus*, the *Bull*, containing 48 Stars, whereof one in the Bull's Eye is of the First Magnitude, called *Aldebaran*, and by the *Romans*, *Palilicium*; and another in the Tip of his Horn is of the Second Magnitude. This great Constellation has two smaller Constellations belonging to it, 1. the *Pleiades*, or Seven Stars, in the Bull's Neck; sometimes they are called *Vergilia*, because of their Cosmical Rising in the Spring: 2. *Hyades*, which are Five Stars near the Bull's Eye, called sometimes *Sucula*.

3. The *Twins*, or *Gemini*, a Constellation of 34 Stars, whereof 3 are of the Second Magnitude, the first preceding in the Head is called *Castor*, that in the Neck following is called *Pollux*, and the third is in the Foot.

4. The *Crab*, or *Cancer*, containing 32 Stars, two of them of the Third Magnitude, the rest of the Fourth, Fifth, and Sixth.

5. The *Lion*, *Leo*, containing 43 Stars, whereof two are of the First Magnitude, *viz.* the Lions Heart or *Regulus*, and that in the Extremity of the Tail called *Cauda Leonis*, very fair Stars; and two of the Second Magnitude, *viz.* that in the middle of the three in his Neck, and that on the top of his Loins.

6. The *Virgin*, *Virgo*, hath 45 Stars belongs to her, and one considerable of the First Magnitude, in the Virgin's-Left-hand, called *Spica Virginis* or *Vindemiator*.

7. The

7. The *Ballance*, *Libra*, containing 14 Stars, whereof two are of the Second Magnitude, *viz.* one in the Southern Scale called *Lanx Meridionalis*, the other in the very End of the Handle called also *Lanx Septentrionalis*.

8. The *Scorpion*, containing 35 Stars, one of the First Magnitude in the Body called *Cor Scorpionis*, and of the Second in the Head.

9. *Sagittarius*, or the *Centaur*, hath 30 Stars in it, two whereof are of the Second Magnitude, one in the Knee of his Right-leg, and the other in the Heel of the same Leg.

10. *Capricornus*, containing 28 Stars, but none of them either of the First or Second Magnitude.

11 *Aquarius*, having 42 Stars in it, but none of any considerable Magnitude.

12. *Pisces*, the *Fishes*, have 36 Stars in them, but none of them either of the First or Second Magnitude.   ➤

The Constellations on the South side of the Zodiack, are,

1. The *Whale*, or *Cetus*, a Constellation of 29 Stars, whereof two are of the Second Magnitude, one near his Mouth, and another near the Tail.

2. *Orion*, a most noted Constellation of 56 Stars, whereof there is one of the First Magnitude in his Left-shoulder of a ruddy colour, and another of the same Magnitude in his Right-foot called *Rigel*; there are four of the Second Magnitude, one in his Right-shoulder, and three in his Girdle in a straight Line called the *Yard Wand*: There are two in the Shoulders before-mentioned, two in his Feet, three in the *Yard Wand*, and three below in the Sword, which fashion this great Warriour, and are very notorious.

3. *Eridanus*, or the *River*, of 44 Stars, in the Extremity whereof one is of the First Magnitude called *Enar*, the rest are small ones.

4. The *Hare*, *Lepus*, of 13 Stars, all small ones.

5. The *Great Dog*, *Canis major*, a Constellation of 19 Stars, whereof that in his Mouth is of the First Magnitude; a great sparkling Star called *Sirius*, and one near his Left-knee is of the Second Magnitude; the rest are small Stars.

6. The *Little Dog*, *Canis minor*, of 10 Stars, whereof one in his Belly called *Procyon* is of the first Magnitude, the rest are small.

7. The *Ship*, or *Argo Navis*, a Constellation of 51 Stars, whereof one in the Rudder called *Canopus* is of the First Magnitude, and there are seven Stars of the Second dispersed in this Constellation.

8. *Centaurus*, or the *Centaur*, a Constellation of 41 Stars, wherein there are two of the First Magnitude, one in his Left-thigh, and another in the Extremity of his Right-foot; there are five of the Second Magnitude, the rest small.

9. *Crater*, the *Goblet*, is a small Constellation of 11 Stars.

10. Corvus,

10. *Corvus,* the *Crow,* another fmall Conftellation of 8 little Stars.

11. *Hydra,* the *Serpent,* containing 29 Stars, whereof one is of the Firft Magnitude called *Alphard* in the third Wreath, and is fometimes called *Cor Hydra;* the reft are fmall.

12. *Lupus,* the *Wolf,* a Conftellation of 20 Stars, all fmall.

13. The *Altar, Ara,* of 6 fmall Stars.

14. The *Southern Crown, Corona Meridionalis,* of 13 fmall Stars.

15. The *Southern Fiſh, Piſcis Notius,* of 12 Stars, whereof one in the Mouth called *Fumahant* is of the Firft Magnitude, the reft are fmall Stars.

16. There are 12 Conftellations more towards the South Pole, *viz.* 1. The *Peacock, Pavo;* 2. *Toucan;* 3. *Grus;* 4. *Phenix;* 5. *Dorado;* 6. *Piſcis volans;* 7. *Hydra;* 8. *Chameleon;* 9. *Apis;* 10. *Apis Indica;* 11. *Triangulum;* and 12. *Indus.*

There is befides to be noted, the *Milky-way,* defcribed upon the Globe round about, and feveral other little Clouds, or white Spots, the which viewed by a good and long Telefcope are found to be very many fmall Stars together, and infinite in number, fo clofe, that with the bare Eye they difappear, and feem to be a fmall Cloud and white Way. So that the Number of the Stars mentioned in the former Conftellations are not all, nor it may be not the Thoufandth part of the Stars; for as fome Eyes may fee more of them than others, fo by Glaffes ftill longer than others more are feen, and may be almoft accompted infinite.

PROP. XXX.

Of the Magnitude of the Stars, and their Proportions to the Earth, how great their viſible Diameters are, and which of them ſhould chiefly be known by Pilots, &c.

S. **H**OW do Aftronomers determin or agree concerning the Magnitude of the Stars?

T. They agree by the Comparifon or Refpect they have one to another, (being impoffible to know it otherwife, becaufe of the exorbitant Diftance they are from us,) and this is their way. They fay that the Stars of the firft Magnitude are thofe which are moft confiderable, amongft which, there is fome which gives a greater light than others; as for Example, *Sirius* in the Mouth of the *Great Dog* is much Brighter than the *Bulls Eye,* call'd by the *Arabians, Aldebaran;* altho both are faid to be of the Firft Magnitude: Next to thefe, thofe which appear a little lefs Bright or give leffer Light, are faid be of the Second Magnitude; then follow thofe yet a little leffer, or one fize inferiour to the fecond for the Third Magnitude; and fo the Stars Gradually decreafe unto the Sixth Magnitude, which is the fmalleft of all, except fome few called *Nebula* or Dark, (becaufe they are hardly feen when the weather is moft Clear,) and fome other which cannot be feen but with a Telefcope or Perfpective Glafs.

S. What fay thofe who compare the Stars with the Earth, to know their Magnitude?

T. They fay that the Stars of the firft Magnitude contain the Globe of the Earth . . . . 107 Times.

Thofe of the Second Magnitude	90 Times.
Thofe of the Third Magnitude	72 Times.
Thofe of the Fourth Magnitude	54 Times.
Thofe of the Fifth Magnitude	36 Times.
And thofe of the Sixth Magnitude . . .	18 Times.

S. How great is their vifible *Diameters?*

T. According to the Obfervations of the Famous *Ticho-Brahe*, the vifible Diameter of the Stars of the Firft Magnitude is but of 2 Minutes, and yet there is fome whofe Diameter wants fifteen Seconds of it.

Diameter of the Second Magnitude . . . . . 1 Minute.

Thofe of the Third Magnitude . . . 1 Minute, and 1 Min. $\frac{1}{12}$ for fome.

Thofe of the Fourth Magnitude . . . . . $\frac{1}{4}$ of a Minute.

Thofe of the Fifth Magnitude half a Minute or 30 Seconds.

Thofe of the Sixth Magnitude the third of a Minute or 20 Seconds.

But this is no Article of Faith, and you are not obliged to believe it. For fince by Telefcopes 'tis found that thofe of the Firft Magnitude do exceed 5 or 6 Seconds in Diameter.

S. What Stars fhould a Pilot know?

T. A good Pilot ought to know moft of the Stars of the Firft and Second Magnitudes, named in the following Fifth Book, or at leaft thefe few:

The Pole Star.
The Brighteft of the Guards.
The lower of the Pointers.
The Bull's Eye, *Aldebaran.*

The

The Left-foot of *Orion, Rigel.*

The Great Dog *Sirius.*

The Little Dog *Procyon.*

The Whales Tail. I mean the Brighteſt and moſt Northern.

The *Hydra's* Heart.

The skirt of *Bootes, Arcturus.*

The Eagles Heart.

The Lions Heart, *Regulus.*

The Virgins Spike.

S. How ſhall I do to come to the knowledge of theſe Stars?

T. You may know them by the help of a Celeſtial Globe, but the ſureſt and beſt way for you is to get ſome pretender to Aſtronomy to ſhow them to you; I mean thoſe Maſters that teach it, which for a ſmall preſent will grant you your requeſt, if they ſee you are Ambitious of Learning, and deſirous to be an Artiſt: This I adviſe you to, if you cannot learn it aboard from your Captain, Pilot, or any other.

S. Is this all you have to ſay to me at preſent concerning the Stars?

T. No, for beſides this you muſt be informed concerning the Croſiers, (or the Croſs) which are Four Stars in the South Hemiſphere, of great uſe to Navigators, in their Voyages to the *Eaſt-Indies,* when they loſe the North Star: And therefore take notice of this their Figure.

$$* \\ B$$

$$* \qquad * \qquad *$$

$$Cox * Foot. \\ A$$

And that the time for Obſervation is when the Star A is Perpendicularly right under the Star B.

S. What Star muſt I then make uſe of, for my Obſervation?

T. You muſt make uſe of the Star A, (called Cox-foot) or the lower-moſt of the Four.

S. Why that Star?

T. Firſt, becauſe it is the neareſt to the Pole; and Secondly, becauſe it is right over it.

Q   *S.* How

S. How many Degrees is it then Diftant from the Pole?

T. It is now Diftant from the South Pole 28 Degrees 43 Minutes, (as the Table of the Stars Declination fheweth) and therefore you will eafily know by it, if you have paffed the Equinoxial or not, for if its Meridional height is greater than 28 Degrees 43 Minutes, you may be fure that you have paffed the Equinoxial towards the South; but if it be lefs than 28 Degrees 43 Minutes, fo much as it wanteth of it, fo much will your Ship be to the North of the Equinoxial; but if it be 28 Degrees 43 Minutes, you may conclude that you are under the Equinoxial.

S. In what Conftellation is this Star?

T. It is in the fame Conftellation that the reft of the Groffers are, to wit, in *Sagittarius* or the *Centaur*.

The End of the Second Book.

THE

THE
Compleat ART
OF
NAVIGATION.

THE THIRD BOOK.

The Practical Part of Navigation.

PROPOSITION I.

Of the Sea Compass, That excellent Instrument of Navigation.

S. WHAT have you to say of the *Sea Compass* ?

T. My design being to show you only the Essentials of Navigation for practice: I shall not trouble you with a long Discourse of the *excellency* and *Antiquity* of the Sea Compass, and who first invented it; but will presently begin with those things which a good Pilot must or ought to know of it, since it is the chief Instrument they have at Sea; and therefore, I say that in a Sea Compass as in other Instruments, there are three things which an Artist should know : First if it be *good*, Secondly its *Use*, and Thirdly its *Defects*, and how to Correct them.

S. How shall I know if the Compass be good ?

T. You are first to examin the Fly, or Card of it, to make sure that the Flower-de-luce points right North, I mean that the Needle or Wyer be

Q 2   exactly

exactly under the Line of North and South, which you may eafily know by thrufting a Pin at the end of the Wyer through the Card, and if it goeth out at the Point of the Flower-de-luce, you may affure your felf that it points right North, (variation excepted) and that all the other Rhumbs are as they fhould be; you are alfo to fee, that the focket be not rufted, and that it be placed right in the Center or very middle of the Card, leaning neither on one fide nor other.

S. Is the Point of the Needle placed always under the Flower-de-luce of the Card, (or Fly ?)

T. No, for in fome Compaffes it is placed more Eafterly, and in fome other more Wefterly, to Correct the Variation of particular places, and for the eafe or conveniency of ignorant Pilots; but he that will be an Artift, muft have a care of thofe things, and muft find out the Variation by his Obfervations, and Correct it himfelf, according as he finds it to increafe or diminifh: Befides, as the Variation Changes, thofe kind of Compaffes that have ferved at one time, may not ferve at another, therefore you cannot be too careful to prevent thofe Errors, by examining the Fly or Card of your Compafs, and Correcting its defects your felf.

S. What is the next thing to be examined?

T. The next is the Pin of your Compafs, for you muft fee that the Point of it be not blunt nor rufted, that the Card may Swim well upon it: The faid Pin is to be Perpendicular in the middle of the Box, (making Right-angles on all fides) and ought to be of Mettal fit for a Sea Compafs, that is to fay, of Brafs, Copper, or Lattin, for if it were of Iron, or Steel, I do not Counfel you to truft to it: And the fame is to be obferved of the Rings, which muft have a free motion; in fhort, you are to take great care there be no Iron at all, neither to the Compafs, nor to the *Abbitacle*, and that your Compafs be not placed near Iron Guns, or other Inftruments of Iron.

S. How fhall I know if the Pin be Perpendicular in the Center of the Box?

T. It is to be known thus, take with an ordinary Compafs the diftance from the Point of the Pin to the Circumference of the bottom of the Box, or the furthermoft Circle drawn upon it, then remove the Compafs fo open upon other Points of the fame Circle, or Circumference, and if the other Foot of the Compafs fall upon the Point of the Pin, you may affure your felf that it ftands Perpendicular as it ought to be; but if it do not fall out fo, you may believe that it is not well, and therefore you are to fet it right by drawing it towards you, or from you, until you find it to be Perpendicular, (this is the way alfo to know if the focket of your Card is right in the middle.)

S. What do you fay of the Needle?

T. As to the Needle, I fhall only advife you to fee it well touched your felf (before you go to Sea) by thofe Compafs-Makers which have the Reputation of having the beft Load-ftone, or that make the beft Compaffes.

S. What

S. What Metal are they to be of?

T. They are to be of the beſt Temper'd Steel, and are to be very bright or clean when they are touched, that they may receive the better the Attractive vertue of the Load-ſtone.

S. What doth the Fly of the Compaſs repreſent?

T. It repreſents the Horizon.

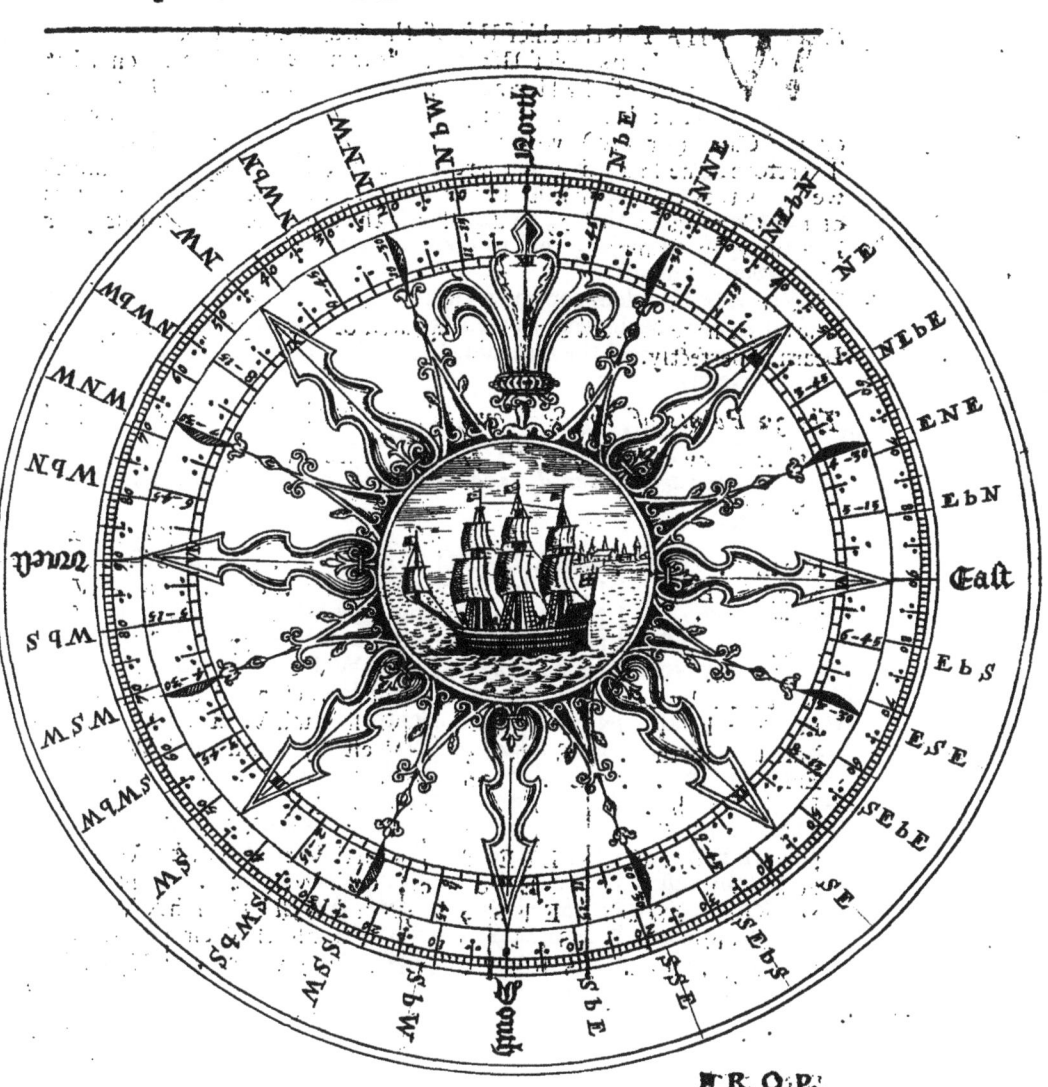

NROE.

PROP. II.

The Use of the Compass

S. WHAT is the chief Use of the Sea Compass?

T. Its chief Use is to show us at any Time, on what part of the Horizon, the *North, South, East, West,* and all other *Rhumbs* are; only by looking on the Lines drawn on the Card (or Fly) which represent to us the several Courses. Therefore, the first thing you are to Learn, is to Box your Compass well, (as they call it) that is, to name by heart all the Rhumbs or Points of the Compass in order, and to know them as soon as you cast your Eye on your Compass.

S. Who is he that does not know that?

T. Many young beginners, as well Sea-men as Gentlemen, for whose sake only I have inserted the foregoing Card of a Compass, that they may Learn it perfectly.

The 32 Points of the Sea Compass, with the Characters by which their Names are shorten'd, and what each Point is worth in respect of Time.

	Charact.	H.	M.		Caralters.
North	N	12	0	South	S
North by East	N b E	00	45	South by West	S b W
North North East	N N E	01	30	South South West	S S W
North East by North	N E b N	02	15	South West by South	S W b S
North East	N E	03	00	South West	S W
North East by East	N E b E	03	45	South West by West	S W b W
East North East	E N E	04	30	West South West	W S W
East by North	E b N	5	15	West by South	W b S
East	E	06	00	West	W
East by South	E b S	00	45	West by North	W b N
East South East	E S E	07	30	West North West	W N W
South East by East	S E b E	08	1	North West by West	N W b W
South East	S E	09	00	North West	N W
South East by South	S E b S	09	45	North West by North	N W b N
South South East	S S E	10	30	North North West	N N W
South by East	S b E	11	15	North by West	N b W

The

The Distance of the Rhumbs, or Points, from the Meridian

North.	South.	D.	M.	South.	North.	Points
		2	49			¼
		5	38			½
		8	26			¾
N by E	S by E	11	15	S by W	N by W	1
		14	4			1 ¼
		16	53			1 ½
		19	41			1 ¾
N N E	S S E	22	30	S S W	N N W	2
		25	19			2 ¼
		28	8			2 ½
		30	56			2 ¾
NEbN	SEbyS	33	45	SW by S	NWbN	3
		36	34			3 ¼
		39	23			3 ½
		42	11			3 ¾
N E	S E	45	00	S W	N W	4
		47	49			4 ¼
		50	37			4 ½
		53	26			4 ¾
NEbE	SEbyE	56	15	SWbW	NWbW	5
		59	4			5 ¼
		61	52			5 ½
		64	41			5 ¾
E N E	E S E	67	30	W S W	W N W	6
		70	19			6 ¼
		73	07			6 ½
		75	56			6 ¾
E by N	E by S	78	45	W b S	W b N	7
		81	34			7 ¼
		84	22			7 ½
		87	11			7 ¾
East	West	90	00			8

Besides this you muſt know, that one only Point of the Compaſs (miſtaken) will bring you 4 Leagues higher, or lower, than the deſired Harbour in 20 Leagues Courſe, 8 Leagues if the Courſe is of 40 Leagues, and 16 Leagues in a Courſe of 80 Leagues : You muſt know alſo, what Point of the Compaſs the Wind bloweth over; as for Example, if the Wind be North, it bloweth over the Flower-de-luce towards the South, and ſo of the reſt: You muſt know beſides how to ſet the Sun, and any Point of Land by your Compaſs.

S. What do you mean by ſetting the Sun, and any Point of Land?

T. I mean to ſee upon what Point of the Compaſs they bear from you.

S. Is this all?

T. No, for beſides that, you muſt know how the Ship Capes, that is to ſay, what Point of the Compaſs looks ſtrait forward to the Head of the Ship.

S. This is eaſie enough, and therefore I wiſh now that you would ſhow me by ſome Example, how I muſt chiefly make uſe of the Sea Com-paſs?

T. It is chiefly to be uſed thus : Suppoſe that you would Sail from one Harbour to another; as for Example, from *Dover* to *Diepe,* (in *France*) and that having conſulted your Card, you find that you muſt direct your Courſe South; you are accordingly to Steer away South, that is, upon that Point which is oppoſite to the Flower-de-luce, which Point you muſt always keep before you, in a Straight-line with the Head of the Ship, (as much as is poſſible,) for that Courſe will bring you to your deſired Haven, if you make it good.

PROP. III.

The Errors of the Sea Compaſs.

S. IS the Compaſs ſubject to any Errors when the Wyer is well touched with the Load-ſtone, and all its other Parts made and placed exactly according to your Directions?

T. Yes, for although it be made as well as Art can make it, and as well touched, the Needle is ſubject to Variation, which cauſes an Error of great Conſequence in Navigation; and therefore muſt be Corrected by the Pilot or any other that undertakes to guide a Ship.

S. Is the Compaſs (or rather the Needle of it) ſubject to any other Error beſides the Variation?

T. Yes, it is ſubject to incline to the Pole which is neareſt to it, but that is not of ſuch conſequence as the firſt, for you may ſoon Correct it, only by dropping a little Sealing Wax under your Card, on that ſide which is higheſt, (and that only as much as will ſerve to put it in its
Equilibrium,

Equilibrium, as it was at firft,) but it is not fo eafie with the Variation, which is the grand Error or Defect of the Compafs; and of fuch confequence, that you cannot deferve the Conduct of a Ship, except you underftand it very well: Therefore mind what followeth, and put it in Practice, whenever you go to Sea in long Voyages.

PROP. IV.

Of the Variation of the Compafs.

S. WHAT do you call the *Variation?*
 T. The Variation is nothing elfe, but the difference betwixt the true North of the World, and the North which the Wyer or Needle fhoweth; that is, the Degrees and Minutes which the North of the Needle (or Point of the Flower-de-luce) is diftant from the true North of the World.

S. What do you call the *true North* of the World?
 T. It is a Point in the Azimuth which paffeth through the North-pole of the World; I mean that Point of it, which is at the Horizon; which Point is always fix'd and immovable, in what Horizon foever we be, fince it is in the Azimuth which paffeth through the North Pole of the World, which Pole is immovable.

S. Is the Pole of the Magnetick (or North of the Load-ftone) fix'd, and immovable as well as the Pole of the World?
 T. No, the North of the Load-ftone, that is, the North, which the Needle of your Compafs fhoweth, is inconftant and variable, fince it is fome time on one fide of the North Point of the World, and fome time on the other, fome time more, and fome time lefs Diftant from it.

S. Can the Variation be called by any other Name that may be more intelligible?
 T. Yes, it may as properly be called the *Declination* of the Needle; fince the Variation of the Compafs is the fame in refpect of the Azimuth, or Point of it, which fhoweth the true North of the World; as the Declination of the Sun, is in refpect of the Equinoctial Line; to wit, its Diftance from it, fometime more, and fome time lefs; and fome time on one fide, and fome time on the other.

S. Is the Declination of the Needle, or Variation, of *two Denominations*, as the Declination of the Sun is?
 T. Yes, for when the Needle declines, or varies on the Eaft of that Azimuth, which fhoweth the true North of the World, we call it *Eafterly Variation*, but when it declines on the Weft fide of it, we call it *Wefterly Variation*.

R

S. I

S. I underſtand now very well what is meant by *Variation*, therefore, pray teach me next how to *find out*, or *obſerve* the Variation at Sea.

T. It is very juſt you ſhould know it, ſince without it, you can never be an Artiſt in Navigation, and therefore in order to that, I will firſt ſhow you the Uſe of the Azimuth Compaſs.

THE AZIMUTH COMPASS.

A D the great broad graduated Braſs Circle.

b c the Index, movable on the Point *b*.

b a is the ſight erected.

d e the Hypothenuſal Lute-ſtring or Thread.

B B the Braſs or Copper-hoops or Rings, which the Round Box hangs in.

C C the great Square Box that contains all the reſt.

P R O P.

PROP. V.

The Ufe of the Azimuth Compafs.

S. **B**EFORE you fhow me the Ufe of the Azimuth Compafs, pray tell me in what it differs from the Compafs by which we Steer?

T. It differs from it, only in thofe few things which you fee in the foregoing Figure; to wit, a broad Brafs Circle added to it, whofe half is divided (on the Limb) into 90 Degrees, numbred from the middle of the faid Divifions on both fides, with 5, 10, 15, 20, &c. unto 45 Degrees, which Degrees are alfo Subdivided into Minutes, by Diagonal Lines and Excentrick Circles, drawn as from A to D: There is alfo an Index, with a Sight erected on it, from the top of which, to the middle of the Index, is faften'd a filk Thread, or fmall Lute-ftring, whofe fhadow is to fall (in time of Obfervation) upon the Line on the middle of the faid Index; and befides that, there are two Strings, which by croffing one another at Right-angles, divide the broad Brafs Circle into 4 equal Parts, or Quadrants, and from the Termination of thefe Strings, are drawn 4 fmall Black-lines on the infide of the Box, which ferve to rectifie the Inftrument by the 4 Lines that are alfo drawn at Right-angles, on the Superficies of the Fly or Card.

S. To what is this broad Brafs Circle faften'd?

T. It is faften'd upon the fame round Box, wherein the Needle and Fly of your Compafs are.

S. Doth the Index move upon the Center or middle of the Compafs?

T. No, it moves on that part of the Brafs Circle, or Limb of it, from which the Degrees were drawn, becaufe that by it the Degrees come to be as large again as they would be, if it moved upon the Center of the Compafs.

S. What do you fay as to the ufe of the Azimuth Compafs, in time of Obfervation?

T. I fay, that you muft firft rectifie the Brafs Limb on the edge of the Box by the Needle and Fly within the Box, according as the Obfervation requires it; for if the Obfervation be in the Fore-noon, you muft put the Center of the Index upon the Weft Point of the Fly, fo that the four Lines on the edge of the Fly, and the four Lines on the infide of the Box do meet together: Your Compafs being thus rectified, turn the Index towards the Sun, untill the fhadow of the Thread falls exactly upon the Line, in the middle of the Index; and likewife, into the very flit of the Sight erected on it, and at the fame time the inner edge of the Index will fhow you the Degree and Minute of the Suns Magnetical Azimuth from the South, or North part of the Meridian, as you will better underftand by this

Example.

Suppose that your Compass being rectified for an Observation in the Fore-noon, the Index should cut 15 Degrees 30 Minutes, from the East Southerly. You may conclude, that the Magnetical Azimuth of the Sun is 74 Degrees 30 Minutes from the South Part of the Meridian, or else 105 Degrees 30 Minutes from the North Part thereof. But if the Index had cut 15 Degrees 30 Minutes from the East Northerly, then would the Azimuth be 74 Degrees 30 Minutes from the North, and 105 Degrees 30 Minutes from the South Part of the Meridian.

S. What must I do, if when the Compass stands in this position, the Azimuth of the Sun be less from the South Part of the Meridian, than the Degrees or Graduation reaches; that is to say, less than 45 Degrees?

T. Because that, in that case, the Compass would be useless, as it now stands; you must turn the Instrument just a quarter of the Compass, by placing the Center of the Index on the North (or South) Point of the Card, (according to the Suns position from you) and then the edge thereof will show the Degrees and Minutes of the Suns Azimuth as before: This is so plain that it needs no more Example; and I think he that understands well the use of the Azimuth Compass, when the Sun is on the East side of the Meridian, must needs understand it when the Sun is on the West side thereof, there being no more in it than to place the Center of the Index upon the East Point of the Fly, for in every thing else it is the same.

S. What must I do when I observe the Amplitude by the Azimuth Compass?

T. If you would observe the Amplitude at Sun Rising, turn first the Center of the Index exactly over the West Point of the Fly or Card, and then rectifie your Compass by the Lines within the Box, to the Lines on the Fly. Your Compass being thus prepared, stay 'till the Sun be half of his (appearing) Diameter above the Horizon, then looking thorow the Sight, turn the Point of the Index towards the Sun, until you cut the Body of the Sun with the Thread, (that is, until the Thread be so placed between your Eye and the Sun that it seems to divide the Body of the Sun into two equal Parts, from top to bottom,) and at the same time, the inside of the Index will show you the Degrees and Minutes of the Suns Magnetical Amplitude from the East, either Southerly, or Northerly. But when you will observe the Amplitude at Sun Setting, you must place the Center of the Index exactly over the East Point of your Compass, and proceed to observe as before, when the lowest part of the Suns Limb is yet half of the Suns Diameter above the Horizon: But more of this when I show you how to observe the Variation by the Amplitude.

PROP.

P R O P. VI.

How to obferve the Variation at Noon by the fhadow of the Sun.

S. WITH what Inftrument muft I obferve the Variation at Noon?

  T. You may obferve it with the Sea Compafs, by faftening a Thread over the Glafs and Center of the Fly, thus: When by your Obfervation you find that it is 12 of the Clock, or Noon, turn your Compafs untill the fhadow of the Thread falls exactly upon the Point of the focket, and if at the fame time, the fame fhadow falls upon the Point of the Flower-de-luce there is no Variation, but if it falls afide of it, you may conclude that there is Variation, of as many Degrees as the fhadow is Diftant from the Point of the Flower-de-luce.

 S. How fhall I know what fide the Variation is on?

 T. If the fhadow of the Thread falls on the Eaft fide of the Flower-de-luce, the Variation is Wefterly, but if it falls on the Weft fide of the Flower-de-luce, then the Variation is Eafterly; for if the Beam of the Sun, and the fhadow of the fame Beam, make but one and the fame Right-line, it is plain, that the Sun being exactly South, his fhadow will be exactly North. I acknowledge that the hour of 12, or Meridian Altitude of the Sun is known at Sea, only by the greateft Altitude of the Sun above the Horizon, and this might practically caufe fome error, becaufe that about Noon, one cannot for fome time perceive that the Sun Rifes or falls, or be fenfible that he changes place, altho it is certain he doth; wherefore, to obferve more exactly, you are to mind (or caufe to be minded) how much the fhadow all the time that the Sun feems to ftand, changes place upon your Compafs, and to take the middle betwixt the two extreams, for the true Meridional Line, (or North and South) and truft to the Variation that it marks. This is fo eafie that it needs no Example, but yet not fo certain as the ways which follow.

P R O P.

PROP. VII.

How to obferve the Variation by the North Star.

S. **WHAT** time muft I make my Obfervation at the *North-ftar*, for the Variation?

T. The beft time to obferve the Variation by the *North-ftar*, is when the Star is in the Meridian above or under the Pole, for then you will be fure to find the true North without any danger of miftake, and thefe are the affured marks, by which you fhall know when the *North-ftar* is at the Meridian.

The *Firft*, is when the Star called the *Knee* of *Caffiopea*, is exactly between your Zenith and the Pole-ftar, fo that holding up a Thread with a Lead at the end, the faid Thread will cut thofe two Stars, and fall upon the *firft* of the *Great Bear* or *Third-horfe* of *Charles's-Wain*, for then the *North-ftar* is in the Meridian, and above the Pole: This is alfo the time to obferve its Altitude, from which Subtracting 2 Degrees 4 Minutes, the Remainder will be the height of the Pole, which being the fame as the Diftance of your Zenith from the Equinoxial, fhoweth your Latitude.

The *Second* mark to know whether the North-ftar is in the Meridian, is when the Star in the *Knee* of *Caffiopea* is right under it, and the Third Horfe of *Charles's-Wain* above it, fo that your Thread meets or cuts thofe Three Stars, as before; for, then the North-ftar will be in the Meridian under the Pole, and if at that time you obferve what Degree of your Compafs the North-ftar is at, that Degree will be the true North, and the Arch Comprehended between the Flower-de-luce and that Degree, will be the Declination or Variation of your Compafs.

S. I can now eafily find out when the *North-ftar* is in the *Meridian*, above or under the Pole, but how fhall I know by it, the *Variation* of the Compafs?

T. You may eafily know it, with the Azimuth Compafs, thus: Look through the fight, and turn the Index towards the North-ftar, until you cut that Star with the Thread, and at the fame time the edge of the Index will fhow you the true North, and as many Degrees as you find the Index on the Eaft fide of the Flower-de-luce, fo many Degrees is the Variation Wefterly; or as many Degrees as the Index is Diftant from the Flower-de-luce Weftward, fo many Degrees is the Variation Eafterly.

Example.

Having obferved the Azimuth of the North-ftar when it was on the Meridian, and you find the Index upon 5 Degrees 30 Minutes from the Flower-de-luce Eaftward, and would know what that fignifies. I Anfwer

fwer, that it fignifies, that where you made your Obfervation, the Variation is of five Degrees 30 Minutes Wefterly, fince the Flower-de-luce of your Compafs declines 5 Degrees 30 Minutes to the Weftward of the true North of the World, which the edge of the Index fhoweth; but if the Flower-de-luce had been on the Eaft fide of the Index, then the Variation had been Eafterly as many Degrees, as the Flower-de-luce had declined from the edge of the Index Eaft-ward.

S. Suppofe I had no Azimuth Compafs, could I obferve (for a need) the Variation by our ordinary Sea Compafs?

T. Yes, you may, but then you muft divide firft the Limb or outmoft Circles of your Fly into 360 Degrees, (if it be not fo already) or only a quarter of it, to wit, 45 Degrees from the point of the Flower-de-luce Eaftward, and as many from the Flower-de-luce Weftward, which you may eafily do, by dividing each half quarter into three equal Parts, and each of thofe Parts into three equal Parts more, which will be of 5 Degrees each, and each of thefe being divided into five equal Parts more, make the 45 Degrees; your Compafs being thus made fit for your Obfervation, and a light held to it by fome of your Company, hold a Plummet fo between your Eye and the Compafs untill the Thread cover or hides from you the tip of the focket, and cuts the North-ftar; and the Degree which the Thread covers in the fame time fhows the Variation, which is Wefterly, if on the Eaft fide of the Flower-de-luce, or Eafterly, if on the Weft fide of it, as in the firft Example, but this is only in cafe of neceffity.

P R O P. VIII.

How to obferve the Variation by the Suns Altitude, (or his Diftance from your Zenith.)

S. **I**S it very difficult to obferve the Variation by the Suns Altitude?

T. No, not at all, for it is rather eafier to thofe that are ufed to obferve the Latitudes, their being nothing more in it, than to obferve the Suns height (or his Diftance from your Zenith) a little before Noon; for Example, an hour; and at the fame time look (as you were before taught) on what Degree of your Compafs the fhadow of the Thread falls, and fet it down upon your Slate, or Paper; after 12 of the Clock (or Noon) obferve again the Suns Altitude, untill you find it exactly as in the Morning; then mind in the fame time upon what Degree the fhadow of the Thread falls, and take the middle of thofe two fhadows or Points obferved upon your Compafs, (to wit, one before Noon, and the other after) for that will fhow you the true North of the World: Therefore if the Flower-de-luce declines (or varies) on the Weft fide of it, the Variation is Wefterly, but if it declines on the Eaft fide, the Variation is Eafterly.

Eaſterly. Now to know how much Variation there is, you muſt ſub-tract the Degrees which the ſhadow (of the Thread) ſhoweth in the Morning from thoſe Degrees which the ſhadow cuts in the Afternoon, and half of the Remainder will be the Variation, which will be Eaſterly, if the ſhadow in the Morning is nearer the Flower-de-luce than in the Afternoon; but on the contrary, the Variation will be Weſterly, if in the Afternoon the ſhadow is nearer the Flower-de-luce than in the Morning.

It may happen, when there is much Variation, that the two ſhadows obſerved before and after Noon, will fall both on the ſame ſide of the Flower-de-luce: In this caſe you muſt ſubtract as before, the leſſer Obſervation (of the ſhadow) from the greater, and the half of the Re-mainder being added to the leaſt of the two Obſervations will ſhow the Variation, which is always on that ſide that the Flower-de-luce is of.

S. What Compaſs do you uſe moſt for this Practice?

T. The ſame Compaſs with which you were taught to obſerve the Variation at Noon (which hath a Thread or Wyer over the Glaſs) and ſometime with the Azimuth Compaſs; as to the firſt, this is the way you muſt go to work.

In the ſame time that you are preparing to make your Obſervation for the Latitude your Ship is in, you are alſo to prepare your Compaſs, and before you begin to obſerve, ſee that the ſhadow of the Thread fall exactly upon the tip of the ſocket, (which is in the Center of the Fly) and let ſome body keep it ſo while you are obſerving, and when you have obſerved look in the ſame time what Degrees of your Compaſs the ſhadow cuts, (towards the North) minding how far it is from the Flower-de-luce, and on which ſide it is of.

Example.

Suppoſe that about 11 of the Clock in the Morning, you find by your Obſervation the Sun 50 Degrees above the Horizon, (which is his Altitude) and that in the ſame time the ſhadow of the Thread falls upon 30 De-grees from the Flower-de-luce Weſtward, which two Obſervations you write down to remember; then after Noon, when the Sun declines, having obſerved again untill you find your Croſs or Vane upon the ſame Point of your Staff or Quadrant, as in the Fore-noon; to wit, at 50 Degrees; and in the ſame time the ſhadow of your Thread upon 18 De-grees from the Flower-de-luce Eaſtward. If by that you will know how much Variation there is: Subtract the 18 Degrees obſerved in the After-noon, from the 30 Degrees in the Morning, and there will remain 12 De-grees, the half whereof 6 Degrees is the Variation. Now for to know which ſide it is on, ſince the 30 Degrees Weſtward are greater than the 18 Degrees Eaſtward, the middle of the ſhadows will be Weſtward, and the Flower-de-luce Eaſtward from it, and therefore the Variation is Weſter-ly 6 Degrees.

Example 2.

Suppose that by my Obfervation made alfo in the Fore-noon, fometime before 12 of the Clock, I find the Sun Diftant from my Zenith 29 Degrees 30 Minutes, and at the fame time the fhadow upon 9 Degrees 30 Minutes from the Flower-de-luce Weftward ; and that in the After-noon when the Sun is at the fame Diftance from my Zenith (that is to fay, at 29 Degrees 30 Minutes) the fhadow falls upon 27 Degrees 30 Minutes from the Flower-de-luce Eaftward, and would know by it how much Variation there is, and on what fide. Firft I fubtract the 9 Degrees 30 Minutes of the Fore-noon from the 27 Degrees 30 Minutes in the After-noon, and there remains 18 Degrees, whofe half 9 Degrees, is the Variation, which is Wefterly, becaufe the difference of the two Ob-fervations upon the Compafs is Eafterly, there being more Degrees Eaftward than Weftward, and therefore the Flower-de-luce being Wefterly from it we know that the Variation is Wefterly, fince the Variation is always on that fide that the Flower-de-luce is of.

Example. 3.

Suppose that in the Fore-noon my Obfervation had been of 42 Degrees 30 Minutes upon the Crofs Staff, (or Quadrant) and of 16 Degrees 45 Minutes upon the Compafs; to wit, from the Flower-de-luce Weftward ; and that in the After-noon I obferve the Sun at the fame Diftance from my Zenith, and the fhadow upon 3 Degrees 15 Minutes of my Compafs, likewife from the Flower-de-luce Weftward ; for to know by that how much Variation there is, I fubtract the 3 Degrees 15 Minutes of the Obfervation made in the After-noon from the 16 Degrees 45 Minutes ob-ferved in the Fore-noon, and the remainer is 13 Degrees 30 Minutes, the half whereof is 6 Degrees 45 Minutes, to which, I add the 3 Degrees 15 Minutes obferved in the After-noon, which makes 10 Degrees for the Variation, which is Eafterly, fince both Obfervations were Wefterly, and the Flower-de-luce Eafterly from it ; but if the Flower-de-luce (or North of my Compafs) had been Wefterly of both my Obfervations, (or both my Obfervations Eafterly from the Flower-de-luce which is the fame) then the Variation had been Wefterly as many Degrees as I found it Eafterly.

S. What muft I do, if I would make ufe of the Azimuth Compafs for this Practice?

T. You muft take care in preparing it, that the fhadow of the Thread falls directly upon the ftreight Line drawn in the middle of the Index, and alfo upon the very flit of the Sight, keeping it fo all the time that you are obferving the Suns Altitude, (or Diftance from your Zenith) and when you have obferved, look in the fame time what Degrees of the Compafs the edge of the Index cuts from the Flower-de-luce, minding on what fide it is of to write it down, and fo forth; obferving the fame Rules in every thing elfe as you did with the firft, in the three precedent Examples.

And *Note*, that thofe Obfervations are moft to be trufted to, which are made when the Sun is fartheft from the Meridian, provided he be not fo near to the Horizon as to fuffer any very confiderable refraction.

PROP. IX.

To find the Variation by the Scale and Compafs.

S. WHAT muft I do to find the Variation by the Scale and Compafs?

T. You muft firft obferve (with your Azimuth Compafs) the Magnetical Amplitude of the Sun, and by the Tables of Amplitude you muft find the Sun's true Amplitude for that Day: For thefe two things being known, you may with eafe find the Variation (and which way it is) thus:

With the Chord of 60 Degrees defcribe a Circle, and quarter it with two Diameters at Right-angles, at the Extremities of which write N for *North*, S for *South*; E for *Eaft*, and W for *Weft*; as in this following Figure: Then take with your Compafs the true Amplitude, and prick it from the Eaft or Weft Point as the occafion requireth, prick off alfo the Magnetical Amplitude, then take the Diftance between the Magnetical Amplitude and the North, and fet it from the true Amplitude Northward, and the Foot of the Compafs will fall in a certain Point, from which, if you draw a Line through the Center that Line will reprefent the North and South Line of the Compafs or Magnetical Meridian; and that way, that this Point lyeth from the North, that way is the Variation, which is juft fo much as is the Arch Comprehended between the faid Point and the North.

Example. 1.

Admit that in Latitude 46 Degrees North, the Sun having 20 Degrees North Declination: I obferve the Magnetical Amplitude at Sun Rifing to be 42 Degrees from the Eaft, Northward; and then would know the Variation of the Compafs.

Firft, with the Chord of 60 Degrees I defcribe a Circle, and having quarter'd it as before directed, with the 20 Degrees Declination, and 46 Degrees Latitude; I enter the Table of Amplitude, and find the true Amplitude to be 29 Degrees 30 Minutes, which I prick from E, the Eaft Point towards the North to G; and taking the Magnetical Amplitude 42 Degrees, I prick that alfo from E, Northward to H; then taking the Diftance between H and N, the North Point, I fet one Foot of my Compafs in G, the Point of the true Amplitude, and the other

being

being turned toward the North reaches to L , the Point to which the North Point of the Compaſs is directed, ſo that L, N, 12 Degrees 30 Minutes, is the quantity of Variation, which is *Eaſterly* ; becauſe L falls towards E , that is, between the North and the Eaſt.

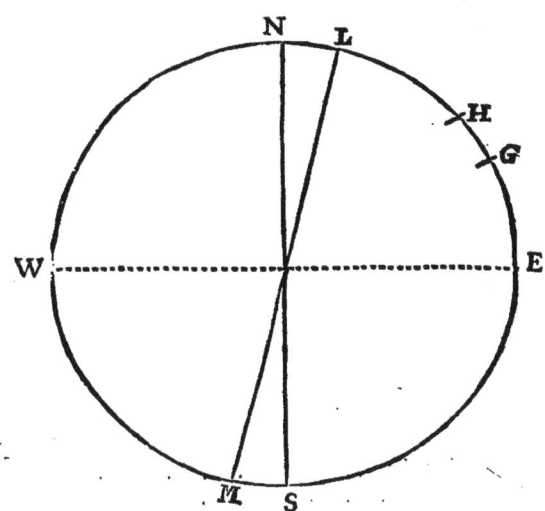

	D.	*M.*
Magnetical Amplitude	42	00
True Amplitude	29	30
Variation	12	30

Example. 2.

Admit that in Latitude 38 Degrees North , the Sun having 11 Degrees Declination South , the Magnetical Amplitude at Sun Set , is obſerved to be 35 Degrees from the Weſt Southward: *I demand what is the Variation, and which way it is?*

Firſt enter the Table of Amplitudes with the Latitude and Declination, as before directed ; and you will find the true Amplitude to be 14 Degrees 1 Minute, which being pricked from W, to G ; and alſo the Magnetical Amplitude to H, as before directed ; the Diſtance between L, N, ſhows the Variation to be 20 Degrees 59 Minutes to the Eaſtward, from the North.

Mag-

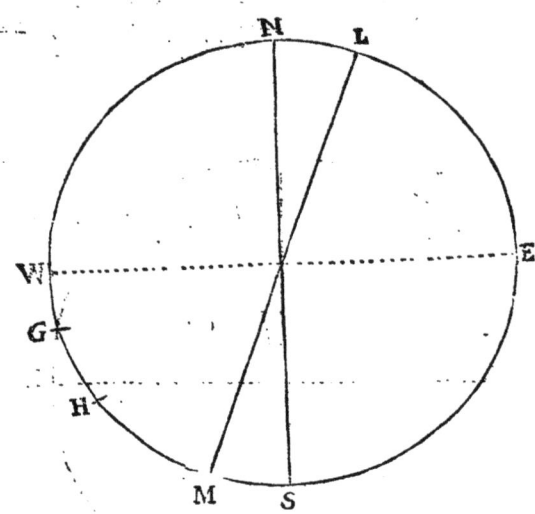

	D.	*M.*
Magnetical Amplitude	35	00
True Amplitude	14	01
Variation	20	59. *Easterly.*

Example. 3.

Admit the Latitude to be 43 Degrees North, the Suns Declination 18 Degrees North, and the Magnetical Amplitude at Sun Rifing 8 Degrees 30 Minutes South, and it be required to find the Variation of the Compafs.

Firft defcribe a Circle as before directed, and having quartered it, prick the true Amplitude (which by the Table of Amplitudes is found to be 25 Degrees) from E (the Eaft Point) towards the North to G; and the Magnetical Amplitude 8 Degrees 30 Minutes from E, Southward (according to the Obfervation) to H, then taking the Diftance between H the Magnetical Amplitude, and N the true North, fet it from G Northward, and it will fall on the Point L, then meafure the Diftance L, N, and you will find it to be 33 Degrees 30 Minutes for the Variation, which is Weftward, becaufe L falls between the North and the Weft.

Mag-

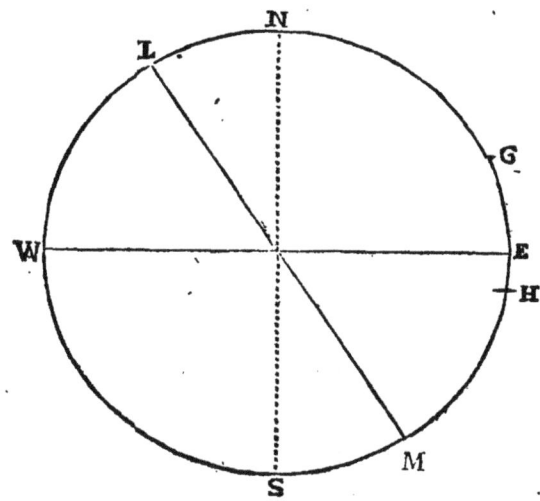

	D.	*M.*	
Magnetical Amplitude	8	30	
True Amplitude, add	25	00	
Variation	33	30	*Westerly.*

Example. 4.

Admit the Suns Declination to be 8 Degrees South, the Latitude 50 North, and the Magnetical Amplitude at Sun Rising 6 Degrees North, the Variation is required.

The true Amplitude by the Table is 12 Degrees 30 Minutes from the East Southward, because the Declination is South.

	D.	*M.*
Magnetical Amplitude	6	00
True Amplitude, add	12	30
Variation is Easterly	18	30

Example.

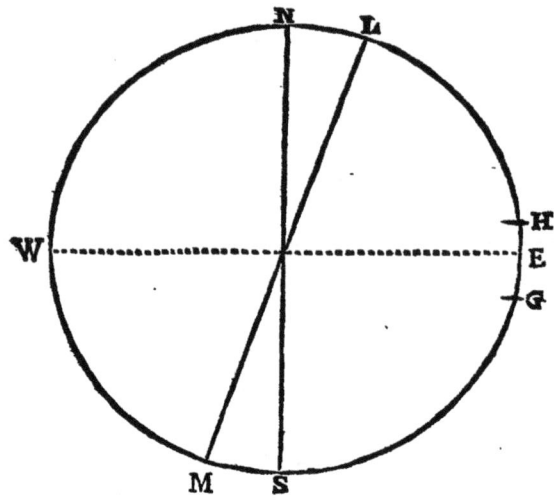

Example. 5.

Admit the Suns Declination to be 22 Degrees North, the Latitude 45 Degrees North, and the Magnetical Amplitude at Sun Setting 46 Degrees from the West Northward; the Variation is required.

First by the Table of Amplitudes, you will find the true Amplitude to be 31 Degrees 59 Minutes, which must be pricked from the West Point Northward, as in the next Figure: You must also prick down the Magnetical Amplitude, as before directed, and it will fall on the Point H, then take the Distance H, N, and placing one Foot of your Compass in G, (the Point of true Amplitude) the other Foot will fall on L; whose Distance from N being measured, (upon the Line of Chords) will show that the Variation is 14 Degrees Westerly, because L falls between the North and the West.

	D.	M.
Magnetical Amplitude	46	00
True Amplitude Subtracted	31	59
The Variation is Westerly	14	01

P R O P.

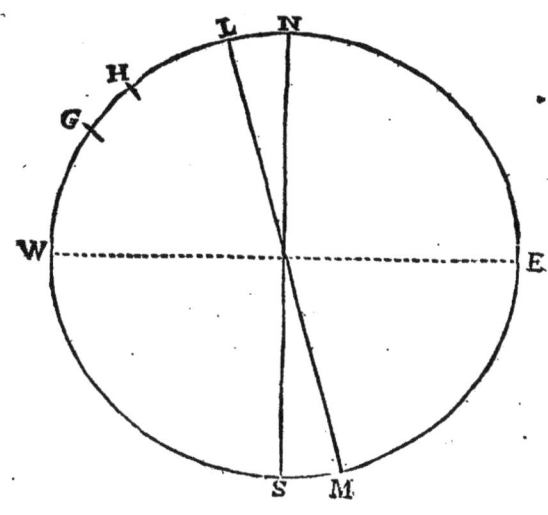

P R O P. X.

How to obſerve the Variation by the Suns Amplitude
Ortive and Occaſive

S. **H**OW ſhall I obſerve the Variation by the Suns Amplitude at
his Riſing?

    T. The way to obſerve the Variation by the Amplitude
Ortive of the Sun, (that is to ſay, at his Riſing) is firſt to prepare your Azimuth Compaſs by turning the Center of the Index, right
over the Weſt Point of the Fly, and rectifying it by the Lines within
the Box, to the Lines on the Fly, as its uſe ſhows you; then when the
Sun is one third part of his Diameter above the Horizon, (that is to ſay,
when the Sun is ſo high above the Horizon, that the Diſtance between
the Horizon and the lower Limb of the Sun, is equal to one third part of
his Diameter,) look through the Sight and turn the Index to the Sun,
untill you cut the Body of the Sun in the very midſt with the Thread
into two equal parts, and at the ſame time the edge of the Index will
ſhow you the Degrees of the Suns Magnetical Amplitude, either Northerly, or Southerly : Then with the Latitude of the Place, and Declination of the Sun that Day, find out the true Amplitude of the Sun, thus :
Look in the firſt Column of your Table of Amplitudes, for the Latitude
of

of the place you are in, which we will suppose 40 Degrees; follow that Line to the Right-hand, untill you are under the Suns Declination, for that Day, which we suppose 18 Degrees, (to be found on the top of your Table) and there you will find 23 Degrees 47 Minutes of Amplitude Ortive, that is to say, that in the Latitude of 40 Degrees the Suns true Amplitude at his Rising, is 23 Degrees 47 Minutes, when his Declination is 18 Degrees.

S. I underſtand now well enough how to find out the Amplitude, both Magnetical and by Calculation, but how ſhall I know by it how much Variation there is, and whether it be Eaſterly or Weſterly?

T. You may eaſily know it if you obſerve well theſe following Rules.

Firſt Principle.

ARTICLE I.

When you find by the Azimuth Compaſs, that the Sun Riſes or goes down on the ſame ſide that the true Amplitude is of, you muſt Subtract the Magnetical Amplitude (or Degrees and Minutes obſerved with the Azimuth Compaſs) from the true Amplitude found by your Table, or rather the leſſer from the greater, and the Remainder will be the Variation.

S. WHAT do you mean by the Sun's Riſing or going down on the ſame ſide that the true Amplitude is of?

T. I mean by the ſame ſide when (for Example) you find with the Azimuth Compaſs that the Sun Riſes or goeth down from the Eaſt, or Weſt Northerly, when his Declination is North; or elſe when he Riſes or goeth down from the Eaſt, or Weſt Southerly, when his Declination is South: But it is called the contrary ſide, when, for Example, the Sun Riſes or Sets from the Eaſt, or Weſt Southerly, when his Declination is North; or elſe when he Riſes or goeth down from the Eaſt or Weſt Northerly, when his Declination is South.

Example.

The 20th. of *May* 1684. my Ship being then in 50 Degrees of Latitude North, I obſerve the Suns Magnetical Amplitude, as before taught, (that is to ſay, with the Azimuth Compaſs, when the Body of the Sun is one third part of his Diameter above the Horizon) and find it to be 37 Degrees 54 Minutes from the Eaſt Northerly, the Queſtion is how much Variation there is?

To find by this Obſervation, how much Variation there is, I firſt look for the Declination of the Sun in one of the following Tables of Declination, &c. (in the 5th. Book) Calculated for that Year, and find that that Day at Noon, the Suns Declination is 22 Degrees North, then with the Latitude and Declination, I find by my Table of Amplitudes

that

that the Sun's true Amplitude is 35 Degrees 39 Minutes, which signifies that the Sun is to Rise at 35 Degrees 39 Minutes from the East Northerly, and neverthelefs my Compafs fhows that he Rifes at 37 Degrees 54 Minutes from the East Northerly: Therefore there muft be fome Error in the Inftrument, which did not fhow the true place the Sun did Rife at, becaufe of the Variation; which I find out, by Subtracting the 35 Degrees 39 Minutes of the true Amplitude from the 37 Degrees 54 Minutes obferved with my Azimuth Compafs, and there remains 2 Degrees 15 Minutes for the Variation.

A R T I C L E   II.

When you find by your Compafs, that the Sun Rifeth or goeth down on the fide contrary to his Declination, you muft add the Degrees and Minutes of your Obfervation, to the Degrees and Minutes of the Suns true Amplitude, and the whole will be the Variation.

Example.

THE Second of *October*, 1684. my Ship being then in 46 Degrees 20 Minutes of North Latitude; I obferve at Night the Suns Magnetical Amplitude, (when the lower part of him was about one third part of his Diameter above the Horizon) and find it to be 3 Degrees 45 Minutes from the Weft Northerly: The Queftion is, how much Variation there is? I Anfwer, fince on the Second of *October* the Suns Declination is 7 Degrees 48 Minutes South, in the Latitude of 46 Degrees 20 Minutes North, the Sun is to go down at 11 Degrees 20 Minutes from the Weft Southerly; I muft then add both Amplitudes together, (fince they are on contrary fides) and there will come 15 Degrees 5 Minutes for the Variation.

S. This I underftand, but I have often heard that it is difficult to know what fide the Variation is on, therefore you will oblige me to inftruct me well in that particular, that I may not fall into the fame Errors with thofe that take one fide for another?

T. Your requeft is very juft, and therefore I will give fome Rules concerning it, which muft needs make it very plain and eafie to you.

Second Principle.

A R T I C L E   I.

When by your (Azimuth) Obfervation you find that the Sun Rifes nearer the Flower-de-luce, than the true Amplitude, it fhows the Variation is Eafterly.

Example.

THE Twenty-ninth of *May*, 1684. my Ship being then in 42 Degrees of Latitude North, I obferved in the Morning the Sun's Magnetical Am-

T

Amplitude, (when the lower part of him was about one third part of his Diameter above the Horizon) and found it 35 Degrees 45 Minutes from the Eaſt Northerly; the Queſtion is, how much Variation there is, and what ſide it is on?

Anſwer, Since the Sun's Declination is 23 Degrees in the Latitude of 42 Degrees North, he is to Riſe at 31 Degrees 43 Minutes from the Eaſt Northerly, which muſt be Subtracted from the 35 Degrees 45 Minutes obſerved, (becauſe they are both on the ſame ſide,) and there remains 4 Degrees 2 Minutes for the Variation, according to the firſt Article of the firſt Principle. Now to know what ſide the Variation is on, I conſider that the 35 Degrees 45 Minutes, obſerved with the Compaſs from the Eaſt Northerly, are nearer the Flower-de-luce than the 31 Degrees 43 Minutes of Amplitude, and therefore I conclude according to this preſent Article, that the Variation is Eaſterly, and conſequently the reſolve of this preſent Queſtion is, that the Variation is 4 Degrees 2 Minutes Eaſterly.

ARTICLE II.

When by your Azimuth Compaſs you find that the Sun Riſes further off from the Flower-de-luce, than the true Amplitude, it ſhews the Variation is Weſterly.

Example.

THE Firſt of *September*, 1684. my Ship being then in 33 Degrees 30 Minutes North Latitude; I obſerve in the Morning the Sun's Magnetical Amplitude, (when the lowermoſt part of him was one third part of his Diameter above the Horizon,) and find it 4 Degrees 15 Minutes from the Eaſt Southerly; the Queſtion is, how much Variation there is, and what ſide it is on?

Anſwer, The Suns Declination the firſt of *September*, 1684. was 4 Degrees 13 Minutes North, ſo that his true Amplitude was 5 Degrees 19 Minutes from the Eaſt Northerly, which I add to the 4 Degrees 15 Minutes obſerved from the Eaſt Southerly, (becauſe they are on contrary ſides) and there comes 9 Degrees 34 Minutes for the Variation, according to the Second Article of the firſt Principle.

Now to know what ſide it is on, I conſider that the 4 Degrees 15 Minutes, obſerved from the Eaſt Southerly, are further from the Flower-de-luce, than the 5 Degrees 19 Minutes of (true) Amplitude from the Eaſt Northerly, and therefore I ſay, that the Variation is 9 Degrees 34 Minutes Weſterly.

Third Principle.

ARTICLE I.

When by your (Azimuth) Compass, you find that the Sun goeth down nearer the Flower-de-luce than his Amplitude, it shows the Variation is Westerly, (quite contrary to his Rising.)

Example.

THE Twenty eighth of *October*, 1684. my Ship being then in the Latitude of 46 Degrees North, I observed at Night the Sun's Magnetical Amplitude, (when his lowermost edge was one third of his Diameter above the Horizon,) and found it 13 Degrees 20 Minutes from the West Southerly; the Question is, how much Variation there is, and what side it is on?

Answer, The Sun's Declination being that Day 16 Degrees 40 Minutes South, I find by it and the proposed Latitude, that the Sun's true Amplitude is 24 Degrees 23 Minutes from the West Southerly; from which I Subtract the 13 Degrees 20 Minutes of Magnetical Amplitude, (because they are both on the same side) and the Remainder 11 Degrees 3 Minutes is the Variation.

Now to know what side it is on, I consider that the 13 Degrees 20 Minutes observed with the Compass from the West Southerly, are nearer the Flower-de-luce than the 24 Degrees 23 Minutes of Amplitude from the West Southerly; and so I conclude that the Variation is 11 Degrees 3 Minutes Westerly.

ARTICLE II.

When by your Compass you find that the Sun goeth down further from the Flower-de-luce than his true Amplitude, the Variation is Easterly.

Example.

THE Twenty sixth of *August*, 1684. being in the Latitude of 30 Degrees North, I observe in the Evening the Sun's Magnetical Amplitude (when he was yet one third of his Diameter above the Horizon) and find it to be 3 Degrees 15 Minutes from the West Southerly; the Question is, how much Variation there is, and what side it is on?

Answer, The Sun's Declination being that Day 6 Degrees 30 Minutes North, I find by it, and the proposed Latitude, that the Sun's (true) Amplitude is 7 Degrees 30 Minutes from the West Northerly, which I add to the 3 Degrees 15 Minutes of Magnetical Amplitude, (since they are on contrary sides) which makes 10 Degrees 45 Minutes, for the Variation.

Now

Now to know what fide it is on, I confider that the 3 Degrees 15 Minutes obferved from the Weft Southerly, are further from the Flower-de-luce than the 7 Degrees 30 Minutes of true Amplitude from the Weft Northerly, and therefore the Variation is Eafterly 10 Degrees 45 Minutes.

S. What do you mean by *Magnetical Amplitude ?*

T. I mean that Amplitude which I did obferve with the Azimuth Compafs; but I call true Amplitude that which is found out by your Table of Amplitudes, becaufe by it is found the true Point or Degree and Minute of the Horizon that the Sun is to Rife or Set at.

S. I underftand now how to find the Variation by the Sun's Amplitude, but I am not fatisfied why I muft not obferve the Sun's Magnetical Amplitude, untill the Sun is one third of his Diameter above the Horizon, therefore pray give me fome Reafons for it, that I may be convinced of the Error of thofe who obferve the Amplitude when the Sun is at the Horizon without allowing for the Refraction.

T. It is very fit that you fhould be fatisfied in it, fince it is fo contrary to the Practice of many Pilots, whofe Error is eafie to be proved; for if it is certain, that the Sun appears at the Horizon when he is yet under it, (as we are well affured, not only by the Obfervations of all the famous Aftronomers, but by the *Hollanders* own experience in their Voyage to *Nova Zembla*, as you have already read:) It is as certain alfo that there is an Error in the Amplitude obferved when the Sun is juft at the Horizon, (without allowance for the Refraction) becaufe the Sun then appears in another Azimuth or Vertical Circle, than that he fhould Rife at, which

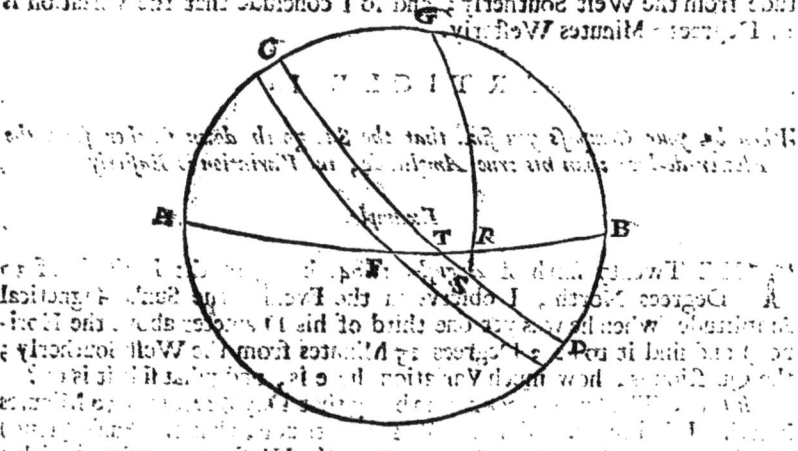

to Demonftrate to you, fuppofe that the Line A B is the Horizon, and C D the Parallel the Sun is in, and fhould Rife at T, fo that F being the true East Point, his Amplitude should be F T, but the Sun being yet under the Horizon in S, he will appear in R, because of the Refraction, and his Rising Amplitude will be F R, that is to say, the Refraction will make him appear nearer the North: Therefore to Correct that Error which is sufficiently proved, you must observe the Magnetical Amplitude when the Sun is about one third of his Diameter above the Horizon, except you be near the Equinoxial, for then you need not, because he Rises so Perpendicularly above the Horizon that the Refraction (in those parts) cannot cause any sensible Error, though the Amplitude be observed when the Sun is just at the Horizon; but it is not so where the Sphere is Oblique, or at any Distance from the Equinoxial, for that Error becomes the more Considerable the further we are from it, and therefore must be Corrected either by observing your Amplitude when the Sun is one third of his Diameter above the Horizon, or by the Supputation of the Arch T R, which may easily be done by those who understand Trigonometry and the use of Signs or Logarithms, for the little Triangle T R S, may be taken for a Right-lined Triangle, its sides being but of half a Degree; now in the Triangle T R S, the Angle R, is a Right-angle, since the Azimuth makes Right-angles with the Horizon, the Angle S T R, is equal to that which the Equator makes with the Horizon, that is to say, equal to the Complement of the height of the Pole, and therefore the Angle T S R, will be equal to the height of the Pole; the Arch S R, is according to *Ticho* of 34 Minutes for the Sun, but we will take but 30 in part because of the Parallax, (which must be subtracted from it) and so it may serve as well for the Stars as for the Sun, say then; *As the Sine Complement of the height of the Pole, is to the Sine of the height of the Pole: So is the Sine of 30 Minutes, to the Sine of the Arch T R:* But this being too hard for a beginner, I give you here a Table of the Error which the Refraction causes in the Amplitude both at Sun Rising, and Setting; the use of it is easie, for you see by it how many Minutes the Refraction increases the Magnetical Amplitude, when observed on that side of the Equinoctial the Sun is on, and diminisheth it when observed on the contrary side, as you will better understand by this

Example.

Admit that the Sixth of *May*, 1684. your Ship being in the Latitude of 45 Degrees North, you observe in the Morning the Suns Magnetical Amplitude, when the Sun is just at the Horizon (that is to say, when the Sun is divided by the Horizon into two equal parts, one half being above it and the

Latitude.	Amplitud. Refraction
Degree	Minutes.
5	3
10	5
15	8
20	11
25	14
30	18
35	21
40	25
45	30
50	35
55	43
60	51
65	D. 57
70	1 3
75	1 52

other

other under) and find it 22 Degrees from the East Northerly; to Correct the Error which the Refraction causes in your Observation, look in the first Column of the precedent Table for 45 Degrees, and in the same Line you will find in the Column of the Refraction 30 Minutes, which being Subtracted from 22 Degrees, the remainer 21 Degrees 30 Minutes is the true Magnetical Amplitude, since the Sun's Declination is North, and your Ship is in North Latitude; but if the Sun's Declination had been South, you must have added it to your Observation, because the Refraction then would have leffen'd your Magnetical Amplitude 30 Minutes; that is to say, while you are in North Latitude or the contrary fide of the Declination; for when the Latitude you are in, and the Sun's Declination are of the fame Denomination, you must always subtract what your Table showeth, but when they are of contrary Denominations, you must add it to your observed Amplitude. Thus, the Amplitude must be Corrected, if your Observation be made when the Sun is one half above, and the other half below the Horizon. But if the Observation be made when his lower Limb is about one third part of his Horizontal Diameter, or a very little more, suppose eleven or twelve Minutes above the Horizon, there needs no Correction, because he is raifed to that height above the Horizon, by the excefs of his Refraction above his Parallax, fo that his true place is then in the Horizon. For his lower Limb (being not in the Horizon, but a little above it) is then raifed almoft 30 Minutes, or about 28 or 27 Minutes above its true place; and confequently, if there were no Refraction nor Parallax would then be about 16 Minutes below the Horizon, and fo the Center of the Sun which is about 16 Minutes higher, would appear juft in the Horizon. If you will make your Observation by the Sun's Perpendicular Diameter, (which may perhaps be thought fomething more eafie) you muft make it when the Sun is almoft half or about half that Diameter above the Horizon, because that Diameter is a little fhortened by the Sun's Refraction.

A

A
TABLE
OF
AMPLITUDES

ORTIVE and OCCASIVE:

SHOWING

Every Day what Degree and Minute the
Sun is to rise or go down from the East
or West, either Northerly or Southerly, according to his Declination; to
wit, from one Degree of Latitude, to
thirtyfive Degrees of Latitude.

A Table of AMPLITUDES

Latitude	DECLINATION											
	1	2	3	4	5	6	7	8	9	10	11	12
	D.M.	D.M.	D.M.	D.M.	D.M.	D.M.	D.M.	D.M.	D.M.	D.M.	D.M.	D.M.
1	1.0	2.0	3.0	4.0	5.0	6.0	7.0	8.0	9.0	10.0	11.0	12.0
2	1.0	2.0	3.0	4.0	5.0	6.0	7.0	8.0	9.0	10.0	11.0	12.0
3	1.0	2.0	3.0	4.0	5.0	6.0	7.0	8.1	9.1	10.1	11.1	12.1
4	1.0	2.0	3.0	4.1	5.1	6.1	7.1	8.2	9.2	10.2	11.2	12.2
5	1.0	2.0	3.1	4.1	5.1	6.1	7.2	8.2	9.3	10.3	11.3	12.3
6	1.0	2.1	3.1	4.2	5.2	6.2	7.3	8.4	9.4	10.4	11.4	12.4
7	1.0	2.1	3.1	4.2	5.2	6.3	7.4	8.5	9.5	10.5	11.5	12.5
8	1.0	2.1	3.2	4.3	5.3	6.4	7.5	8.6	9.6	10.6	11.7	12.7
9	1.1	2.2	3.3	4.3	5.4	6.5	7.6	8.7	9.7	10.8	11.9	12.9
10	1.1	2.2	3.3	4.4	5.5	6.6	7.7	8.8	9.8	10.10	11.10	12.11
11	1.1	2.2	3.4	4.4	5.6	6.7	7.8	8.9	9.10	10.12	11.12	12.14
12	1.1	2.3	3.4	4.5	5.7	6.8	7.10	8.11	9.13	10.14	11.14	12.16
13	1.1	2.3	3.5	4.6	5.8	6.9	7.11	8.13	9.14	10.16	11.17	12.19
14	1.2	2.4	3.5	4.7	5.9	6.11	7.13	8.15	9.17	10.18	11.20	12.22
15	1.2	2.4	3.6	4.8	5.11	6.13	7.15	8.17	9.19	10.21	11.24	12.26
16	1.2	2.5	3.7	4.10	5.13	6.15	7.17	8.19	9.22	10.24	11.27	12.30
17	1.2	2.6	3.8	4.11	5.14	6.17	7.19	8.22	9.25	10.28	11.31	12.34
18	1.3	2.6	3.9	4.13	5.16	6.19	7.22	8.25	9.28	10.31	11.35	12.38
19	1.3	2.7	3.10	4.14	5.17	6.21	7.24	8.28	9.31	10.35	11.39	12.42
20	1.4	2.8	3.11	4.16	5.19	6.23	7.27	8.31	9.35	10.39	11.43	12.47
21	1.4	2.9	3.13	4.17	5.22	6.26	7.30	8.35	9.39	10.43	11.48	12.52
22	1.5	2.10	3.14	4.19	5.24	6.29	7.33	8.38	9.43	10.48	11.53	12.58
23	1.5	2.11	3.16	4.21	5.27	6.32	7.38	8.42	9.48	10.54	11.59	13.4
24	1.6	2.12	3.17	4.23	5.29	6.34	7.40	8.46	9.52	10.58	12.3	13.10
25	1.6	2.13	3.19	4.25	5.32	6.37	7.44	8.50	9.57	11.3	12.5	13.16
26	1.7	2.14	3.21	4.27	5.34	6.41	7.48	8.54	10.1	11.9	12.15	13.23
27	1.7	2.15	3.22	4.30	5.37	6.44	7.51	8.59	10.7	11.14	12.22	13.30
28	1.8	2.16	3.24	4.32	5.44	6.48	7.56	9.4	10.12	11.21	12.29	13.37
29	1.8	2.17	3.26	4.34	5.45	6.52	8.1	9.9	10.13	11.27	12.35	13.45
30	1.9	2.19	3.28	4.37	5.46	6.55	8.5	9.15	10.24	11.34	12.44	13.53
31	1.10	2.20	3.30	4.40	5.50	7.0	8.10	9.21	10.31	11.41	12.52	14.2
32	1.11	2.22	3.33	4.43	5.54	7.5	8.16	9.27	10.38	11.49	13.0	14.11
33	1.12	2.23	3.35	4.46	5.58	7.10	8.21	9.33	10.45	11.57	13.9	14.21
34	1.12	2.25	3.37	4.50	6.2	7.15	8.27	9.40	10.52	12.5	13.18	14.32
35	1.13	2.27	3.40	4.53	6.6	7.20	8.33	9.47	11.1	12.14	13.28	14.42

ORTIVE and OCCASIVE.

Latitude.	DECLINATION.											
	13	14	15	16	17	18	19	20	21	22	23	23 30
	D.M.	D.M.	D.M.	D.M.	D.M.	D.M.	D.M.	D.M.	D.M.	D.M.	D.M.	D.M.
1	13. 0	14. 0	15. 0	16. 0	17. 0	18. 0	19. 0	20. 0	21. 0	22. 0	23. 0	23.30
2	13. 0	14. 0	15. 0	16. 1	17. 1	18. 1	19. 1	20. 1	21. 1	22. 1	23. 1	23.31
3	13. 1	14. 1	15. 1	16. 2	17. 2	18. 2	19. 2	20. 2	21. 2	22. 2	23. 2	23.32
4	13. 2	14. 2	15. 2	16. 3	17. 3	18. 3	19. 3	20. 3	21. 3	22. 5	23. 4	23.34
5	13. 3	14. 3	15. 3	16. 4	17. 4	18. 4	19. 5	20. 5	21. 5	22. 7	23. 6	23.36
6	13. 4	14. 5	15. 5	16. 6	17. 6	18. 6	19. 7	20. 7	21. 7	22. 9	23. 8	23.38
7	13. 6	14. 7	15. 7	16. 8	17. 8	18. 9	19. 9	20.10	21.10	22.11	23.11	23.41
8	13. 8	14. 9	15. 9	16.10	17.10	18.11	19.12	20.12	21.13	22.13	23.14	23.46
9	13.10	14.11	15.12	16.12	17.13	18.14	19.15	20.16	21.16	22.17	23.18	23.49
10	13.12	14.13	15.14	16.13	17.16	18.17	19.18	20.19	21.20	22.21	23.23	23.53
11	13.15	14.16	15.17	16.19	17.20	18.21	19.22	20.23	21.24	22.26	23.27	23.58
12	13.18	14.19	15.21	16.22	17.24	18.25	19.27	20.28	21.30	22.31	23.33	24. 4
13	13.21	14.23	15.24	16.26	17.28	18.29	19.31	20.33	21.35	22.37	23.39	24. 9
14	13.24	14.26	15.29	16.30	17.32	18.34	19.36	20.38	21.41	22.42	23.45	24.16
15	13.28	14.30	15.33	16.35	17.37	18.40	19.42	20.44	21.47	22.49	23.52	24.23
16	13.32	14.34	15.37	16.40	17.43	18.46	19.48	20.51	21.53	22.57	23.59	24.31
17	13.35	14.40	15.44	16.45	17.48	18.51	19.54	20.57	22. 1	23. 4	24. 7	24.39
18	13.41	14.44	15.47	16.51	17.54	18.58	20. 1	21. 5	22. 8	23.12	24.15	24.47
19	13.46	14.50	15.53	16.57	18. 1	19. 5	20. 9	21.12	22.16	23.21	24.24	24.57
20	13.51	14.55	15.59	17. 4	18. 8	19.12	20.16	21.21	22.25	23.39	24.34	25. 7
21	13.56	15. 1	16. 6	17.10	18.15	19.20	20.24	21.30	22.34	23.40	24.45	25.17
22	14. 2	15. 8	16.13	17.18	18.23	19.28	20.33	21.39	22.44	23.50	24.56	25.28
23	14. 9	15.16	16.22	17.25	18.34	19.39	20.43	21.49	22.55	24. 1	25.11	25.44
24	14.15	15.22	16.28	17.34	18.40	19.48	20.53	21.55	23. 6	24.13	25.21	25.53
25	14.22	15.29	16.36	17.42	18.49	19.56	21. 3	22.10	23.18	24.25	25.33	26. 6
26	14.30	15.37	16.44	17.52	18.59	20. 7	21.14	22.22	23.30	24.38	25.46	26.20
27	14.37	15.45	16.53	18. 1	19. 9	20.18	21.26	22.34	23.45	24.52	26. 1	26.35
28	14.46	15.54	17. 3	18.11	19.20	20.29	21.38	22.47	23.57	25. 6	26.16	26.51
29	14.54	16. 4	17.13	18.22	19.32	20.41	21.51	23. 1	24.11	25.22	26.32	27. 7
30	15. 3	16.13	17.23	18.34	19.44	20.54	22. 5	23.16	24.27	25.38	26.49	27.25
31	15.13	16.24	17.35	18.45	19.57	21. 8	22.19	23.31	24.43	25.55	27. 7	27.43
32	15.23	16.34	17.46	18.58	20.10	21.22	22.35	23.47	25. 0	26.13	27.26	28. 3
33	15.34	16.46	17.59	19.11	20.24	21.37	22.51	24. 4	25.18	26.32	27.46	28.23
34	15.45	16.58	18.12	19.25	20.38	21.53	23. 7	24.22	25.37	26.52	28. 7	28.45
35	15.56	17.11	18.25	19.40	20.55	22.10	23.25	24.41	25.57	27.13	28.29	29. 8

V.

A Table of AMPLITUDES

Latitude	DECLINATION											
	1	2	3	4	5	6	7	8	9	10	11	12
	D.M.	D.M.	D.M.	D.M.	D.M.	D.M.	D.M.	D.M.	D.M	D.M.	D.M.	D.M.
35	1.13	2.17	3.40	4.53	6. 6	7.20	8.33	9.47	11. 1	12.14	13.28	14.42
36	1.14	2.28	3.43	4.57	6.11	7.25	8.40	9.54	11. 9	12.24	13.39	14.54
37	1.15	2.30	3.45	5. 1	6.16	7.31	8.47	10. 2	11.18	12.34	13.49	15. 5
38	1.16	2.32	3.48	5. 5	6.21	7.37	8.54	10.11	11.27	12.44	14. 1	15.18
39	1.17	2.34	3.52	5. 9	6.27	7.44	9. 1	10.19	11.37	12.55	14.13	15.31
40	1.18	2.37	3.55	5.14	6.32	7.51	9. 9	10.28	11.47	13. 6	14.25	15.45
41	1.19	2.39	3.59	5.19	6.38	7.58	9.18	10.38	11.58	13.18	14.39	16. 0
42	1.20	2.41	4. 2	5.24	6.44	8. 5	9.26	10.48	12. 9	13.31	14.53	16.15
43	1.22	2.44	4. 6	5.29	6.51	8.13	9.35	10.58	12.21	13.44	15. 7	16.31
44	1.23	2.47	4.10	5.34	6.58	8.21	9.45	11. 9	12.34	13.58	15.23	16.48
45	1.25	2.50	4.15	5.40	7. 5	8.30	9.55	11.21	12.47	14.13	15.39	17. 6
46	1.26	2.53	4.19	5.46	7.12	8.39	10. 6	11.33	13. 1	14.31	15.57	17.25
47	1.28	2.56	4.24	5.52	7.21	8.49	10.18	11.46	13.16	14.45	16.15	17.45
48	1.30	2.59	4.29	5.59	7.29	8.59	10.30	12. 0	13.28	15. 2	16.34	18. 6
49	1.32	3. 3	4.35	6. 6	7.38	9.10	10.42	12.15	13.48	15.21	16.55	18.29
50	1.33	3. 7	4.40	6.14	7.48	9.21	10.56	12.30	14. 5	15.40	17.16	18.52
51	1.35	3.11	4.46	6.22	7.58	9.34	11.10	12.47	14.24	16. 1	17.39	19.17
52	1.37	3.15	4.52	6.30	8. 8	9.47	11.25	13. 4	14.43	16.23	18. 3	19.44
53	1.40	3.20	5. 0	6.40	8.20	10. 0	11.41	13.22	15. 4	16.47	18.29	20.13
54	1.42	3.24	5. 6	6.49	8.32	10.15	11.58	13.42	15.26	17.11	18.37	20.43
55	1.45	3.29	5.14	6.59	8.44	10.30	12.16	14. 3	15.50	17.37	19.26	21.15
56	1.47	3.35	5.22	7.10	8.58	10.46	12.35	14.25	16.15	18. 6	19.57	21.50
57	1.50	3.41	5.31	7.22	9.13	11. 4	12.56	14.48	16.42	18.36	20.30	22.27
58	1.53	3.47	5.40	7.39	9.28	11.23	13.18	15.13	17.10	19. 8	21. 6	23. 6
59	1.57	3.53	5.50	7.47	9.45	11.43	13.41	15.41	17.41	19.42	21.45	23.49
60	2. 0	4. 0	6. 1	8. 1	10. 2	12. 4	14. 6	16.10	18. 9	20.19	22.26	24.34
61	2. 4	4. 8	6.12	8.16	10.21	12.27	14.34	16.41	18.50	20.59	23.11	25.24
62	2. 8	4.16	6.24	8.34	10.42	12.52	15. 3	17.15	19.28	21.43	23.59	26.17
63	2.12	4.25	6.37	8.50	11. 4	13.19	15.37	17.51	20. 5	22.29	24.51	27.15
64	2.17	4.34	6.51	9. 9	11.28	13.48	16. 8	18.31	20.52	23.20	25.48	28.19
65	2.22	4.45	7. 7	9.30	11.54	14.19	16.46	19.14	21.43	24.16	26.50	29.28
66	2.28	4.55	7.24	9.53	12.22	14.54	17.26	20. 1	22.37	25.16	27.55	30.44
66½	2.31	5. 1	7.33	10. 4	12.38	15.12	17.48	20.26	23. 6	25.45	28.35	31.26

ORTIVE and OCCASIVE.

Latitude	DECLINATION.											
	13	14	15	16	17	18	19	20	21	22	23	23 30
	D.M.	D.M.	D.M.	D.M.	D.M.	D.M.	D.M.	D.M.	D.M.	D.M.	D.M.	D.M.
35	15.56	17.11	18.25	19.40	20.55	22.10	23.25	24.41	25.57	27.13	28. 2	29. 8
36	16. 9	17.24	18.40	19.55	21.11	22.27	23.44	25. 1	26.18	27.35	28.53	29.32
37	16.22	17.38	18.55	20.11	21.28	22.46	24. 3	25.21	26.40	27.58	29.17	29.57
38	16.35	17.53	19.19	20.26	21.47	23. 5	24.24	25.43	27. 3	28.23	29.44	30.24
39	16.50	18. 8	19.27	20.46	22. 6	23.26	24.46	26. 7	27.28	28.49	30.11	30.52
40	17. 5	18.25	19.45	21. 5	22.26	23.47	25. 9	26.31	27.54	29.17	30.40	31.22
41	17.20	18.42	20. 3	21.25	22.48	24. 1	25.33	26.57	28.23	29.46	31.11	31.54
42	17.37	19. 0	20.23	21.46	23.10	24.34	25.59	27.24	28.50	30.16	31.43	32.27
43	17.55	19.19	20.43	22. 8	23.34	25. 0	26.26	27.53	29.20	30.49	32.18	33. 2
44	18.13	19.39	21. 5	22.32	23.59	25.26	26.55	28.23	29.53	31.23	32.54	33.40
45	18.33	20. 0	21.28	22.57	24.25	25.55	27.25	28.56	30.27	31.59	33.33	34.20
46	18.54	20.23	21.53	23.22	24.53	26.25	27.57	29.30	31. 3	32.38	34.14	35. 2
47	19.17	20.47	22.18	23.50	25.23	26.57	28.31	30. 6	31.42	33.19	34.57	35.47
48	19.39	21.12	22.45	24.16	25.55	27.30	29. 7	30.44	32.23	34. 3	35.44	36.35
49	20. 3	21.38	23.15	24.51	26.28	28. 6	29.45	31.25	33. 7	34.49	36.33	37.26
50	20.29	22. 7	23.45	25.24	27. 7	28.44	30.26	32. 9	33.53	35.39	37.26	38.21
51	20.57	22.37	24.17	25.59	27.41	29.24	31. 9	32.55	34.43	36.32	38.23	39.19
52	21.26	23. 8	24.52	26.36	28.21	30. 6	31.57	33.45	35.36	37.29	39.23	40.22
53	21.57	23.42	25.28	27.16	29. 4	30.53	32.45	34.39	36.33	38.30	40.49	41.30
54	22.30	24.18	26. 7	27.58	29.50	31.43	33.38	35.35	37.34	39.36	41.40	42.43
55	23. 5	24.57	26.49	28.43	30.39	32.36	34.35	36.37	38.40	40.47	42.56	44. 3
56	23.43	25.38	27.34	29.32	31.32	33.33	35.36	37.42	39.51	42. 4	44.19	45.30
57	24.24	26.22	28.29	30.24	32.28	34.34	36.43	38.54	41.10	43.27	45.50	47. 4
58	25. 7	27.10	29.14	31.20	33.29	35.40	34.54	40.12	42.33	44.59	47.30	48.48
59	25.58	28. 5	30.10	32.21	34.35	36.52	39.12	41.37	44. 6	46.40	49.21	50.44
60	26.44	28.56	31.10	33.27	36.25	38.10	40.38	43.10	45.47	48.31	51.24	52.53
61	27.39	29.56	32.16	34.39	37. 5	39.56	42.11	44.52	47.40	50.36	53.42	55.20
62	28.38	31. 1	33.27	35.57	38.31	41.10	43.54	46.46	49.46	52.56	56.20	58. 9
63	29.42	32.12	34.46	37.23	40. 5	42.54	45.19	48.53	52. 8	55.36	59.24	61.26
64	30.52	33.30	36.15	38.58	41.50	44.49	47.57	51.17	54.50	58.43	63. 3	65.27
65	32.10	34.55	37.46	40.43	43.46	46.59	50.23	54. 2	58. 0	62.26	67.36	70.39
66	33.35	36.30	39.31	42.40	45.58	49.27	53.10	57.14	61.47	67. 5	73.53	78.38
66½	34.20	37.21	40.29	45. 1	47. 9	50.48	54.44	59. 4	64. 0	69.58	78.30	90. 0

PROP. XL.

The Use of the Tables of Amplitude.

S. I Believe I underſtand already the Uſe of this Table, by what you have already ſaid of it, however to be the more perfect in it, you will do well to give me ſome more Examples.

T. Before I give you any Example, I muſt remember you that on the top of your Tables is ſet down the Sun's Declination, and in the firſt Column on the Left-hand of each page the Degrees of Latitude, for that being known there will be no difficulty to find out the Amplitude.

Example.

The Sun's Declination being 16 Degrees South, and your Ship being in the Latitude of 48 Degrees North; the Queſtion is, how many Degrees and Minutes the Sun will Riſe (or Set) Diſtant from the Equinoxial, or from the true Eaſt (or Weſt) Point of the World, which is the ſame; firſt look in the firſt Column for 48 Degrees of Latitude, then ſeek for your 16 Degrees of Declination on the Top of your Table, and under 16 your Declination, and againſt 48 your Latitude, you will find 24 Degrees 16 Minutes, and ſo much will the Sun Riſe (in that Latitude and Declination) from the true Eaſt Southerly; and Set alſo at the ſame Diſtance from the true Weſt Southerly, becauſe his Declination is South.

S. How ſhall I find out the Amplitude when the Latitude is of Degrees and Minutes?

T. In that caſe you muſt find the Amplitude of the Latitude given (without Minutes) as before, and likewiſe the Amplitude of the Latitude that follows next under it, then Subtract the leſſer Amplitude out of the greater, and from their difference (found out by the Subtraction) add to the laſt Amplitude proportionably to the Minutes that are over and above the Degrees of Latitude; to wit, Three Quarters if it be 45 Minutes, Two Thirds if it be 40 Minutes, Half if it be 30 Minutes, One Third if it be 20 Minutes, a Quarter if it be 15 Minutes, a Fifth if it be 12 Minutes, a Sixth if it be 10 Minutes, and ſo of the reſt: Or elſe ſay, (by the Rule of Three) *If 60 Minutes that a Degree contains giveth ſo many Minutes: (Found by the Difference of the two Amplitudes:) What will the Minutes found with the Degrees of Latitude give?* And thereby you will ſee what muſt be added to the leaſt Amplitude.

Example.

Example.

Suppose that the Sun's Declination is of 15 Degrees North, and your Ship is in the Latitude of 48 Degrees 30 Minutes North; the Question is, how many Degrees and Minuutes the Sun will Rise from the East, or Set from the West, (which is what we call Amplitude?)

Answ. Look first for the Amplitude of 48 Degrees, in the Column of 15 Degrees of Declination, which according to the precedent Direction you will find to be 22 Degrees 45 Minutes, look also for the Amplitude of the Latitude that follows next under it; to wit, that of 49 Degrees; which you will find to be 23 Degrees 15 Minutes, from which Subtracting the least Amplitude 22 Degrees 45 Minutes, there remains 30 Minutes for the Difference, now because that besides the Degrees of Latitude, there is 30 Minutes, which are the half of a Degree, take half of the 30 Minutes of Difference, which is 15 Minutes; or else say, (by the Rule of Three) *If* 60 *Minutes* (contained betwen 48 and 49 Degrees) *gives* 30 *Minutes of Difference: How many will the* 30 *Minutes* (which are over and above the 48 Degrees of Latitude) *give?* And you will find it 15 Minutes, which being added to the least Amplitude, *viz.* 22 Degrees 45 Minutes, you have 23 Degrees for the Amplitude of 15 Degrees of Declination, and 48 Degrees 30 Minutes of Latitude.

S. How shall I know what side the Amplitude is on?

T. You may easily know it by the Sun's Declination, for the true Amplitude is always on that side that the Sun's Declination is on, and therefore in this last Example the Amplitude is Northerly, because the Declination is North.

S. What must I do when the Minutes are with the Degrees of Declination, and not with the Latitude?

T. In that case you must not take the Amplitude of two different Degrees of Latitude as in the precedent Example, but you must take the Amplitude of two Different Degrees of Declination; to wit, that next to the first on the Right-hand, as you will easily understand by this

Example.

Admit, that the Sun's Declination is 20 Degrees 45 Minutes South, and that my Ship is in the Latitude of 36 Degrees North; the Question is, how much Amplitude the Sun hath, and what side it is on?

Answ. I look first for the Amplitude of 36 Degrees of Latitude and 20 Degrees of Declination, which I find to be 25 Degrees 1 Minute, and the next to it (in the Column of 21 Degrees of Declination) is 26 Degrees 18 Minutes, from which I Subtract the least Amplitude 25 Degrees 1 Minute, and there remains 1 Degree 17 Minutes or 77 Minutes for the Difference; now because the 45 Minutes which are over and above the 20 Degrees of Declination, are Three Quarters of a De-

gree;;

gree; I take the Three Quarters of 1 Degree 17 Minutes, or 77 Minutes, of Difference which is 57 Minutes $\frac{1}{4}$, or else I say, by the Rule of Three; *If 60 Minutes giveth 77 Minutes Difference, What will 45 Min. give?* And there will come 57 Minutes $\frac{1}{4}$, which must be added to the least of the two Amplitudes 25 Degrees 1 Minute, and there will come 25 Degrees 58 Minutes $\frac{1}{4}$, for the required Amplitude which is Southerly, because the Declination is South.

S. You have made me understand this pretty well, but pray give me an Example with Minutes both to the Latitude and Declination.

T. I will, for an Example will be much more intelligible than any discourse I could make about it, and therefore mind well this Question, because it is harder than the former, and yet as necessary.

Example.

Admit that the Sun's Declination is 22 Degrees 40 Minutes South, and my Ship in the Latitude of 46 Degrees 35 Minutes North; the Question is, how many Degrees and Minutes the Sun will Rise or Set from the East or West?

Answ. First, I look for the Amplitude of 46 Degrees and 47 Degrees of Latitude, in the Column of 22 Degrees of Declination, where I find it to be 32 Degrees 38 Minutes, and 33 Degrees 19 Minutes, which being Subtracted one from another, (the lesser from the greater) there will remain for the Difference 41 Minutes, of which I take the Third and the Fourth, because of the 35 Minutes over and above the 46 Degrees of Latitude, which are the Third and Fourth part of a Degree, which being taken out of the 41 Minutes of Difference, comes 23 Minutes 55 Seconds; or else I say, by the Rule of Three, *If 60 Minutes give 41 Minutes Difference, What will the 35 Minutes (which are over and above the 46 Degrees of Latitude) give?* And there will come 23 Minutes 55 Seconds, which I add to the lesser Amplitude 32 Degrees 38 Minutes, and there will come 33 Degrees 1 Minute 55 Seconds, but because the Seconds are above 30, I neglect the Seconds and reckon a Minute for them, and so I make it up 33 Degrees 2 Minutes. Next I look for the Amplitude of 46 and of 47 Degrees of Latitude, under 23 Degrees of Declination, where I find it to be 34 Degrees 14 Minutes, and 34 Degrees 57 Minutes, which I subtract one from another; to wit, 34 Degrees 14 Minutes from 34 Degrees 57 Minutes, and there remains for the Difference 43 Minutes, of which I take the Third and the Fourth, which comes to 25 Minutes 5 Seconds, or else I say, by the Rule of Three, *If 60 Minutes give 43 Minutes Difference, How many Minutes will the 35 Minutes (of Latitude) give?* And there will come 25 Minutes 5 Seconds as before; but because the Seconds are under 30, I neglect them, and so there remains 25 Minutes, which being added to the least of these two last Amplitudes, *viz.* 34 Degrees 14 Minutes, there will come 34 Degrees 39 Minutes. Then I subtract one from another; that is to say, the two

Am-

Amplitudes found for the Latitude of 46 Degrees 35 Minutes, and for 22 and 23 Degrees of Declination; to wit, 33 Degrees 2 Minutes from 34 Degrees 39 Minutes, and there will remain for the Difference 1 Degree 37 Minutes or 97 Minutes, whose half 48 Minutes must be added to the least of these two Amplitudes; to wit, that of 33 Degrees 2 Minutes, and there will come 33 Degrees 50 Minutes, that the Sun will Rise and Set from the East or West Southerly, as was required.

P R O P. XII.

How to observe the Variation at any Time of the Day.

S. HOW may the Variation be known at any Time of the Day. *T.* This is done by the Sun's Azimuth, which you may find out by the Sun's Declination, the Complement of the Latitude, and the Sun's height.

S. Is it to be found by Numbers?

T. Yes, and chiefly by the Logarithms, and therefore I would have passed it by, as being too hard for a beginner, but that did consider the excellency of the Azimuth to find out the Variation, for which use I can recommend it to you above all the precedent Practices, when you come to understand the use of the Logarithms, which in a short time I design to render very easie to you.

S. since the use of the Azimuth is of such excellency for the Variation, you will do well to set it down however, for although I do not at present understand the use of the Logarithms, I may hereafter, and in the mean time it will be necessary to those who understand it already; therefore pray tell me what is the first thing I must know, to find out the Sun's Azimuth?

T. You must first know the Sun's Distance from the nearest Pole to you, which you may easily do by his Declination; as for Example, suppose that your Ship is in Latitude North, and the Sun's Declination is 14 Degrees 30 Minutes North, and you will know his Distance from the nearest Pole to you, (which is the North Pole,) Subtract 14 Degrees 30 Minutes of North Declination, from 90 Degrees the Distance of the Pole from the Equinoxial, and the Remainder 75 Degrees 30 Minutes, is the Sun's Distance from the North Pole.

S. Must I always Subtract the Sun's Declination from 90 Degrees to have his Distance from the Pole?

T. Yes, when the Latitude and the Sun's Declination are on the same side of the Equinoxial, but when they are of contrary sides, (as for Example, when your Ship is in Latitude North, and the Sun's Declination is South,) in that case you must add the Sun's Declination to 90 Degrees, and the whole will be the Sun's Distance from the nearest Pole to you, which

which you muſt ſet down with the Complement of the Latitude, and
the Complement of the Sun's height, and theſe Three Sums being added
together, take half of their Sum, from which half Subtract the Diſtance
of the Sun from the Pole, and the Remainder will be what we call the
Difference : Then ſay,

   1. *As the Radius,*
      is to the Complement of the Latitude;
   So the Complement of the Sun's height
      to a fourth Sine.

   2. *As this fourth Sine,*
      is to the Sine of the half Sum ;
   So is the Sine of the Difference
      to a ſeventh Sine.

Unto which ſeventh Sine, if you add the Sine of 90 Degrees, half that
Sum will be the Sine of an Arch, the double of whoſe Complement is
the Azimuth from the neareſt Pole to you, but if you would reckon the
Sun's Azimuth from the furthermoſt Pole from you, or rather from that
part of the Meridian which marks our Noon, as we will do here, you
muſt Subtract it from 180 Degrees, and the Remainder will be the
Diſtance from Noon, or from the further Pole from you, as you will
better underſtand by the following

Example.

In the Latitude of 46 Degrees 30 Minutes North, the Sun's Declination being 13 Degrees 50 Minutes North, and his height in the Morning 42 Degrees 30 Minutes: I deſire to know the Sun's Azimuth; Therefore I Subtract the Sun's Declination 13 Degrees 50 Minutes from 90
Degrees, and there Remains 76 Degrees 10 Minutes for the Sun's Diſtance
from the neareſt Pole, which I ſet down thus,

Complement of the Sun's Declination   76 Deg. 10 Min.

Next, I Subtract the Latitude 46 Deg. 30 Min. from 90
Degees, and the Remainder 43 Deg. 30 Min. is the Com- . 43 .30
plement of the Latitude . . . .

In like manner I find out the Complement of the Sun's height 47   30

The Sum . . . 167 10

The half Sum . . . 83 : 35

The Difference between the half Sum, and the Complement of the Sun's Declination . . . . is 07 25

Analogy

Analogy by the Logarithms.

(1.) *As the Radius (or Sine of 90 Degrees)* 10,0000000

  is to the Sine Complement of the Latitude 43° 30′ . 9,8378122
  So is the Sine Complement of the Suns height 47 30 . 9,8676309

  to the fourth Sine . . . . ✗9,7054431

(2.) *As that fourth Sine* 9,7054431

  is to the Sine of the half Sum 83° 35′ 9,9972708
  So is the Sine of the Difference 7 25 9,1108726

                                              ✗9,1081434

  to a seventh Sine 9,4027003

To which I add the Radius 10,0000000

The Sum whereof is . . . . ✗9,4027003
The Half of the Sum is 9,7013501

Which is the Sine of 30° 11′, whose Complement is 59° 49′, the double whereof is 119° 38′, for the Suns Azimuth from the North.

Example 2.

In the Latitude of 38 Degrees 30 Minutes South, the Suns Declination being 18 Degrees 48 Minutes South, and his height in the Morning 45 Degrees 44 Minutes; I desire to know the Sun's Azimuth? According to the precedent Direction, I set down the

	D.	M.
Complement of the Sun's Declination	71	12
Complement of the Latitude	51	30
Complement of the Sun's height	44	14
The Sum . . .	166	56
The half Sum	83	28
The Difference . . .	12	16

Analogy by the Logarithms.

(1.) *As the Radius* 10,0000000

  is to the Sine Complement of the Latitude 51° 30′ . 9,8935444
  So is the Sine Complement of the Sun's height 44 15 . 9,8437250

  to a fourth Sine . . . . ✗9,7372694

Take

Take notice, that I cut off the Unit for the Radius, which must be Subtracted from this last Number, and the Remainer is the fourth Sine required, which is 9,7372694.

(2.) *As that fourth Sine* . . . , ...	9,7372694
is to the Sine of the half Sum 83° 28′	9,9971704
So is the Sine of the Difference 12 16	9,3272811
	19,3244515
to a seventh Sine	9,5871821

To which I add the Radius by setting down an Unit before it, thus 19,5871821

Whose half is the Sine of 38° 26′ 30″ 9,7935910

Whose Complement is 51° 33′ 30″
Which being doubled 51 33 30

The Sum is . . 103 07 00   that is to say, 103 Degrees 7 Minutes for the Azimuth from the South, which being Subtracted from 180 Degrees, there Remains 76 Degrees 53 Minutes, that the Sun is Distant from the North, where now it is Noon, contrary to those who are in Latitude North, whose Noon is Southerly: But if you would know the Azimuth from the East or West, Subtract 90 Degrees from 103 Degrees 7 Minutes, the Azimuth from the South, and the remainer 13 Degrees 7 Minutes, is the Azimuth from the East or West Northerly.

Example 3.

In Latitude of 48 Degrees 20 Minutes North, the Sun's Declination being 21 Degrees 42 Minutes South, and his height in the Afternoon 34 Degrees 40 Minutes: I desire to know the Sun's Azimuth? I add the Sun's Declination 21 Degrees 42 Minutes, to 90 Degrees, (because the Latitude and Declination are on contrary sides of the Equinoxial) which makes 111 Degrees 42 Minutes, for the Sun's Distance from the North Pole, which is the nearest Pole to me: So I set it down in this Order:

	D.	M.
The Sun's Distance from the North Pole	111	42
The Complement of the Latitude	41	40
The Complement of the Sun's height	55	20
The Sum . .	208	42
The Half Sum . .	104	21
The Difference	7	21

Analogy

Analogy by the Logarithms.

(1.) *As the Radius* 10,0000000

is to the Sine Complement of the Latitude 41° 40' . . 9,8226883
So is the Sine Complement of the Sun's height 55 20 . . 9,9151228

to a fourth Sine *9,7378111

(2.) *As that fourth Sine* 9,7378111

is to the Sine of the half Sum 104° 21' 9,9862340
So is the Sine of the Difference 7 21 9,1069729

19,0932069

to a seventh Sine 9,3553958

To which I add the Radius (or Sine of 90 Degrees)
by setting an Unit before it, thus 19,3553958

Whose half is the Sine of 28 Deg. 26 Min. 9,6776979

Whose Complement is, 61° 34'
Which I double . . 61 34
The Sum . . 123 08   for the Azimuth from

the North, which being Subtracted from 180 Degrees, (the Distance of
the North from the South) there remains 56 Degrees 52 Minutes, that
the Sun is Distant from the South Westerly, which is the required
Azimuth. But if you would know the Sun's Distance from the West,
(or East) Subtract 90 Degrees from 123 Degrees 8 Minutes, the Azimuth from the North, and the remainer 33 Degrees 8 Minutes, is the
Azimuth from the West (or East) Southerly.

S. How do you know the *Variation* by the Azimuth?

T. The Variation is known by the *Difference* between the *Sun's Azimuth*,
and the *Magnetical Azimuth*, which having been explained to you already
I shall only give you these few Directions, to render it the more intelligible.

First Principle.

When the Sun's Azimuth, is on the same side or Denomination of the Mag-
netical Azimuth, you must Subtract the lesser Azimuth from the greater,
and the Remainder will be the Variation.

X 2   *Example.*

Example.

Admit the Sun's Azimuth to be 32 Degrees from the South Easterly, and the Magnetical Azimuth to be 43 Degrees 15 Minutes, also from the South Easterly. You muſt Subtract the 32 Degrees of the Sun's Azimuth (becauſe it is the leſſer) from the 43 Degrees 15 Minutes of the Magnetical Azimuth, and the 11 Degrees 15 Minutes remaining, is the *Variation*.

The ſame is to be done when both Azimuths are from the South. Weſterly; or from the North Eaſterly; or from the North Weſterly.

Second Principle.

When the Sun's Azimuth and Magnetical Azimuth are of contrary ſides, or Different Denominations, you muſt add them both together, and the whole will be the Variation.

Example.

Admit that the Calculated Azimuth of the Sun is 4 Degrees from the South Weſterly, and the Azimuth by Obſervation is 5 Degrees from the South Eaſterly, to find out the Variation you muſt add them both together, (ſince they are of Different Denominations) and there will come 9 Degrees for the required Variation.   The ſame muſt be done if the Sun's Azimuth was from the South Eaſterly, and the Magnetical Azimuth from the South Weſterly; and the ſame is to be done from the North, for if one was from the North Eaſterly, and the other from the North Weſterly, you are to add them both together, and their Sum will be the Variation.

S. How ſhall I know what ſide the Variation is on; that is to ſay, when it is Eaſterly or Weſterly.

T. Well enough, if you mind well theſe following Directions.

ARTICLE I.

Your Ship being in Latitude North, if you obſerve (in the Morning) the Magnetical Azimuth to be nearer the South than the Sun's Azimuth; the Variation is Weſterly.

Example.

ADmit the Magnetical Azimuth in the Fore-noon to be 5 Degrees from the South Eaſterly, and the Sun's Azimuth 10 Degrees 38 Minutes alſo from the South Eaſterly: And I would know the Variation and what ſide it is on?

First

First I Subtract the 5 Degrees from 10 Degrees 38 Minutes, (since both Azimuths are on the same side) and the Remainder 5 Degrees 38 Minutes is the Variation, then considering that the Magnetical Azimuth is nearer the South than the Sun's Azimuth. I conclude that the Variation is Westerly 5 Degrees 38 Minutes, (or half a Point of the Compass) as was required.

ARTICLE II.

If you be in Latitude North, and you observe in the Morning the Magnetical Azimuth to be further from the South than the true Azimuth, the Variation is Easterly.

ARTICLE III.

Being in Latitude North, if you observe in the After-noon the Magnetical Azimuth to be nearer the South Westerly than the Sun's Azimuth, the Variation is Easterly.

ARTICLE IV.

If in the After-noon being in Latitude North, you observe the Magnetical Azimuth further from the South Westerly, than the Sun's Azimuth, the Variation is Westerly.

ARTICLE V.

If you be in Latitude South, and you observe in the Morning the Magnetical Azimuth to be nearer the North Easterly than the Sun's Azimuth, the Variation is Easterly.

ARTICLE VI.

If in the same Latitude, you observe in the Morning the Magnetical Azimuth to be further from the North Easterly, than the Sun's Azimuth, the Variation is Westerly.

ARTICLE VII.

Being in Latitude South, if you observe in the After-noon the Magnetical Azimuth to be nearer the North (Westerly) than the Sun's Azimuth, the Variation is Westerly.

A. R

ARTICLE VIII.

If in the Latitude South, you observe in the After-noon the Magnetical Azimuth, to be further from the North Westerly than the Sun's Azimuth, the Variation is Easterly.

S. What Compass do you Counsel me to make use of, to observe the (Magnetical) Azimuth?

T. The same Compass with which you were taught to observe the Variation at Noon: First, because it is very easie, and Secondly, because you may Steer by it, it being much better to Steer by the same Compass with which you have observed the Variation, than by any other; because of many accidents that commonly happen to one Compass, more than to another.

S. How shall I know by this Compass the Magnetical Azimuth?

T. You may know it by the same directions that I gave you to observe the Sun's Altitude, which are, that at the same time that you observe the Sun's height, to find out by it its true Azimuth you must turn your Compass so, that the shadow of the Thread which lies over the Glass falls exactly upon the Point or tip of the socket; and the shadow of the same Thread that falls upon the Degrees on the Limb of the Fly, will show you the Magnetical Azimuth.

S. Pray show me now how to Correct the Variation?

T. Although my design is to show you fully how to Correct the Variation of your Compass; I shall say for the present but little concerning it, because it shall be made very plain and easie to you in the Use of the Chart and the Sinical Quadrant; therefore 'till then, be satisfied to know that the Variation is the same to the Compass, that the Lee-way is to a Ship, and therefore must be allowed for, if you will make good your Course, by taking as much on the contrary side; as for

Example.

Admit that your Compass varie a Point Westerly, and that with a good Wind you Steer away *W* b *N*; it is certain, since your Compass doth varie a Point Westerly, that your true Course is not *W* b *N*, but only *W*; because of the Variation, which makes it fall off.

But on the contrary, if to go to such a place, it is necessary to make good a *W* b *N*, you must Steer a Point more Northerly, since the Variation is a Point Westerly, that is to say, you must Steer away *W N W*; and by so doing you will make good the proposed Course, because of the Variation of your Compass which makes you fall off a Point more Westerly.

PROP.

PROP. XIII.

How to make a Plain Chart.

s. ADMIT I have ready a Sheet of the beſt *Dutch* Paper, what muſt I do to deſcribe a *Plain Chart* on it?

T. You muſt draw on one ſide of it a Line repreſenting the firſt Meridian; as for Example, M N upon the two extremities, of which you ſhall raiſe two Perpendiculars of the ſame length; to wit, M O for the Equinoxial, and N P for a Parallel, then from P to O, you muſt draw another Meridian Parallel to the firſt. This done, you ſhall divide the ſaid Lines into as many equal parts for Degrees as you have accaſion for, *viz.* the two Meridians for Degrees of Latitude, and the two Parallels for Degrees of Longitude; (but if you will you may divide but one of the Parallels into equal parts; to wit, the Southermoſt;) and then ſubdivide each Degree into 12 equal parts, of five Minutes each, and each Meridian into two parts exactly, by drawing through the very midſt of them the Parallel R S, which muſt be croſſed at Right-angles, by drawing in the ſame manner a third Meridian T V, from the midſt of the graduated Parallels, and at the Point where theſe two Lines intercept or croſs one another in the midle of your Chart, draw an occult Circle as great as you can, (that may be rub'd out) and it will be divided by the Lines R S, and T V, into 4 quarters; then divide each quarter of it into 8 equal parts, and draw Lines through each Point of Diviſion and the Center of your Circle, and ſo it will be divided into 32 Rhumbs. This they call the *Mother Compaſs,* upon the Center of which, having deſcribed two very little Circles, mark with a Flower-de-luce that Rhumb which ſhoweth the North, (which you may eaſily know by the graduated Meridian or Equinoxial) and with a Croſs that which ſhoweth the Eaſt, as you ſee in the Figure: Then upon the Circumference of your great Circle, *viz.* Upon each of the 16 Points of Diviſion, through which the chief or Principle Rhumbs are drawn; deſcribe another ſecret Circle leſſer than the firſt, and divide its Diameter into two equal parts, by drawing a Line at Right-angles through the middle or Center of it, (as you were taught in the firſt Book) and ſo your Circle will be divided 4 quarters; then divide each quarter into 8 equal parts as before, and into draw Lines through the Center and each Point of Diviſion, and it will be divided into as many parts as the Mother Compaſs which it repreſents, and doing as much upon each Point at two Rhumbs diſtant one from another, you will deſcribe as it were 16 little Compaſſes, which is the moſt that any Chart hath how great ſoever it be. The next thing will be to make a Scale of Leagues, which is thus done, taking upon the graduated Meridian 5 Degrees, and dividing them into 5 equal parts

of

of 20 Leagues each, (that a Degree contains) then subdivide each part into 4 parts more of 5 Leagues each, and the first of these parts into 5 more, and your Scale will be ready for use.

S. What is next to be done?

T. The next thing is to place exactly the *Harbours, Capes, Islands,* &c. in their proper Situations, taking their Latitudes and Longitudes out of the best and most approved Charts; and in doing this, you must endeavour as much as you can to describe each Harbour, Cape and Bay, in their proper Form and Figure: Chiefly the Havens Mouth, and its Sea-marks, (if you make it a particular Chart for the Coast) as *Steeples, Light-houses* or the like; as for *Rocks,* you must Mark them with little *Crosses* or *Blots;* the *Banks* or *Sands,* with *Points;* the *Roads* and places fit for Anchorage with an *Anchor,* and the *Depth* of the Water with *Numbers.*

S. Why do you not describe as many little Compasses in your Chart, as you desire me to do when I make one?

T. It is because my Chart don't require it, it being too little; however the thing is so plain, that you cannot but understand it.

PROP. XIV.

How to know if the Plain Chart be well made.

S. HOW shall I know if my Chart be well made?

T. The way to know it, is first to examin with an ordinary Compass if all the Lines of *Rhumbs* which are of the same Denomination be Parallel one to another; and if the *Harbours, Capes,* and *Islands,* are placed in their proper *Latitudes,* which you may know by comparing it, with an approved Table of Latitudes; as also by your own observations with the *Astronomical Ring,* or *Astrolabe* in all the places you come to: But most commonly it is examined by another approved Chart, as that of *Bleau, Colon,* or *Guerard,* or else by the Charts inserted into the Great Wagoner or Sea Column.

You are likewise to mind if the *Sands, Rocks,* or *Showls,* are Marked or set down in it, as you have found them by experience, and if the Degrees are all equal, and contain each 20 *English* Leagues, taken upon the Scale of Leagues.

PROP

P R O P. XV.

The Use of the Plat or Plain Chart.

S. FOR what Use is this Chart?

T. Its first Use is to show by what Rhumb or Point of the Compass you are to Sail by, to go from one place to another; as for Example, if you would go from the Harbour at A, (in this plain Chart) to another at B. To do this, fancy to your self the Line A B, and find to what Line of Rhumb it is Parallel: Thus, Set one Foot of your Compass in the very place from whence you depart at A, and the other Foot upon the nearest Line or Rhumb which you think Parallel to the imagined Line A B; (as the Compass C showeth,) then draw it forward towards B, keeping still one Foot thereof upon the Line of the Rhumb, and if the other Point describe the imagined Line A B; you may conclude that the said Rhumb (which is *W S W*) being Parallel to the Line A B, is the course you must Steer to go to B.

S. If one Rhumb alone could not serve to bring me to the desired Harbour, What must I do then? For Example, if I would go from the Harbour at A, to that D.

T. In that case you must find out as before, that Rhumb which will Clear you from the *Sands*, *Rocks*, or other dangerous places, (and bring you nearest to your Harbour) which you will find to be a *NW* b *W*, but when you are Clear of the Sands and other Impediments, and your Ship is arrived at E; then you must change your Course, making choice (as before) of that Rhumb which will bring you from E to D, which you will find to be a *W* b *N*.

P R O P. XVI.

How to know by the Plain Chart the Distance between two Places.

S. HOW shall I know the Distance between two places? As for Example, from the Harbour at A to that at B.

T. You must take with your Compass the Line A B, or the just Distance between the two places so marked, and apply the said Compass so open to the Scale of *English* Leagues, (20 to a Degree) and the Scale will show you 50 Leagues that the Harbour at A, is Distant from that at B, as was required.

Y

S. What

S. What muft I do when the Diftance between the two places is greater than the Scale of Leagues?

T. Then you muft take the length of the Scale of Leagues, and look how many times that is contained in the Space between the two places, and if there remains any odd meafure, then having taken that odd meafure with your Compafs, apply it to the firft part of the Scale of *Englifh* Leagues (if it be a *Dutch* Chart) fo fhall you know the whole Diftance, which from A to D is 59 Leagues or 177 Miles, (for a League is 3 Miles.)

PROP. XVII.

How to know by the Plain Chart, the Latitude of any Place.

S. HOW fhall I know by my Chart in what Latitude a *Point* of Land, an *Ifland*, or any other place is Situated?

T. To know that by you Chart, fet one Foot of your Compafs in the very place whofe Latitude you would know, and the other Foot on the neareft Parallel or Line of Eaft and Weft, and then draw your Compafs forward, until it touch the graduated Meridian, (that is to fay, that Meridian where the Degrees are marked) and the Foot which was on the place will fhow its Latitude; as for

Example.

If you would know the Latitude of the *Ifland* at F: Set one Foot of your Compafs in the very middle of it, and the other Foot in the neareft Parallel at G, and keeping that Foot ftill upon that Line, draw it forward until it touch the graduated Meridian; and the Foot that was at F, will fhow you that the faid *Ifland* is in 49 Degrees 30 Minutes of North Latitude, (fince the propofed *Ifland* is on the North fide of the Equinoxial.)

PROP. XVIII.

How to find the true Point of the Ship.

S. HOW can I determin by a Point, the place my Ship is arrived at?

T. To make it very eafie to you, it is fit that I fhould give you an Example: Suppofe then that your Ship being departed from

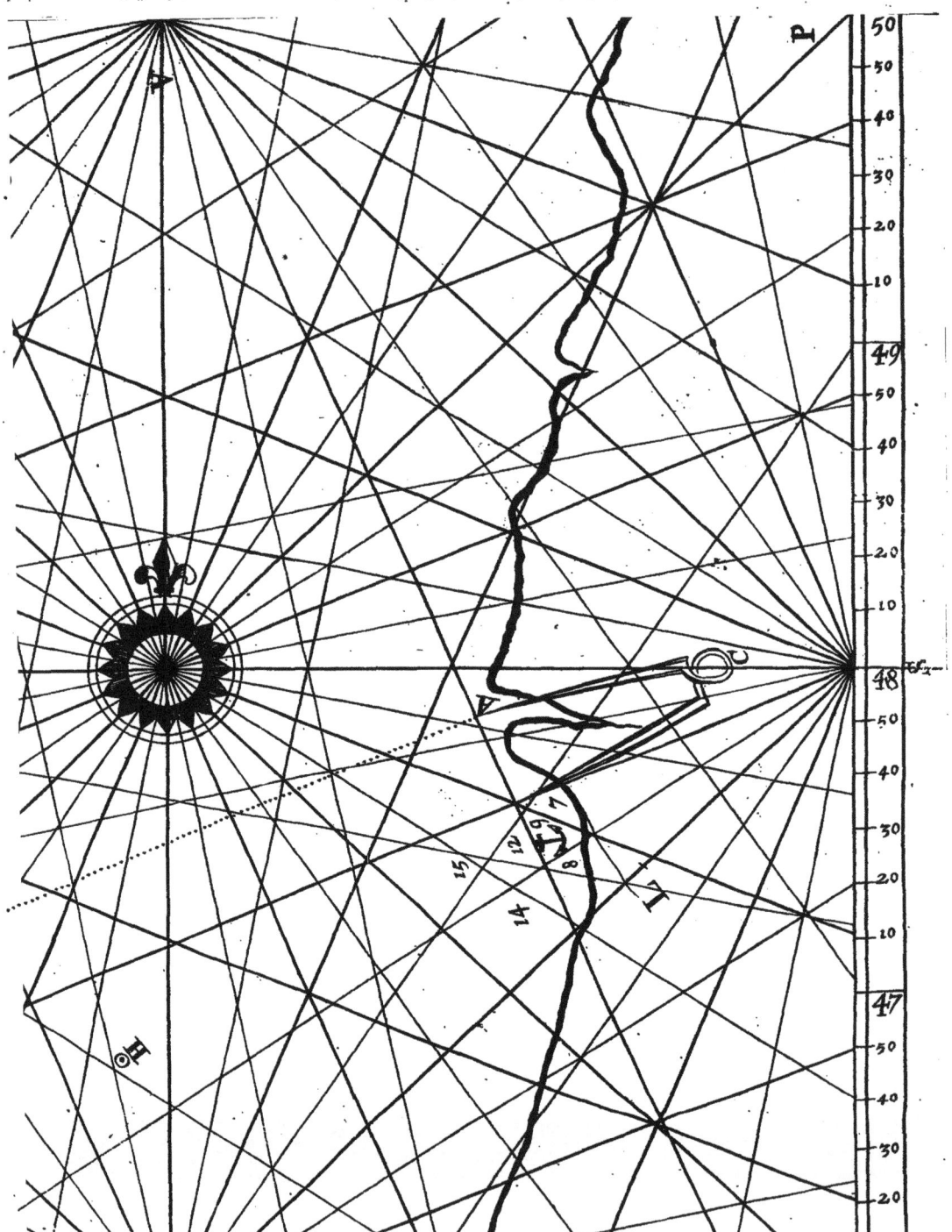

from the Point A, hath Sailed *S W* 30 Leagues, and you would fee in your Chart the place it is arrived at. Firft, take with the Compafs 30 Leagues, (upon the Scale of *Englifh* Leagues) and with another Compafs take the neareft Diftance between the place of your departure at A, and the neareft *S W* Line at L; then take in your Left-hand the firft Compafs which contains the Leagues, and fetting or placing one Foot thereof in the Point of your Departure at A, turn the other Foot towards that part your Ship hath Sailed, then fet one Foot of the Compafs which contains the Diftance of the Rhumb upon the fame Line of *S W*, (whofe Diftance you took) holding it fo with your Right-hand, that it doth not fwerve from the faid *S W* Line, draw your Compafs forward to the *N E*, untill the other Point of it, meet exactly with the Point of the firft Compafs at H, and there mark a Point, for that is the place where your Ship is at that inftant, (fuppofing you have allowed for Lee-way, &c.) and doing the fame for any other Courfes, you may fee in your Chart the place your Ship is arrived at, and from whence you may direct your Courfe again to the place whereunto you would go, or as near it as you can, in cafe the Wind be contrary.

S. Doth not this Point H, which fhoweth the place my Ship is arrived at, fhow alfo my Difference in Latitude and Longitude?

T. Yes, it doth, for if you Subtract the Latitude of your Point from the Latitude of the place from whence you departed, or rather the lefs Latitude from the greater, the remainder will be your Difference in Latitude; and the fame is to be done for your Longitude, when they are both of the fame Denominations, as you may better underftand by what has been already faid in the Second Book.

S. How do you call this manner of keeping an account of the Ships way?

T. It is called the *Dead Reckoning.*

PROP. XIX.

Proving the Plain Chart falfe, and not to be trufted to, but in very fhort Voyages.

S. HOW can the *Plain Chart* be falfe when moft Pilots make ufe of no other, and for all that bring their Ships as well to the defired Port, as thofe who Sail by that of *Mercators Chart?*

T. If any fucceed in long Voyages, by Sailing by the *Plain Chart*, you may be affured that it is more by *chance* than *skill*, except their Courfe were (Eaft or Weft) under the Equinoxial, or (North or South) under a Meridian, being certain that upon any other Point there is a confiderable Error, by which many have loft their Ships and Lives, as

many

many more will do, if not prevented by leaving the Plain Chart, and making use of *Mercator's*.

S. If Sailing by the *Plain Chart* is fo dangerous, I wifh you would prove it by fome convincing Demonſtration, for until now I have not been fenfible of it.

T. To make it very plain to you, it muſt be by fome Figure which reprefents the Plain Chart, of which I know no better than that which *Fournier* hath fet down in the 14th. Book of his *Hydrography*, which for your Inſtruction I fhall near hand defcribe to you (with his opinion) it being impoſſible to render it more intelligible by any other.

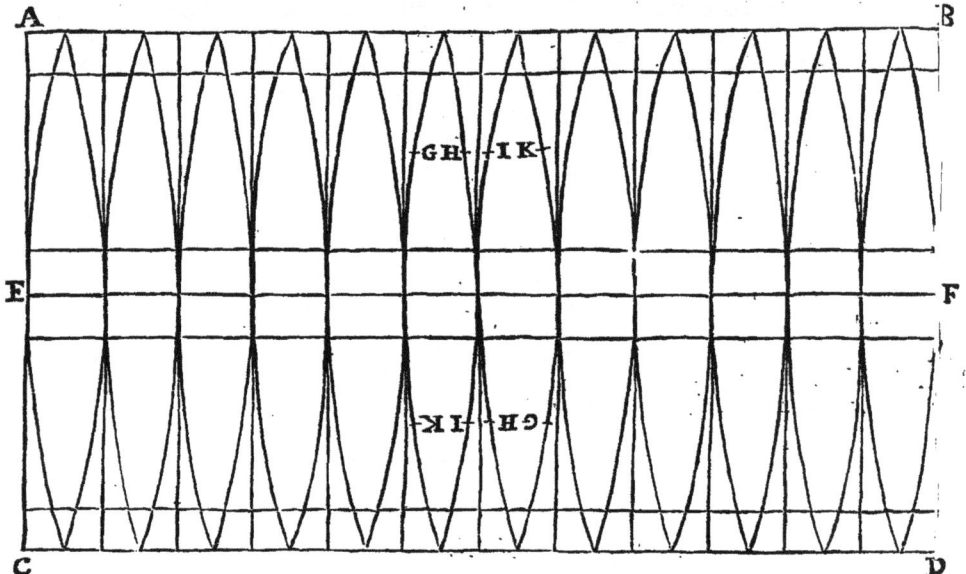

Admit then the Plat ABCD to be a plain Chart, and that the Lines AB and CD are Parallels, and are joyned to two other Lines alfo Parallels, *viz.* AC and BD, and to make a Right-angle draw threw the middle of the Chart the Line EF, which may reprefent the Equinoxial and be Parallel to CD, and double of AC; divide EF into 12 equal Parts, and opening your Compafs until it contains almoſt 9 of thofe Parts; fet one Foot thereof upon any Point of the Divifion, and the other upon the fame Line EF, from whence draw an Arch, and fo on from every Point of Divifion, until your Chart is divided as you fee, and reprefents as it were 12 pieces of Musk-Million: Now if you were t joyn (in round) the Point E with the Point F, and the extremities o ends of all thofe pointed Ribbs together, the two Points where they fhoul

joy

joyn would represent the Poles of the World, and the whole would represent the Terrestial Globe. Now if you imagin that Globe open, and displaid in such a manner that those Circular Lines become Parallels one to another, all those crooked or bowing pieces which did joyn and unite together in the Poles, will remove and be set at a Distance one from another, and their extremities or ends will fall upon a Line of as great extent or length, as the Equinoxial, and by that means those parts of the World which are at any Distance from the Equinoxial, will appear at a greater Distance one from another than they really are, and so much the more the further off they are from it.

S. Can you prove that this error is of such consequence to Navigation?

T. Yes, and that you may not doubt in the least of it, I will make it plain to you by an Example: Admit then that a Pilot is Sailing from the *Rock* of *Lisbon* to the Island *Tercera*, and that the said *Rock* is at the Point K, and the Island *Tercera* at G. I say that his Ship in going from K to G, shall have Sailed of that Parallel, but the Part K I, G H, and not the part H I, which however is marked in his Chart (although it is a *Vacuum.*) That Pilot then who sees by his Chart that the Distance between K and G, is of 330 (*English*) Leagues, and by his reckoning finding that he hath Sailed but 250 Leagues, will be perswaded that he is arrived but at H, and will believe that G is yet 80 Leagues Distant from him; and therefore Sailing with all his Sails out, if it be dark, or misty weather, he will infallibly strike against it, and make Ship-wrack, or at least will miss it, passing aside and leaving it behind him, since it is certain that the parts of that Parallel, to wit, G H and I K contain but 250 (*English*) Leagues.

S. What is the Cause of this mistake?

T. It is because this Pilot believeth that the Degrees of the 39 Parallel (which passeth through the Rock of *Lisbon*, and the Island *Tercera*) are equal to those of the Equinoxial, because they appear so by his Chart, not minding that the Meridians on the Globe are not Parallel as in his Chart, but draw nearer to one another proportionably to their Distance from the Equinoxial, and that in the 39 Parallel a Degree is almost 12 Minutes shorter than a Degree of the Equinoxial, for a Degree of the Equinoxial is 20 Leagues, and a Degree of the 39 Parallel is but 15 Leagues and a half. Likewise he will find by the Plain Chart the Distance from *Rochel* to *Canada* in the 46 Parallel to be of 1050 Leagues, although there are really but 730, each Degree of that Parallel being but of 14 Leagues, which is 6 Leagues or 18 Miles in every Degree less than appeareth by his Chart.

S. How many Miles do you think that a Pilot would mistake that should sail (by the Plain Chart) 2000 Miles in the Parallel of 60 Degrees?

T. He

T. He would commit a miftake of a 1000 Miles (except he knew how to Correct the error of his Chart) becaufe a Degree of that Parallel is lefs by half than it appeareth by his Chart; to wit, 10 Leagues; and therefore it is no wonder that fo many Ships are loft, when there is fuch great Number of ignorant Pilots, that Sail by a Chart which fhoweth them that they are a great way from the Land, when (if the fight of the Land do not rectifie the error) they will certainly ftrike on it.

S. Are there any more defects or faults in the Plain Chart?

T. Yes, there is a confiderable one in the Rhumbs, but I fhall not trouble you with the Demonftrations of it, fince all thefe errors are Corrected in the following or next Chart, which I recommend to you as the beft for your Practice, and the only one I defire you to ufe at Sea.

PROP. XX.

How to make the true Sea Chart, commonly called Mercator's Chart.

S. SINCE this Chart is the beft for practice, pray fhow me how to make it?

T. There is no neceffity for your making your Chart, fince you may buy them better made than you can make them your felf, however becaufe that by doing it, you will better underftand how the Degrees of Latitude come to be unequal; I think fit to fatisfie you, and in order to that, you are firft to take care that the skin of Parchment defigned for your Chart, be of the largeft fize, and very white and fmooth. But if you cannot get one fmooth and white, you may make it fo your felf; thus, Firft rub it with fome Cerufs, or white Lead, which is to be had at any Colour-fhop, and wipe it with a clean linnen Cloth, then boil the fhreds of your Velom in fair Water, untill it comes to be of a Glüifh or Clammy fubftance, and then having well ftretcht your skin upon a board, rub it with a linnen Cloth or Spunge dipped in this water, and when it is dry rub it again with white Lead, and your Velom will be very white all over, and fo fmooth that nothing fhall ftop your Pen, provided you rub it well after this fecond time of rubbing it with white Lead; for elfe the Rhumbs, and what elfe you fhould defcribe on your Chart would foon wear out and difappear.

S. When my Velom is thus prepar'd, how muft I draw my Chart?

T. You muft firft defcribe upon it a Square as directed for the plain Chart, then divide the Equinoxial Line, into as many equal Parts or Degrees as you have occafion for; then draw upon a piece of Paper or piece of Board, the Line A B equal to one Degree of the Equinoxial, at the end of which at A, raife a Perpendicular of the fame length A C, then

draw

draw the Quadrant or quarter of a Circle C B, and divide it into 90 Degrees; then from the Point B, draw the Line B D Parallel to A C, and draw the Lines called Secants through every Degree or Point (of Divifion) of your Quadrant, as A F, A G, A H, &c. And fo on untill the 70 Parallel, if you have occafion to make your Chart fo large, that done, graduate your firft Meridian: Thus,

For the firft Degree of Latitude, take the length of the Line A B; (which here we fuppofe to be but one Degree of the Equinoxial,) for the fecond Degree the next Line or Secant A F, for the third Degree the Secant A G, for the fourth A H, for the fifth A I, and fo forth, every Degree increafing as the reft of the Secants do, to which they are to be equal: Then draw Parallels through every (fifth or) tenth Degree of the Meridian, and Meridians through every (fifth or) tenth Degree of the Equinoxial; as for the Rhumbs, Harbours, Capes, Iflands, Rocks, Shoalds, &c. to be defcribed on it, they are to be done as you have been taught in the plain Chart, you muft place them by their true Latitude and Longitude, and then your Chart will be fit for ufe. But here take notice, that if you would make a particular Chart; as for Example, from 30 Degrees of Latitude to 50, then the Secant 30, at H, muft be the firft Degree from the graduated Parallel, A I the fecond Degree, and fo increafing onward from 30 to 50, as the Secants do.

S. Why do you make the Degrees of Latitude to increafe more and more, the further they are from the Equinoxial?

T. There is a neceffity (for the conveniency of Pilots) to make the Meridians of their Chart Parallel one to another, for then if a Right-line cuts them, it will make equal Angles (*Euclid,* Prop. 28.) according to the nature and definition of the Loxodromick Lines reprefented by the Rhumbs upon the Terreftial Globe, it is alfo neceffary that thofe Degrees of the Meridian fhould keep the fame Proportion with the Degrees of the next Parallel as they do upon the faid Globe, then becaufe we make the Degrees of all the Parallels equal to thofe of the Equinoxial, (by drawing the Meridians Parallel one to another) we muft alfo increafe the Degrees of the Meridians (or Latitude) over and above the Degrees of the neareft Parallels according to the Proportion that is between them, for (as Mathematitians know) a Degree of a Parallel hath the fame Proportion to a Degree of the Equinoxial or Meridian, as the Sine Complement of its Latitude to the Total Sine; (or Radius;) Trigonometry teaching us, that there is the fame reafon or proportion between the Sine Complement of an Arch and the Total Sine, as the Total Sine to the Secant of that Arch. And therefore to obferve the fame Proportion, as we increafe the Degrees of the Parallels by the Parallelifm of the Meridians; fo we muft alfo increafe the Degrees of the Meridians by the Addition of the Secants.

S. Why do you make no Scale of Leagues for this Chart?

T. Becaufe the Degrees of the Meridian is the true Scale, which how great foever it appeareth, it containeth but 20 Englifh Leagues or 60

Miles;

Miles (or Minutes;) for if your Voyage had been between two Parallels; as for Example, between the 40 and 50, the Degrees of the Meridian Comprehended between the two places must serve for a scale to measure your Course, and so making use of a Scale which represents the Leagues or Miles by a greater Line, the further we shall be from the Equinoxial, the fewer Leagues we shall find in the Parallel nearest to the Pole, by which you may easily understand that although we have increased the Degrees of the Parallels and Meridians, we have not done it in value of Leagues, because we represent also the Leagues or Miles by a greater Line, and we make use of a greater measure.

PROP. XXI.

How to know by Mercator's *Chart what Rhumb (or Point of the Compass) you must Steer to go from one Harbour to another.*

S. HOW must I know (by *Mercator's* Chart) upon what Rhumb to Sail from one Harbour to another?

T. This is to be found out in the same manner as by the plain Chart: thus, place a Ruler upon the two determined Harbours, (or imagin a Line drawn from one Harbour to another) and look what Rhumb is Parallel to the side of your Ruler, (or imagined Line) and that shall be the Course you must Steer, as for

Example.

If you would go from A to C, you will find that a *S S W* Rhumb will be Parallel to the imagined Line, or edge of your Ruler placed on A and C; by which you know that your Course from A to C is *S S W.*

PROP. XXII.

How to find by Mercator's *Chart, the Distance of two places.*

S. PRAY show me how to find the Distance of two places that are under the same Meridian, (or North and South one from another.)

T. To measure the Distance between two places in the same Longitude, you must only look how many Degrees of the Meridian is

com-

comprehended between the two propofed places, which being Multiplied by 20, (each Degree containing 20 Leagues or 60 Miles) will fhow you the Diftance in Leagues: Example, The Diftance between A and B is 10 Degrees of Latitude, which being Multipled by 20, gives 200 Leagues for the Diftance between A and B.

S. What if there be odd Minutes over and above the Degrees ?

T. You muft add a League for every three Minutes.

S. How muft I find the Diftance of two places that are in the fame Parallel ?

T. When two places are in the fame Latitude, you muft meafure their Diftance by the Degree of their Latitude, as for Example, if you would know how far the Ifland at D, is from that at B; you muft take upon the Meridian the length of the Degree R S, with which you are to meafure the Diftance B I, for fo often as you fhall find that length between the two propofed Iflands, as here 18 times, fo many times 20 Leagues is D Diftant from B, *viz.* 360 Leagues.

S. Muft I always take 30 Minutes on each fide of the Parallel of the two places propofed, as you have done in taking the Degree R S ?

T. Yes, if you will be the more exact.

S. I underftand this well enough, but how fhall I meafure the Diftance between two places that differ both in Latitude and Longitude ?

T. You muft meafure it by the Degrees of the Meridian comprehended betwixt the two places, as for Example, if you would know the Diftance from A to C, you muft firft meafure the Diftance A F, by the 10 Degrees of the Meridian comprehended between the two Parallels F M, *(viz.* between 60 and 70,) and the Diftance F G, by the 10 Degrees from 50 to 60, and G C, by the 10 Degrees from 40 to 50, counting every Degree for 20 Leagues, and every third Minute for a League: Or elfe thus, Look how many Degrees the Difference of Latitude contains, and take with your Compafs as many Degrees upon the Equinoxial Line, then laying a Ruler on the two places, apply one point of your Compafles fo to the Edge of the Ruler, that the other point (defcribing an Arch of a Circle) may juft touch (but not cut) one of the Parallels or Lines of Eaft and Weft, that pafs between the two Places, and then meafure the Diftance from the Point (or Center) where your Compafs was fixed by the fide of your Ruler, to that place of the Parallel (fo touched) where the Ruler croffes it, and that Extent being meafured in the Equinoxial, will fhow you the Degrees or Leagues between the two places.

Example: *To find the Diftance between* A *and* C.

A, being in 70 Degrees of North Latitude, and C in 39 Degrees, their Difference is 31; which being taken off from the Equinoxial graduated Line, and your Ruler laid over the two Places, A and C, one Foot of your Compafs being placed by the fide of the Ruler at I; the

Z

other

other will juft touch the Parallel of 60 Degrees at L, which the Ruler croffes at F. Then take with your Compaffes the Diftance from I to F, which being meafured on the Equinoxial Line, you will find 34 Degrees, which being Multiplied by 20, gives 680 Leagues for the Diftance between A and C.

PROP. XXIII.

How to find (by Mercator's *Chart) the Latitude and Longitude of any Place.*

S. IS the Latitude and Longitude of places to be found by *Mercator's* Chart, in the fame manner as by the Plain Chart?

T. Yes, the very fame. To find the Latitude, take with your Compafs the Diftance betwixt the propofed place and the neareft Parallel, (or Line of Eaft and Weft;) then remove your Compafs fo open to the graduated Meridian, (or Line of North and South,) and fetting one Foot of it on the fame Parallel, (as before) the other Foot will fhow you upon the Meridian the Latitude of the place: And to find the Longitude, take the Diftance betwixt the propofed place and the neareft Meridian, then removing your Compafs upon the Equinoxial Line, fet one Foot of it upon the fame Meridian, (as before) and the other Foot will fhow you upon the Equinoxial the Longitude required; and in this manner you will find that the Ifland at N is in 25 Degrees of Latitude North, and 15 Degrees of Longitude Eaft.

PROP. XXIV.

The Latitude and Longitude of your Ship being known how to prick it down in your Chart.

S. I Fancy this very eafie?

T. So it is, for there is no more in it, than to take the Diftance between the Degree of Latitude your Ship is in, and the neareft Parallel; holding your Compafs fo open with your Left-hand, with your Right-hand and another pair of Compaffes, take alfo the Diftance between the Degree of Longitude your Ship is at, and the neareft Meridian, and holding your two Compaffes flat, draw them forward untill

the

the two Feet of the Compaffes which mark the Degree of Latitude and of Longitude meet, and there make a Point, for that is the place your Ship is at, at that inftant; but take care in drawing your Compaffes that the Foot of the firft Compaſs do not ſwerve from the Parallel, whoſe Diftance you took, nor the Foot of the laft Compaſs from the Meridian; and do not forget to make a little Circle about your Point, that you may find it the better when you have a mind to look on it.

PROP. XXV.

How to prick down your Reckoning by Mercator's *Chart.*

S. I S it harder to prick down my Reckoning by *Mercator's* Chart, than by the Plain Chart?

T. No, it is almoft the ſame, all the Difference being only in the Leagues, which in *Mercator's* muft be taken upon the Degrees of the Meridian, and not upon a Scale of equal parts as in the Plat, which to make plainer to you, I fhall give you an

Example.

Admit then that departing from D, in the Latitude of 60 Degrees North, you have Sailed *E N E* 50 Leagues, or 150 Miles; to wit, the firft 4 Hours 15 Miles, the next 4 Hours 20, and ſo forth, and that having added all thoſe Miles Sailed in 24 Hours, you find them to a-mount to 150; you muft with your Compaſs take upon the Meridian (to wit, from 60 Degrees towards 70) the 150 Miles Sailed, then remove your Compaſs, ſetting one Foot of it in D, and turning the other Foot towards that part you have Sailed, with another Compaſs mark the Rumb of your Courſe, (as you were taught by the 15th. Propoſition of this Book,) and where the two Feet of the Compaffes meet mark a Point at H, for the place of your Ship at that time, and the ſame is to be done the next Day if you alter your Courſe, or Sail upon any other Point.

S. Is it as eaſie to take any Number of Miles, as Leagues upon the Meridian?

T. Yes, becauſe commonly the Degrees are divided into 12 equal parts of 5 Minutes each, which Minutes may alſo be taken for Miles, ſince a Degree contains 60 Miles as well as 60 Minutes, and therefore it is as eaſie to take your Miles or Leagues upon the Meridian, as upon a Scale made on purpoſe for it.

Z 2

PROP.

PROP. XXVI.

The Latitude of two places, and their Distance given, to find the Course and Difference in Longitude.

S. IS this Proposition of any use in Navigation?

T. Yes, for by it you may Correct your Course, as I shall make it plain to you; suppose then that by my Observation I know the Latitude my Ship is arrived at, and by the Log-line the Miles Sailed, but am in doubt of my Course, because I could not observe the Variation or any thing else, as for Example, Admit that departing from A, which is in 70 Degrees 0 Minutes of Latitude North, and 5 Degrees 0 Minutes of Longitude: I Sail several Days in a Storm between the South and the West, upon the same Point of the Compass, until by Observation I find I am arrived in 60 Degrees of Latitude North, and that by my Reckoning (to which I trust more than to my Course) I have Sailed 270 Leagues: I take upon the graduated Meridian the 270 Leagues Sailed, *viz.* upon the Degrees from 70 to 60, then setting one Foot of my Compass in A, with the other Foot I describe an Arch which toucheth the Parallel of 60 Degrees in L, which is the place my Ship is at at that instant, then with my Compass I look what Rumb is Parallel to the imagined Line A L, and I find that it is a *S W*, by which I conclude that my Course hath been *S W*. Now as to the Difference of Longitude, I add 360 Degrees to the Longitude of A, (since the Longitude of both places is of Different Denominations) and there will come 365, from which I Subtract the 338 Degrees of the Longitude of my Ship at L, and the remainder 27 Degrees is my Difference in Longitude.

PROP. XXVII.

The Latitude of two places and the Course being known, how to find the place of your Ship, and how many Leagues you have Sailed, and the Difference in Longitude.

S. IS this as easie as the precedent Proposition?

T. Yes, and as necessary, therefore I must give an Example of it, admit then that I depart from N in 25 Degrees of Latitude North, and 14 Degrees of Longitude, and Sail away *S S W*, until by Observation I find I am arrived in 10 Degrees of Latitude North,

and

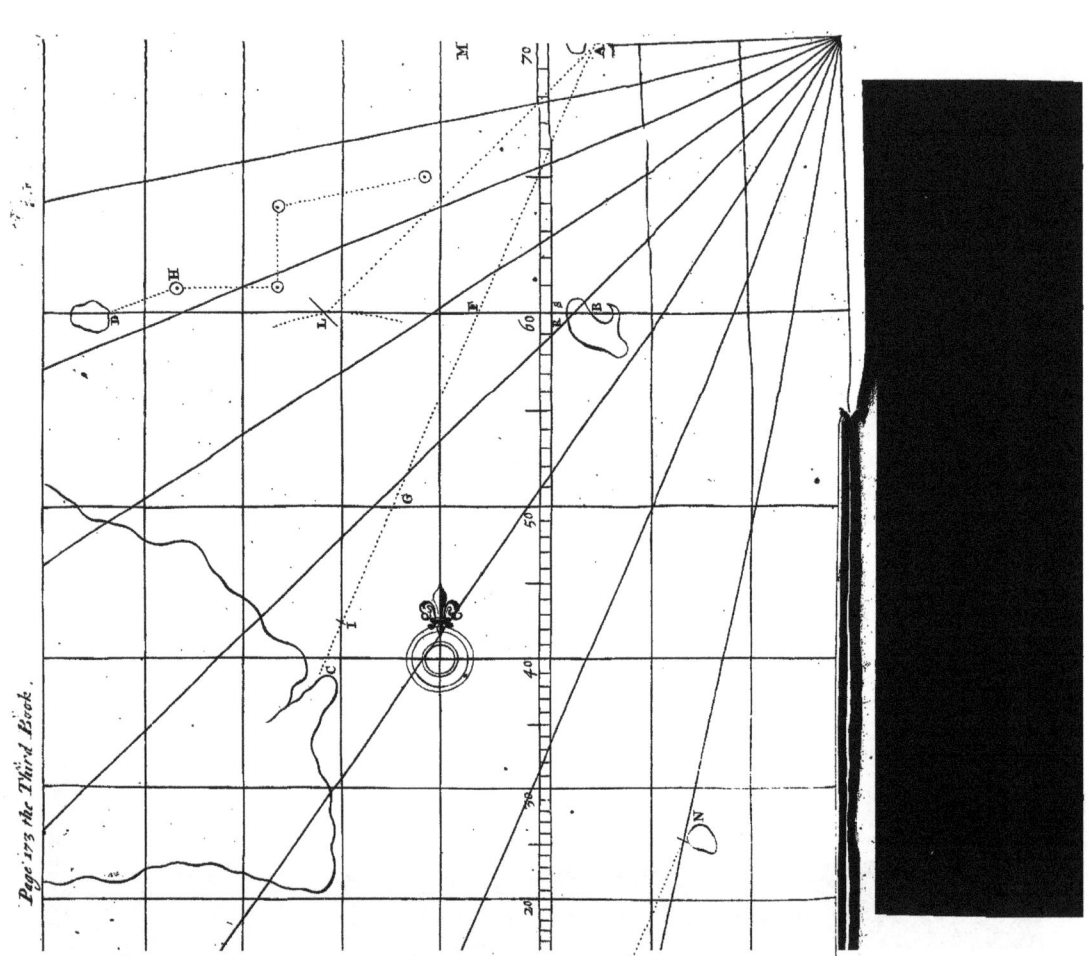

Page 173 the Third Book.

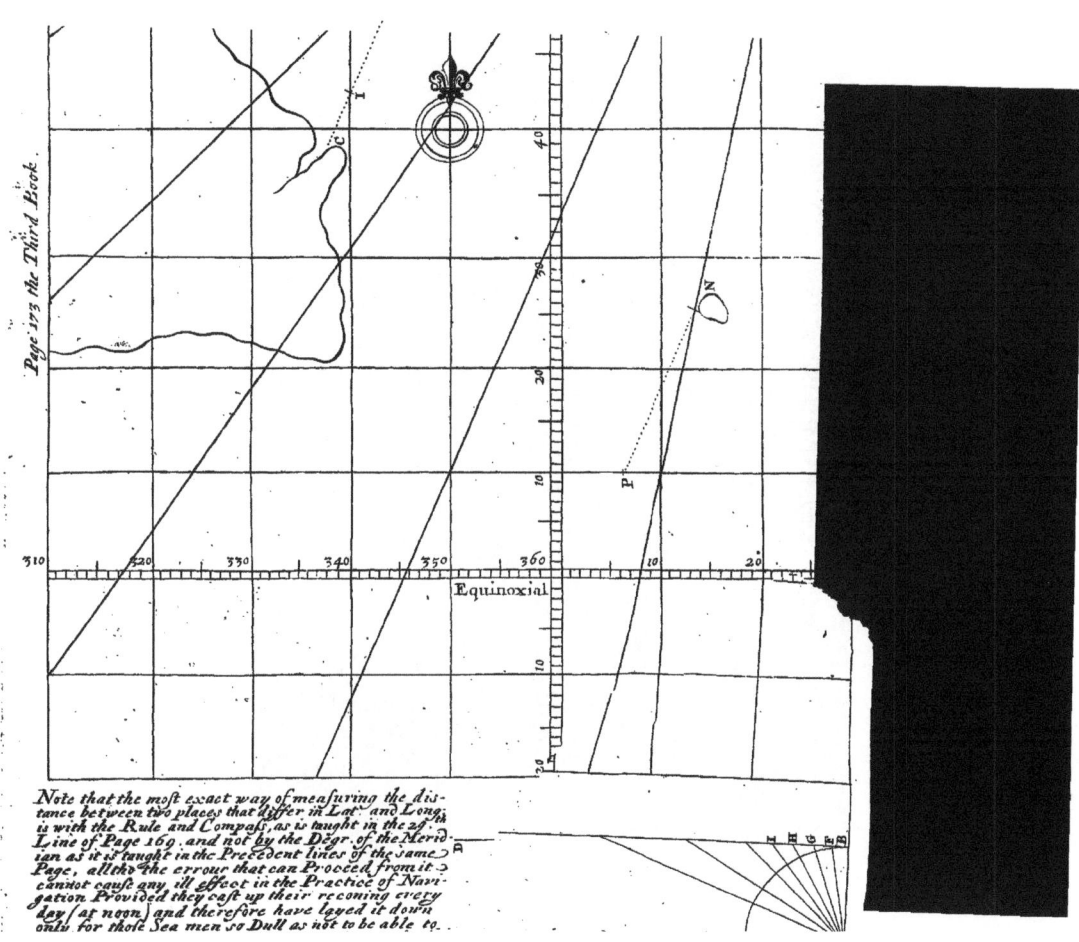

Equinoxial

Note that the most exact way of measuring the distance between two places that differ in Lat. and Long,
is with the Rule and Compass, as is taught in the 29th
Line of Page 169, and not by the Degr. of the Meridian as it is taught in the Precedent lines of the same
Page, altho the errour that can Proceed from it
cannot cause any ill effect in the Practice of Navigation Provided they cast up their reconing every
day (as noon) and therefore have layed it down
only for those Sea men so Dull as not to be able to.

and then would find by my Chart the place of my Ship; the Leagues Sailed, and my Difference in Longitude: First I set one Foot of my Compass upon the Point of my Departure at N, and the other upon the nearest *S S W* Line, and drawing my Compass forward as before taught, I observe where the Foot (that was) at N, toucheth the Parallel of 10 Degrees, and there I prick my Point at P, for the place of my Ship, *viz.* 10 Degrees North Latitude, and 6 Degrees East Longitude; then if I measure the Distance N P, upon the Degrees of the Meridian, to wit, from 25 Degrees towards 10 Degrees. I shall know how many Leagues (or Miles) I have Sailed, *viz.* 320 Leagues or 960 Miles: As for the Longitude, I subtract the lesser out of the greater, and the remainder is my Difference in Longitude, which in this Example I find to be 8 Degrees.

P R O P. XXVIII.

The Course, Distance, and Variation, being known, how to find the place of your Ship.

S. **C**AN this be done by *Mercator's* Chart?

T. Yes, it can, but not by every Pilot, for there is few that know it, and therefore I will teach it you, being willing to serve any that incline to that Noble Art of Navigation; and I hope that Examples will make it very easie and plain to you: Note, this is not Mathematical, and is only laid down for those who know but the use of the Chart, and yet would Correct the Variation by it. Admit then that with a Compass whose Variation is 7 Degrees Westerly, I have Sailed *N N W* 99 Miles, and then would see in my Chart the place of my Ship. First I consider that since the Course and Variation are of the same side, the true Course hath been further from the North than a *N N W*; to wit, 7 Degrees more Westerly; I must then divide the 99 Miles Sailed by the 11 Degrees, which each Point of the Compass contains, and the Quotient will be 9, and because of the 7 Degrees of Variation, I take 7 of those parts which amounts to 63 Miles, for 7 times 9 makes 63, which being Subtracted from 99, there remaineth 36; by which I know that I must prick my Course at twice; to wit, 36 Miles upon the *N N W*, and 63 Miles upon the *N W* b *N*, and, this last Point will show (in my Chart) the place of my Ship, as was desired.

S. What must I do when the Variation is both of Degrees and Minutes?

T. In that case it is necessary to know the *Rule of Three*, for admit that the Variation of your Compass is of 8 Degrees 20 Minutes Westerly, and your Course hath been *N N W*; you must say, *If* 11 *Degrees*

15 Minutes contained between the N NW, and N W b N giveth 91 Miles. What will the 8 Degrees 20 Minutes of Variation give? And you will find 73 Miles; which you must prick upon the *N W b N*, and the rest of the 99 Miles; to wit, 26 Miles upon the *N NW*, and this last Point will mark the place of your Ship.

S. How must I prick down my Reckoning if I had Sailed *S W* 120 Miles, with a Compass that varies 17 Degrees Westerly?

T. Since the Course and Variation is of the same Denomination, or both Westerly; your true Course hath been further from the North, than the *S W*, as many Degrees as your Compass did vary; that is to say, 17 Degrees more Southerly: Now because one Point of the Compass contains 11 Degrees 15 Minutes; I first Subtract 11 Degrees 15 Minutes for the *S W b S*, from 17 Degrees of Variation, and there remains still 5 Degrees 45 Minutes from the *S W b S* to the *S S W*; I must then prick my Reckoning at twice; to wit, one part upon the *S W b S*, and the other upon the *S S W*, and to know how many Miles (or Leagues) upon each Rumb; I say, (by the Rule of Three) *If 11 Deg. 15 Min.* (the Distance of the *S S W* from the *S W b S*) *give 120 Miles, How much will 5 Degrees 45 Minutes give?* and you will find it to be 61 Miles, which I prick upon the *S S W*, and the remaining 59 Miles upon the *S W b S*.

S. Must I always retire from the North to know what my true Course hath been?

T. Yes, when your Course and Variation are on the same side; that is to say, when they are both Westerly, or both Easterly.

S. What must I do when the Variation and my Course are on contrary sides?

T. You must then approach nearer the North; as for Example, Admit that I Sail *W S W* 66 Miles, with a Compass that varies 6 Degrees Easterly, I must retire from the South, for my true Course hath been 6 Degrees nearer the North than a *W S W*, because the Variation is 6 Degrees Easterly; therefore I divide the 66 Miles by 11, and Multiply the Quotient by 6, and that gives 36 Miles, which being Subtracted from 66, there remaineth 30 Miles, which I prick upon the *W S W*, and the 36 Miles upon the *W b S*, but when the Variation is both of Degrees and Minutes, you must do as in the precedent Example.

S. I understand now very well how to find out what my true Course hath been; but what must I do to make good my Course notwithstanding the Variation?

T. When the Variation and your Course is on the same side, you must approach or draw nearer the North, as many Degrees as the Variation is, but when the Variation and Course differ, (that is to say, when one is Westerly and the other Easterly) then you must retire from the North as many Degrees as there is Variation, as you will better understand by this Example. Admit then, that to go to an Harbour Distant from me 180 Miles, my Course is *N W b W* 3 Degrees 30 Minutes more Westerly, and that the Variation of my Compass is of 13 Degrees Easterly,

Eafterly; the Queſtion is, how muſt I prick down this Courſe, and upon what Point of the Compaſs I muſt Steer to make my way good a *NW* b *W* 3 Degrees 30 Minutes Weſterly? As to the Reckoning, there is no doubt but it muſt be pricked upon the propoſed Rumb, ſince I deſign to make it good by my Courſe, but then it is of neceſſity to Cape or direct my Courſe ſo, that I make good a *NW* b *W* 3 Degrees 30 Minutes Weſterly; I muſt then contrary to the precedent Example retire from the North, or rather from the *NW* b *W*, (3 Degrees 30 Minutes Weſterly) as many Degrees as there is Variation, and therefore I muſt direct my Courſe *W N W* 5 Degrees 15 Minutes more Weſterly, for 3 Degrees 30 Minutes, and 13 Degrees of Variation comes to 16 Degrees 30 Minutes, from which I Subtract 11 Degrees 15 Minutes Comprehended between the *NW* b *W* and the *W N W*, and ſo there remaineth 5 Degrees 15 Minutes; now to know how many Miles I muſt prick upon the *NW* b *W*, and upon the *W N W*; (becauſe of the 5 Degrees 30 Minutes) I ſay by the Rule of Three; *If 11 Degrees 15 Minutes give 180 Miles, What will 5 Degrees 15 Minutes give?* And there will come near 84 Miles, which being Subtracted from 180, remaineth 96 Miles, which I prick upon the *NW* b *W*, and the 84 Miles upon the *W N W*. Admit I were to Sail to an Iſland Diſtant from me 255 Miles, my Courſe is *S E* b *S* half a Point Southerly, (that is to ſay, 5 Degrees 37 Minutes more to the Southern) and that I am forced to make uſe of a Compaſs whoſe Variation is of 11 Degrees 15 Minutes Eafterly; the Queſtion is, how my Reckoning muſt be prickt down, (in my Chart) and upon what Point of the Compaſs I muſt Steer to make my Courſe good a *S E* b *S* 7 Degrees 37 Minutes Southerly?

Anſw. Since the Variation is on the ſame ſide with the Courſe; to wit, both Eafterly; I muſt approach to the North as many Degrees as my Compaſs varieth, and therefore I muſt direct my Courſe *S E* 5 Degrees 37 Minutes Southerly, now to prick this Courſe upon my Chart; I ſay by the Rule of Three, *If 11 Degrees 15 Minutes give 255 Miles, What will 5 Degrees 37 Minutes give?* And you will find it to be 127 Miles, which being Subtracted from 255, remaineth 128 Miles, which I muſt prick upon the *S E*, and the 127 Miles upon the *S E* b *S*, becauſe of the 5 Degrees 37 Minutes.

P R O P.

PROP. XXIX.

*What must be observed by those that will keep an account
of the Ships way.*

§. THIS must be very necessary, and therefore pray give me
your advice concerning it.

T. My advise is, that you be very careful in this Affair,
for it is the essential part of Navigation; you must look often
upon the Compass to be certain of your Course, and that your Men
have Steered as was commanded them; think this of Consequence, and
place a careful Man to look to it in your absence, a person whom you
can trust, that you may have a true account of it; and also if the
Wind have blown fresher or calmer, and what ever hath happen'd
when you were a Sleep, especially when you have a Lee Shore and Sail
near it in the Night time, that being very dangerous, many Ships having
been lost by the negligence of Pilots, and therefore when you Sail near
the Shore in the Night time, you ought to Steer your Course a Point of
the Compass more towards the Sea, (or from the Land) than you would
do in the Day time, to prevent any accident that might happen, since one
cannot be too careful. When we Sail close by a Wind, or as near it as
we can with our top Sails out, we allow commonly a Point of the Compass for the Lee-way; as for Example, if the Wind was *E S E*, and we
should Cape *S*, (which is as near as any cross Ship can well Lay) the Course
would be *S b W*. If with our Courses only, we allow commonly two
Points for the Lee-way; and if with our Main Sail only, (which is called
a Trie) we allow for Lee-way 4 Points of the Compass: Hulling in a
Storm we allow 6, and some times 7 Points of the Compass for Lee-
way, and then the Ship may drive 10 or 12 Leagues, and some times
14 in 24 hours. But these Rules or Directions are not always certain,
for according as the Sea and Currents are you must judge, (of your Course)
besides some times a Sea will take your Ship on the Lee-bow, and then
the Drive (or Lee-ward-way) is lesser, some time also the Sea takes her
on the Weather-bow, and then she drives more, when the Sea is smooth
we commonly allow less for Lee-ward-way. To be certain how much
your Ship hath fallen to Lee-ward, you must look to the wake or smooth
Water which your Ship leaveth at Stern, setting it by the Compass,
for its opposite Point will show you the true Course of your Ship; as for
Example, Admit that Sailing (close by a Wind) exactly *S*, the Wind at
E S E, the wake (or smooth Water) of your Ship falls at *N b E*, its
opposite Point, which is the *S b W*, will be your Course, and not the
South, because of your Lee-ward-way. As to the Distance or Number
of Miles Sailed, because there are several ways by which you may judge
of

of it; I think fit to Treat of it by it self in the following Propofitions, and therefore I fhall only add here, that you muft never fail to make your Obfervations when the weather permits it, and to write down all that happens in your Voyage, as for Example: Your feveral Courfes, how many Miles (or Leagues) you have Sailed upon fuch a Point of the Compaſs, or upon every fhift of Wind; the time of the Day or Night you Tack'd about, and when you faw fuch a Cape, Ifland, and the like, how it did bear from you; what Ships you have feen in fuch and fuch a Traverfe; if it hath been fair weather, if it did blow hard, and with what Wind, in fhort all that happens in 24 hours is to be Written down in the fame manner, as I will fhow you by the Journal, which you will find in the Fifth Book, for an Example of that which you are to keep at Sea, and in the mean time, let me advife you to have a care of the Lee-latch when you Sail Large and it bloweth hard.

PROP. XXX.

How to judge of your Diftance, or Number of Miles Sailed.

S. ARE there not feveral ways to judge of the Diftance Sailed or run?

T. Yes, there are; but the moft practiced by our Pilots, is by the Log and half Minute Glafs, whofe ufe is very well known by moft Sea-men; neverthelefs becaufe there may be many beginners as your felf, that never faw it, or if they did, do not know upon what ground they ufe it, I fhall firft fhow you how to make it, then its ufe, and at laft its defects or errors.

S. How is it commonly made?

T. Firft they take a fmall Line of 100 Fathom, or there about, at one end of which, they faften a piece of board of about 8 Inches in length, and 4 in breadth, to which they faften a piece of Lead at the bottom that it may fwim right up, they faften alfo on the other fide of the Log two wooden quils to fupport it the better upon the water, about 18 or 20 Fathom from this Log-board they make a knot, and from that knot they divide the Line by feveral other knots or marks made at 7 Fathoms Diftance one from another, and then rowling it upon a tower made on purpofe, it is fit for ufe.

S. Upon what ground do they make their knots at 7 Fathoms one from another?

A a

T. It

T. It is grounded upon this, that five of our Feet make a Geometrical pace, and 1000 such (Geometrical) paces a Mile, and 60 such Miles a Degree, by which account a Degree containeth 300000 of our Feet, and one Mile 5000 Feet; now because half a Minute of time is the 120th. part of an hour, they make the Log-line to Answer to that Proportion, by taking the 120th. part of a Mile, which you will find to be 41 Feet $\frac{2}{3}$; but because it wants but little of 42 Feet, (the length of 7 Fathom's) for more convenience they mark their knots at 7 Fathom's Distance one from another.

S. Show me now the use of this Line?

T. The use of it is thus: From the Stern of the Ship the Pilot heaves out the Log, and lets run with it 12 or 15 Fathoms of the Line before he reckons any thing, for fear that the height of the Stern and oddy should cause some error; as soon as he comes to the first knot (where the Division of the Line beginneth) they turn the half Minute Glass, letting the Line run out with ease until the half Minute is all past, and then just at that time, he that holds the half Minute Glass bids aloud *Stop*, which the Pilot doth accordingly, and then counts how many knots hath run out in the time that the half Minute Glass was running, and as many knots as he finds, so many Miles he reckons that his Ship hath Sailed in an hour; and for half a knot, he reckons half a Mile, as you will better understand by the next Proposition.

S. What are the Errors or Defects of the Log?

T. The Error of the Log is in the Division of its Line, for it is the opinion of most Learned Men of our Age, that there ought to be 50 of our *English* Feet Distance from knot to knot, and not 42, having found by good Experience that a Degree of the Earth containeth a great deal more than 300000 *English* Feet.

S. What is the Cause of so great an Error?

T. It is the measure with which we divide the said Line; that is to say, our *English* Foot, which is too little, and less by 2 Inches $\frac{1}{10}$, than the Foot of *Boloign*; for although it be very true that 5 Feet make a Geometrick pace, a 1000 paces or 5000 Feet a Mile, 60 Miles or 300000 Feet a Degree, you must know that the Foot differs according to the diversity of Countries, and that the Foot generally made use of to measure a Mile, is that of *Boloign*, (as they call it) which is 2 Inches $\frac{1}{10}$ part of an Inch greater than our *English* Foot; and therefore Mr. *Norwood* might very well affirm (in his *Sea-man's Practice*) to have found by his own experience, that a Degree of the Circumference of the Earth containeth 367200 of our *English* Feet, when the *Royal Academy of Sciences*, and many more Learned Men of our Age, find yet more, their measure being reduced to *English* Feet, and therefore you may very well follow the said Mr. *Norwood's* advise, which he giveth you in the same Book as followeth: *Because the Ships way is more than really appears by the Log-line, and because it is more safe to have the reckoning to be somewhat before the Ship, together with the evenness of the Numbers, to allow but* 360000 *Feet to be one*

Degree

gree, and consequently 6000 English Feet to be one Minute or 60th part of a Deg. vulgarly call'd a Mile, which Number being divided by 120, giveth 50 Feet betwixt knot and knot upon the Log-line. So that upon this ground if a Ship runneth out one of these knots in half a Minute, she runneth one Mile an hour, and if more accordingly.

S. If the knots should be at 50 Feet Distance from one another upon the Log-line, how cometh it then that most Pilots mark them but at 42 Feet?

T. It is because they do not know that 5 of our *English* Feet maketh the Geometrical pace less than really it is, and by consequence the Miles, for else they would not be so obstinate to follow the ill Custom of their predecessors; since their own experience must needs show them their error, which is to out-run their Ship by their Reckoning, if their half Minute Glass be of 30 Seconds, or just half a Minute.

S. How shall I know if my half Minute Glass be of a true length, and how to measure any small portion of Time?

T. It is very easie, and I do not know any that hath taught it in more intelligible terms, than Mr. *Phillip's* in his *Advancement of Navigation*, where you will find that this Experiment is thus to be performed.

" Take a Bullet of any weight whatsoever, and make fast a piece of
" Thread or Silk to it, being 38½ Inches in length from the Center of
" the Bullet, unto the end of the Thread, where a Noose must be made
" to hang it on a small Pin, which is to be fasten'd to any place where
" the Bullet may swing freely.

" This Pendulum being thus prepared, hang its Noose on the Pin,
" the Thread being exactly 38½ Inches between the Center of Gravity,
" and the Center of Motion, each of the swings of this Bullet (being
" either swift or slow) shall be a true Second of Time; so that 60
" of these swings will be the true length of a Minute, and 30 the true
" length of half a Minute, so by this ingenious Experiment you may know
" which of all your half Minutes is a true Glass, and if you have no
" Glass, you may measure any small portion of time by this Experi-
" ment, for half a Second of Time is discovered every time the Pendu-
" lum doth pass the Perpendicular, that is supposed to fall from the Pin
" whereon the Pendulum doth hang.

Note, that it is best to let the Pendulum vibrate according to the length of the Ship, and to hang it near the Main-mast, where the least Motion of the Ship is. And if the Ship be tossed, it will be best to make the Pendulum but a quarter of the length above assigned, that is, 9 Inches and ⅝ parts of an Inch long, and count two vibrations, that is, one backward and one forward, to a Second.

S. Admit that I find my half Minute Glass of true length, is there no danger at all to make use of a Log-line whose knots are at 50 Feet Distance one from another?

T. No,

T. No, and there is no doubt but you will find your Reckoning more just and exact by making use of a Line so divided, however if you fear any thing, because it is not yet so much in use as the other of 42 Feet, you may make Trial of both together, and that which you find truest by your own Experience, you may make use of afterward in other Voyages: But take notice that in this Practice you must Act with great Prudence, taking care to discover if there is any Current, for to succeed by it the Log must remain where it falls, and therefore if the Current carries it a Stern, or push it forward, there will be some mistake, and likewise if you Sail before the Wind it will be pushed forward according as the waves are, and it will be carried a Stern when you Sail close by a Wind, in all which cases you must give some small allowance, shortening or lengthening your Reckoning according as the occasion requires.

P R O P. XXXI.

How to judge of the Ships way or Run by the Pendulum.

S. HOW is the Ships way to be known by the Pendulum?

T. You must mark 50 Feet (or more) upon the sides of your Ship, then having your Pendulum ready, you shall desire one of your Company to throw a piece of a Stick, Chip, or the like over board, a weather the fore part or Prow of the Ship, and as soon as the piece of wood comes even with the mark on the side of your Ship, where you begun the Division of your 50 Feet, then let go the Bullet of your Pendulum, and reckon the swings of your Bullet, until the piece of wood comes even with the last mark, where the Divisions of the 50 Feet end, and then Stop, and so by the *Rule of Three* you may easily judge of the way or run of your Ship; as for

Example.

Admit that the Stick or Chip thrown into the Sea, had passed from the first mark (on the fore part of the Ship) to the last mark, (on the Stern) in the time that the Bullet of your Pendulum had 15 swings, every swing being a Second. I say by the Rule of Three: *If 15 Seconds give 50 Feet; What will 36000 Seconds* (that an Hour or Degree contains) *give?*

15) 180000 (12000	3600
6000) 12000 (2 Miles	50
	180000 *Answer.*

Anſwer, 2 Miles an hour, which ſet down accordingly upon your Log-board, thus, 2 *knots*. If there had been any Feet remaining you ſhould allow for them, as if there had been 3000, you ſhould ſet down half a knot for half a Mile, and if 855, you ſhould ſet down a Fathom, and thus at any time you may take your Reckoning from the Log-board, in the ſame manner as when you keep your Reckoning by the Log.

S. Why did you divide the Product of the Multiplication 12600 by 6000?

T. Becauſe a Mile containeth 6000 of our Feet, and thereby to reduce it into Miles.

P R O P. XXXII.

Other ways to judge how many Miles or Leagues your Ship Sails in an hour.

S. **A**R E there other ways to judge of the Ships Run as eaſie as the former?

T. Yes, and eaſier too, for they require neither Log, Pendulum, nor Rule of Three. For ſome throw a piece of wood over board a weather, about the fore Caſtle or fore part of their Ship, and then walking towards the Stern, always even with the piece of wood which they threw into the Sea: They judge in the ſame manner as if they were walking upon the Land, how many Miles (or Leagues) they go in an hour, and ſo they judge of the Ships Run according to the ſlowneſs or ſwiftneſs of their pace. Theſe of Experienced judgments look only how ſwiftly the Water paſſeth about the middle of the Ship, conſidering alſo the force of the Wind, (and how it blows, whether from the Stern quarter, or bow,) as likewiſe the Currents and Waves, and by theſe they judge how many Miles their Ship Sails in an hour.

Another way to know your Diſtance Sailed.

Is to obſerve in how many Glaſſes you raiſe or depreſs a Degree, for by it you will know your Diſtance Sailed, as for

Example.

Admit that Sailing North or South, you find by your Obſervation you have raiſed or depreſſed the Pole one Degree, you may be ſure that you have Sailed or run 20 Leagues, (or 60 Miles) which being divided by the Glaſſes which did run in that time will ſhow you how much it is an hour;

and

and although this practice is easier Sailing North and South , than any
other Course, nevertheless there is not any Course how oblique soever
it be , but you may know this way (if you can observe well) how many
Leagues (or Miles) you must Sail upon any Point of the Compass, to raise
or depress a Degree , as the following Table showeth.

	Leagues. Miles.
To raise a Degree your Course being N or S, you must Sail	20 or 60
If N b E, or S b W or N b W or S b E,	20½ or 61
and then your Distance from the Meridian you departed, is of	4 or 12
If NNE or NNW or SSE or SSW	21⅓ or 64
Distance from the Meridian	8¼ or 25
If NE b N or NW b N or SE b S or SW b S	24 or 72
Distance from the Meridian	13⅓ or 40
If NE or NW or SE or SW	28⅓ or 85
Distance from the Meridian	20 or 60
If NE b E or NW b W or SE b E or SW b W	36 or 108
Distance from the Meridian	30 or 90
If ENE or WNW or ESE or WSW	52 or 156
Distance from the Meridian	48¼ or 145
If E b N or W b N or E b S or W b S	102½ or 307½
Distance from the Meridian	100⅓ or 301½

S. Of these several ways to judge of the Distance Sailed , which is
it that our Pilots most practice?

T. The most practiced is by the Log : To judge by the passing of the
Water along the Ships side, the strength of the wind , the Sails the
Ship hath out , and the manner the wind bloweth in them, an experienced
Judgment is very necessary.

PROP. XXXIII.

How to allow for Currents in Judging of the Course and Distance.

S. **W**HAT Instructions will you give me concerning Currents?
T. Only this, that when you Sail against a Current, if
it be swifter than the Ships way you will fall a Stern ; but
if it be slower you will get a head , so much as there is
                                                     difference

difference between the Ships way, and the race of the Current, as you
will better underftand by

Examples.

Admit that by your Reckoning, (by the Log or otherwise) you Sail
Eaft 6 Miles an hour againft a Current that fets Weft 3 Miles in an hour,
and you would know how many Miles an hour your Ship goeth a head
Eaft.

Subtract the 3 Miles that the Current fets W an hour, from the 6.
Miles that you Sail in an hour, and the remainer 3 will be the Miles that
your Ship goeth a head Eaft.

I Sail in an hour Eaft	**6** Miles.
The Current fets W in an hour	3
My Ship goeth a head in an hour	3. Miles.

Example. 2.

The Current fets Weft in an hour	6. Miles.
And I Sail Eaft in an hour	3
My Ship falls a Stern in an hour	3 Miles.

S. Admit that your Ship run W 5 Miles an hour, and that the Current fets alfo W 3 Miles an hour, what muft I do then ?

T. You muft add the Miles Sailed and thofe of the Current together,
and the whole will be what your Ship goeth a head in an hour.

Example 3.

My Ship runs W in an hour	5 Miles.
The Current fets alfo W in an hour . . .	4
My Ship goeth a head in an hour	9 Miles W.

S. Admit my Ship crofs a Current that fets WSW 3 Miles an hour,
and in 8 hours fhe Sails 12 Leagues SSW, and in 12 hours more 15 Leagues
SW. Now, how fhall I know what Courfe and Diftance my Ship hath
made her way good, fuppofing fhe firft fet out from A.

T. You may eafily know it by the Quadrant of Reduction, or the
plain Scale.

First draw the Line A B, upon the extremity of which at A, erect the Perpendicular C, and with the Chord of 60 Degrees (taken from your Scale) describe a Quadrant on it, then set off your Ships first Course two Points from the South from D to E, and draw the Line A E, and from A to G prick off your first Distance 12 Leagues, then lay off the Course of the Current, which is 6 Points, from D to L, and draw the Line A L, for a *W S W*, which is the Course of the Current.

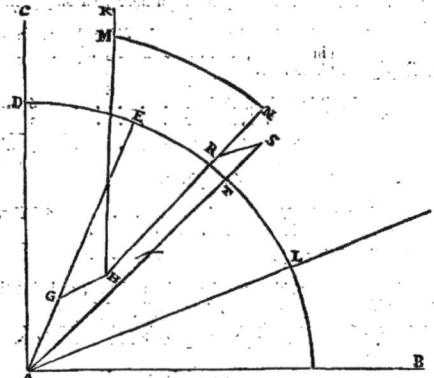

Then draw the Line G H Parallel to A L, and prick on it from G to H, the 6 Leagues that the Current sets forward in 8 hours; and for the second Course draw the Line H K Parallel to A C, then take again with your Compasses the Chord of 60 Degrees, and setting one Foot in H, with the other describe the Arch M N; and from M prick your Ship's second Course, which is 4 Points from the South, then draw N H, and prick down on it the 13 Leagues Sailed, from H to R, then draw R S Parallel to A L, whereon prick down the 9 Leagues that the Current sets in 12 hours from R to S, and draw the Line A S, which being measured upon the Scale of Leagues (Reckoning 10 but for 5) will show you that your Ships right Distance is 40 Leagues.

Then to know what Point your Ship hath made good, measure the Arch D T, and you will find it 47 Degrees, which is 2 Degrees above 4 Points from the South, and therefore your Course hath been *S W*, almost a quarter of a Point Westerly.

                                                      P R O P.

PROP. XXXIV.

How to find which way the Current Sets.

S. HOW do they go to Work to know the Secret Transport of the Currents?

T. The Practice of it, is to have a Log half a Foot longer and three Inches wider than the ordinary one, that sinking deeper by the weight of the Lead, the Current may the have more force to command it the better; this Log they heave out in the wake of the Ship, and letting the Line go (in the same manner as when they will know the Ships way by the half Minute-glass;) they observe if this Log keep in the wake of the Ship; for if it do, they conclude that there is no Current; but if it not, but falls off from it, they know by it which way the Currents Sets. Also when they meet with some Cape, Island, or Rock, they take notice which way the Current runs; these Currents are commonly discovered by their Rippling.

PROP. XXXV.

Of the Log-Board.

S. WHAT do you mean by Log-Board?

T. I mean a Board divided into several Columns (like the following Table) upon which most Pilots use to set down their dead Reckoning, *viz.* The Course and Distance Sailed upon every Shift of Wind, which you are likewise to do in this Order.

First, set down the Time or Hours as you see in the first Column: Then the Ships Course as in the Second: Then the Knots, Half-knots, and Fathoms that the Ship runs, as in the Third, Fourth, and Fifth Columns: Then the Point of the Compass the Wind is at, as in the Sixth Column: Then the force of the Wind as in the Seventh, and the Variation as in the Eighth Column.

B b   The

The Log-Board.

Hours.	Courfe.	Knots.	Half Knots.	Fathoms	Wind at	Force of the Wind	Variation D. M.
2	SW¼W	5	1	o	S S E	F. G.	2 30 W
4							
6							
8							
10							
12							
2							
4							
6							
8							
10							
12							

Note, That when you begin a long Voyage, you muft keep a Log-Book; in which you muft write down every day what is in your Log-board, in the fame order as you fee here; and, befides that alfo, what you have feen, and how it did bear from you; and alfo the Work of your Obfervations.

S. How.

S. How muſt I uſe the Log-Board

T. To underſtand how to uſe it, you muſt remember that ſo many Knots of the Log-Line as your Ship runs in half a Minute Glaſs, ſo many Miles ſhe Saileth in an hour, and that you muſt heave the Log every two hours, and that the beſt way is to ſet down the Courſe made good upon each Point of the Compaſs.

As for Example.

Admit that my Courſe is *S W*, the Wind being at *S S E*, and that after two hours Sailing (with the ſame Gale,) I heave out the Log and find that to run 5 ¼ Knots in half a Minute, and in the ſame time I obſerve alſo (by the wake of my Ship) that there is half a Point of Leeward way, by which I conclude that my Ship makes good but the *S W* half a Point Weſterly; which accordingly I ſet down upon the Log-Board, and ſo continuing as you ſee in the foregoing Table, I ſet down every two hours the Knots, Half-knots, and Fathoms that my Ship runs in half a Minute, with the ſeveral Courſes my Ship hath made good, and what Wind bloweth, the force of the Wind and Variation.

P R O P. XXXVI.

How to keep a Reckoning of 24 hours without the Log.

T. THE way of thoſe that Judge of their Diſtance without a Log is to have a Board of ſix Columns; in the firſt of which they ſet down the Hours; in the ſecond the Courſe; in the third the Miles; in the fourth the Wind that bloweth; in the fifth the force of the Wind; and in the ſixth the Variation; (and ſometime alſo how the Current Sets.)

Example.

Example.

Hours.	Courfe.	Miles.	Wind at	Force of the Wind	Variation D. M.
1	SW ½ W	5 ½	S S E	F G	2 30 W
	SW b W	3	S S E		
2	W S W	6	S b E	F G	
3					
4					
5					
6					
7					
8					
9					
10					

PROP. XXXVII.

*How to take the Reckoning from the Log-Board, thereby to
Compute the true Courfe and Diftance.*

S. **H**OW do you Compute the true Courfe and Diftance of your
Ship, by the feveral Runs expreft upon the Log?
T. Firft I double the Knots, Half-knots, and Fathoms (if
there be any) of each refpective Courfe, by which I know
how many Miles I have Sailed.

S. Why do you double the Knots? &c.

T. I double them to know how many Miles I have Sailed, for if they
exprefs on the Log-Board but the Miles Sailed in an hour, I muft needs
double them to know how many Miles I have Sailed in two hours.

S. What

S. What doth a Fathom expreſs upon the Log-Board?

T. If your Knots be at 42 Foot diſtance one from another it expreſſes the 7th. part of a Mile in an hour, ſo that 2 Fathoms is ſomething more than a quarter of a Mile: But if the Knots are diſtant 50 Foot, 2 Fathoms are ſomething leſs than a quarter of a Mile, for a Fathom contains but 6 Foot.

S. When do you take your Reckoning from the Log-Board?

T. I take it every day at Noon, and commonly after my Obſervation, when I Work my Traverſe in this order. Having made a Table of Six Columns: (as you will ſee in the following Fourth Book:) I firſt ſet down the Courſe which my Ship made good, expreſt in the firſt Line of my Log-Board; to wit, a *S W* by *W*; then the Miles Sailed upon that Courſe, (or Rumb) which (by doubling the Knots) I find to be 11; this done, I Reckon or Count the 11 Miles Sailed upon the Sinical Quadrant, to wit, upon the fifth Rumb from the South, or *S W* by *W*, and the Circles which repreſent the Miles (or Leagues) beginning from the Center of the Inſtrument; and pricking a ſmall pin upon the 11th. Circle and the aforeſaid Rumb; I look how much it gives for Southing, and find it 6 Miles; which accordingly I ſet down in the South Column; I find alſo that it giveth for Weſting 9 Miles, which I likewiſe ſet down in the *W*. Column, and doing in the like manner with the reſt, theſe ſeveral Courſes and Diſtances will be reduced and placed in their proper Columns, which then are to be added together, and after that Subtracting the leſſer Number from the greater, the Remainer will be the difference of Latitude and Departure from the Meridian.

S. I do not underſtand what you mean by Subtracting the leſſer Number from the greater, pray explain your ſelf?

T. I mean that having compared the North and South Columns together, the leaſt of them muſt be Subtracted from the greater, and the like is to be done of the Eaſt and Weſt Columns, but of this more in ſhewing you the uſe of the Sinical Quadrant.

P R O P. XXXVIII.

How to Correct a Dead Reckoning, when the Dead Latitude differs from the obſerved Latitude.

S. **H**OW muſt I Correct my Dead Latitude?

T. There are ſeveral ways, for you muſt Correct your Dead Latitude according as your Courſe hath been, for if your Courſe hath been North and South, and by your Reckoning you find your Latitude leſs than by Obſervation, you muſt Correct the Ships place in the ſame Meridian, and in the obſerved Parallel.

A1

As for Example.

Admit I depart from A in the 50 Parallel (or Latitude of 50 Degr.) and that I Sail directly South 50 Leagues, which is 2 Degr. 30 Min. it will appear by my Reckoning that my Ship is at B, in the Latitude of 47 Degr. 30 Min. but by my Obſervation my Ship is in the Latitude of 47 Degr. therefore to Correct my Dead Reckoning, I prick my Point in C, under the ſame Meridian and in the 47 Parallel according to my Obſervation.

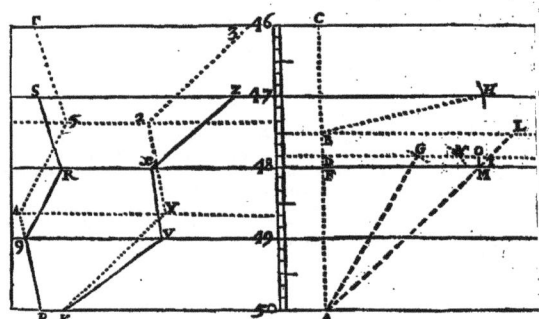

S. What muſt I do when (Sailing North or South) the obſerved Latitude is leſſer than the Latitude by Reckoning?

T. You muſt ſhorten your Diſtance in the ſame Meridian, according to your obſerved Latitude.

As for Example.

Admit that I depart from A, (in the Latitude of 50 Degrees) and Sail South 130 Miles (which is 2 Degr. 10 Min. difference) by my Reckoning, it will appear that I am arrived at D in the Latitude of 47 Degr. 50 Min. but if by my obſerved Latitude I find I am arrived but in the Parallel of 48 Degrees; I ſhall prick my Point in F, and not in D, ſince my Obſervation ſhews that there is an error in my Reckoning.

S. What

S. What is the Caufe of that error?

T. There is two things that may be the Caufe of it, firft the want of Care or of an Experienced Judgment, to judge well of the Ships way (or run) by making it leffer or greater than it really is, or elfe by Stemming a Current, or Sailing with it.

S. Muft I always Correct the place of my Ship under the fame Meridian, when my Courfe hath been either North or South?

T. Yes, when the error comes by not judging well of the diftance Sailed, or when a Current fets with, or contrary to your Courfe; but you muft not do it, when the error proceeds from fome other Caufes, as from the Variation (which by careful Obfervations is eafily prevented) or elfe by croffing a Current, for in that cafe you muft Correct your Courfe otherwife.

As for Example.

Admit that you Sail from A, and that notwithftanding a prudent and careful Reckoning or Judgment of your diftance, you find two or three days together, that your Dead Latitude differs from your obferved Latitude, becaufe of the Currents which your Ship Croffes; in fuch a cafe you muft have great Care not to leave your Dead Reckoning for to ftand to your obferved Latitude, (your Courfe being North or South,) for if you do, you will grofly miftake. You fhall then in fuch a cafe ftand both to your Diftance and Latitude by Obfervation, and with it fhall Correct your Courfe, thus; you fhall take with your Compafs the Miles of your Diftance, and placing one Foot of it upon the place of your Departure, with the other Foot defcribe an Arch upon the Parallel of your obferved Latitude and the Point of interfection will mark the place your Ship is arrived at.

Example.

Admit that by my Reckoning my Ship's Diftance from A, is 150 Miles South, (or 2 Deg. 30 Min.) and fo is arrived at B, in the Latitude of 47 Degr. 30 Min. North : But by my Obfervation I find it to be in the Latitude of 47 Degr. 50 Min. and I know that the Current fets Weftward; therefore to Correct the error of my Courfe, and to find the place of my Ship, I take upon the Meridian 150 Miles or 2 Degr. 30 Min. which is my diftance from A, the place I departed laft, and then placing one Foot of my Compafs in A, with the other I defcribe an Arch, which cuts the obferved Parallel in G, the Point which marks the place that my Ship is arrived at. As for the error that may happen by the Variation, becaufe it may eafily be Corrected when found out by the Obfervations of it : I fhall not trouble you with any more Examples of this kind, but will now fhew you how to Correct the error that may happen by Sailing under a Parallel, (Eaft or Weft.)

P R O P.

PROP. XXXIX.

How to Correct your Dead Latitude by your observed Latitude when you Sail East or West.

S. WHAT muſt I do to Correct my Reckoning when I Sail Eaſt or Weſt, and find that the Dead and obſerved Latitude a-gree ?

T. In that caſe you cannot Correct the error of your Reckoning (if there be any) nor diſcover by your Obſervation if any Current hath hindered your way, or ſet you forward, or made it greater, and therefore when you fear to Stem a Current, you are to alter your Courſe, that by the difference of Latitude you may diſcover the error.

S. How ſhall I Correct my Reckoning when having Sailed ſo many Leagues (or Miles) Eaſt or Weſt, I find by my Obſervation that my Courſe hath not been under that Parallel where my Reckoning places me ?

T. It muſt be Corrected by your difference in Latitude and diſtance as in this

Example.

Admit that departing from B in the Parallel of **47 Deg. 30 Min** after having Sailed **130** Miles Weſt, you find by Obſervation you are in the Parallel of **47** Degrees; to Correct your Courſe wherein there is error, you muſt take with your Compaſs upon the Meridian the **130** Miles (or **2** Degr. **10** Min.) Sailed, and ſetting one Foot in B, with the other deſcribe an Arch which will cut the **47** Parallel, and the Point of Inter-ſection at H will mark the place of your Ship.

S. How muſt I Correct my Reckoning, Sailing obliquely or between a Parallel and a Meridian?

T. When you Sail within **5** or **6** Points of the Meridian and the Dead and obſerved Latitudes differ, you may be ſure that there is ſome error, either by miſ-judging the diſtance run or your Courſe. In this caſe, if you are more ſure of the Rumb or Courſe than of the Diſtance, you muſt ſtand to your Courſe, and the Ship's place muſt be Corrected upon the Rumb by altering your Diſtance, as in this

Example.

Example.

Admit that your Ship Sails from A, *S W* 60 Leagues, and fo by your Reckoning you find you are arrived at I, in the Parallel of 47 Degr. 50 Min. but by Obfervation you are arrived but in the Parallel of 48 Degrees. I fay that you muft fhorten the Ships diftance upon the Courfe, and that you are to prick your Point in M, for the (Corrected) place of your Ship.

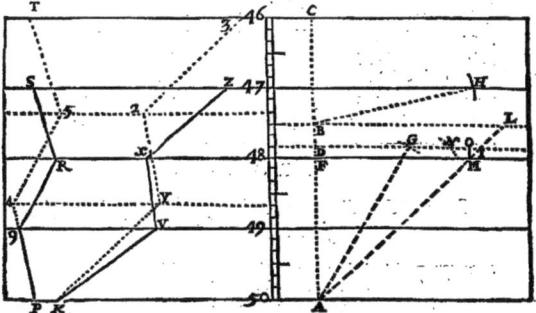

S. What muft I do when my difference in Latitude is greater by Obfervation than my Reckoning?

T. You muft enlarge the Ships diftance upon the Courfe; as for Example, if your Dead Latitude was 47 Degr. 50 Min. and your obferved Latitude 47 Degr. 30 Min. you muft prick your Point in L, and not in I, fince your Obfervation fheweth that your diftance is greater than you did judge it to be, and that your Ship is in the Parallel of 47 Degrees 30 Minutes.

S. Suppofe that I were more fure of my diftance Sailed, than of my Courfe, becaufe of the Currents and Variation which I could not obferve, what muft I do then?

T. In that cafe you muft ftand to your diftance Sailed, and Correct your Courfe.

As for Example.

Admit that departing from A, you Sail away *S W* 167 Miles, and so by Estimation are arrived at M, in the Parallel or Latitude of 48 Degrees; but by the observed Latitude you find you are arrived in the Parallel of 47 Degr. 50 Min. I say that you must stand to your Miles sailed, and Correct your Course thus: You must take with your Compass the distance Sailed; to wit, 167 Miles, (upon the Meridian) and placing one Foot of it in A, with the other describe a secret Arch which cuts the Parallel of 47 Degr. 50 Min. and the Point N of intersection will shew the place of your Ship.

S. What must I do, if I mistrust as much the distance as the Course?

T. You must part the difference; as for Example, if you think to have Sailed by the Rumb A L, which is a *S W*, and to be arrived at M, in the Parallel (or Latitude) of 48 Degrees; and your Observation sheweth that you are in the Parallel D N, you must stand to the Longitude of the Point at M, and drawing the secret Line M O, Parallel to a Meridian; mark your Point in O for the place of your Ship.

PROP. XL.

How to Correct (by the Card) a Composed Course or several Courses together.

S. WHAT do you mean by a Composed Course?

T. I mean several Courses together, as when we have Sailed or Steered upon several Points of the Compass, which commonly happens because of contrary Winds, and different shifts of it, which oblige the Pilot to alter his Course: And sometimes because of Capes, or Points of Land, Rocks, Shoalds, or the like.

S. How shall I Correct my Reckoning, having Sailed upon several Courses, sometime North and sometime South?

T. In this case, if the observed Latitude differs from the Latitude by Estimation, you must stand to the same Longitude, and prick your Point in the observed Parallel.

As for Example.

Admit that being departed from the Point at P, you think to be arrived at S, in the Latitude of 47 Degr. but by your Observation you find to be in the Latitude of 46 Degrees; the Correction of it, is to stand to the same Meridian of S, and to prick your Point in T, the observed Parallel.

S. What must I do when my several Courses have been either to the Southward, or else to the Northward?

T. You must divide to each single Course the difference which you find between the Dead Latitude and observed Latitude.

As

As for Example.

Admit that departing from K, in the 50 Parallel you Sail to the South by the Rumb K V, V X, X Z, if by Eftimation your Ship is at Z, in the 47 Parallel (or Latitude of 47 Degr.) and by Obfervation in the 46th, the difference is of one Degree, which muft be divided according as each Courfe did alter the Latitude by Eftimation, as in this Example; all the difference in Latitude by Eftimation being of 3 Degrees: Admit that the difference in Latitude of the firft Courfe was of one Degree, which is the Third part.   I lengthen my firft Courfe the Third part of my difference, *viz.* 20 Min. (fince a Degree containeth 60 Min.) then placing one Foot of my Compafs in K, and the other in Y, upon the Parallel of 48 Degr. 40 Min. I prick my firft Point 20 Min. more to the Southward than it was at firft; and doing as much from Y and from 2, for the Second and Third Courfe I prick my (corrected) Point in 3, the place my Ship is arrived at according to my Obfervation.

S. But fuppofe that I was more fure of my Courfe than of my diftance, what muft I do then?

T. You muft ftand to your Courfe, but you muft lengthen your diftance.

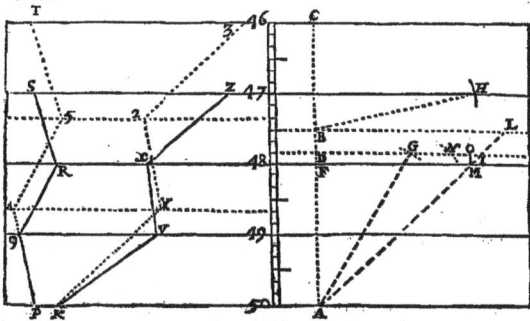

As for Example.

Admit that after thefe three Courfes P 9, 9 R, and R S, you think to be arrived in the Latitude (or Parallel) of 47 Degrees, but by Obfervation you find you are arrived in the Parallel (or Latitude) of 46 Degr. you muft lengthen the diftance upon each Courfe (proportionally, to the difference between the obferved Latitude, and Latitude by Eftimation,) and fo inftead of P 9, prick P 4, and for 9 R, prick 4 5, and for R S, 5 T.

P R O P.

PROP. XLI.

How to Work a Ship at Sea.

S. SINCE you defign to fhew me the Practical part of *Navigation*, I wonder you have not taught me yet, how to Work a Ship at Sea?

T. Although I have not done it ; it is not that I have forgot how neceffary it is to a beginner, but only becaufe Captain *Sturmy* hath Treated of it, fo fully and well in his *Mariners Magazine*, that I think it impoffible to do it better ; however becaufe this Work would be imperfect without it ; I fhall fet it down in his own words, without adding any thing to it, but what I find amifs(may be by the Printers fault) not doubting but a great many who are unwilling to go to the price of his Book ; and others who never faw it nor heard of it, will be glad to find it here in the proper Sea phrafes, as followeth :

The Wind is fair.

The Wind is fair though but little, it comes well, as if it would ftand, therefore up a hand and loofe your Fore-top-fail in the Top, that the Ship may fee we will fail ; bring Cable to the Capfton, heave up your Anchor, loofe your Fore-fail in the Brails, put a broad our Colours, loofe the Mizen in the Brails, is all our Men on board? Thofe that be on Shore may have a Tow, and be bleft with a Ruther ; for we will ftay for no Man : Come my hearts, heave up your Anchor, that we may have a good prize ; come, who fays Amen? One and all ; oh brave hearts, the Anchor is a peck ; heave out your Fore-top-fail, heave out Main-top-fail, hawl home the Top-fail-fheets, the Anchor is away, let fall your Fore-fail, hoife up your Fore-top-fail, hoife up your Main-top-fail ; up and loofe the Main-fail, and fet him ; loofe Sprit-fail, and Sprit-fail Top-fail: A brave Gale, bring the Fore-tack to the Cat-head, and trim our Sails quartering ; hoife up our Small-fails, heave out the Mizen-top-fail and fet him ; now we are clear, and the Wind like to ftand ; hoife in our Boats before it is too much Sea, aboard Main-tack, aboard Fore-tack, a Lee the Helm handfomly, and bring her to eafily, that fhe may not ftay : Brace the Fore-fail and Fore-top-fail to the Maft, and hawl up the Lee bowlines ; that the Ship may not ftay ; pafs Ropes for the Boats on the Lee fide, and be ready to clap on your Tackle, and hoife them in ; ftow them faft, let go the Lee-bowline of the Fore-fail, and Weather-braces : Right your Helm, hale aft the Fore-fheet, trim the Sails quartering as before ; loofe Sprit-fail, and hawl aft the fheets, and hoife up the Sprit-fail Top-fail and other Small-fails: fet the Main-ftay-fail, and Fore-top-
fail

fail-ftay-fail, and Mizen-ftay-fail, and Main-top-fail-ftay-fail, and lace
on your Bonnets, that we may make the moft of our way, to our Station,
clear your Ropes: Come, get up our Steering-fails, the Lee-fteering-
fails of Main-fail, and Main-top-fail, Fore-fail and Fore-top-fail only,
for they will fit faireft, and draw moft. *Thus, you have a brave Ship un-*
der all her Sails and Canvas, in her fwifteft way of Sailing upon the Sea; now
let us have her right before the Wind.

Right afore the Wind and a frefh Gale.

The Wind is vered right aft, take in your fore and fore Top-fail.
Steering-fail, and Fore-top-fail, and Main and Main-top-fail-ftay-fails;
for they are becalmed by the after Sails, and will only beat out; the
Wind blows a frefh Gale, round aft the main fheets, and fore fheets,
Square your Yards, take in your main and main Top-fail Steering-fails;
unlace your Bonnets, take in your main and fore Top-gallant-fails; in
Sprit-fail, and Mizen Top-fail, let go the fheets, hawl home your Clewlines, caft off Top-gallant Bowlines. *Thus, you have all the fmall Sails in,*
and furled, when it blows too hard to bear them.

The Wind vereth forward and fcanteth.

The Wind fcanteth, vere out fome of your fore and main fheets,
and Sprit-fail fheets, and let go your weather braces; Top your Sprit-
fail Yard: The Wind ftill vereth forward; get aboard the fore and
main Tack; caft off your weather fheets and braces; the Sails are in the
Wind, hawl off main and fore fheets; the Wind is fharp, hawl forward
the main Bowline and fore Bowline, and hawl up the main Top-fail and
fore Top-fail Bowline, and fet in your Lee-braces, and keep her as near
as fhe will lie. *Thus, have you all the Sails trim'd fharp, and by a Wind?*

The Wind blows frifking.

The Wind blows hard, fettle our fore and main Top-fails, two thirds
of the Maft down; it is more Wind, come, hawl down both Top-fails
clofe, come, ftand by, take in our Top-fails; let go the Top-fail Bowlines, and Lee-braces; let go the Lee-fheets, fet in your weather braces,
fpill the Sails, hawl home the Top-fail Clewlines, Square the Yard. *Now*
the Top-fails are furled, and you have the Ship in all her Low-fails, or Courfes.

It bloweth a Storm.

It is like to overblow, take in your Sprit-fail, ftand by to hand the
Fore-fail; caft off the Top-fail-fheets, Clewgarnets, Leechlines, Buntlines; ftand by the fheet, and brace; low'r the yard and furl the Sail,
here is like to be very much Wind; fee that your main hallyards be clear,
and

and all the reſt of your Geer, clear and caſt off. (Is all clear?) Low'r the
main Yard, hawl down upon your down hawl; now the Yard is down,
hawl up the Clewgarnets, Lifts, Leachlines, and Buntlines, and furl the
Sail faſt, and faſten the Yards, that they may not Traverſe and gall.
Thus have you the Ship a trye under a Mizen.

A very hollow grown Sea.

We make foul weather, look the Guns be all faſt, come hand the
Mizen; the Ship lies very broad off, it is better ſpooning before the
Sea, than trying or hulling; go reef the Fore-ſail and ſet him; hawl
aft the Fore-ſheet; the Helm is hard a weather, mind at Helm what
is ſaid to you carefully: The Ship wears bravely; ſteady; ſhe is before it;
belay the fore down hall; it is done: the Sail is ſplit; go hawl down the
Yard, and get the Sail into the Ship; and unbind all the things clear
of it: Starboard, hard up, right your Helm, Port, Port hard, more
hands, he cannot put up the Helm; a very fierce ſtorm, the Sea breaks
ſtrange and dangerous; ſtand by to hawl off upon the Lanniard of the
Whip-ſtaff, and help the man at Helm, and mind what is ſaid to you;
ſhall we get down our Top Maſts? No, let all ſtand, ſhe ſcuds before
the Sea very well, the Top-maſt being aloft the Ship is the wholſomer,
and maketh better way through the Sea, ſeeing we have Sea-room. *Thus*
you ſee the Ship handled in fair weather and foul, by and Large; now let us ſee
how we can turn to Windward.

The Storm is over; let us turn to Windward.

The Storm is over, ſet Fore-ſail and Main-ſail, bring our Ship to;
ſet the Mizen, the Main-top-ſail, and the Fore-top-ſail; our Courſe is
E S E, the Wind is at South; get the Starboard tacks aboard, caſt off
our Weather-braces and Lifts; ſet in the Lee-braces, and hawl forward
by the Weather-bowlings, and hawl them taught and belay them, and
hawl over the Mizen-tack to Windward; keep her full, and by as near
as ſhe will lye: How wind you? Eaſt, a quade Wind; no near, hard no
near, the Wind vereth to the Eaſtward ſtill: How Wind you? *N E*, hard,
no near, the Wind is right in our Teeth; no near ſtill: How Wind you?
N W b *N*, the Wind will be Northerly, make ready to go about; we
ſhall lye our Courſe the other way, no near, give the Ship way, that ſhe
may ſtay; ready, ready, a Lee the Helm, vere out the fore ſheet, let
go Fore-top-bowline, caſt off the Lee-braces of your Fore-ſail and Fore-
top-ſail, brace in upon your Weather-braces: The Fore-ſails is a Back-
ſtays, hawl Main-ſail, hawl, let riſe the Main-tack; caſt off your Larboard-braces; let go Main-bowline, and Main-top-ſail-bowline; brace
about the Yard, hawl forward by the Larboard-bowlines; get the Main-
tack cloſe down in the Chefs-tree; the ſheet is cloſe aft, hawl off all;
hawl, get to Fore-tack, let go Fore-bowline, and Fore-top-ſail-bowline;
hawl

hawl aft the Fore-sheet, hawl taught the Main-bowline, and Main-top-
sail-bowline; shift the Mizen-tack, hawl taught Fore-bowline, and Fore-
top-sail-bowline; set in the Lee-braces fore and aft, keep her as near as
she will lye: No near, how Wind you? *NNE*: Thus, ware no more;
no near, keep her full; the Wind is at *NNE*: Thus, ware no more,
(how Wind you?) *ENE*, the Wind is at *N*, keep her away her Course.
ESE: Cast off the Lee-braces, and Weather-bowlines, and set in your
Weather-braces, vere out the Main-sheet, and Fore-sheet, loose the Sprit-
sail, and Sprit-sail Top-sail, and Mizen-top-sail, and Top-gallant-sails;
hoise them up, the Wind veres aft still, let rise the Fore-tack; the Wind's
quartering, hawl aft the Fore-sheet, bring it down to the Cat-head with
a Pass-a-ree; steady in your Weather-braces; the Wind stands. *Thus*
you have the Ship as at first, Steering under all her Canvas, Quarter-wind;
she hath been wrought in all manner of weather, and all sorts of Winds:
Therefore we will draw to a Conclusion with a Man of War in Chase and taking
of her Prize, and so leave this Practick part to your Censure.

The Man of War in her Station.

Now we are in our Station, and a good Latitude, hand your Top-sails and
furl your Main-sail and Fore-sail, and brail up the Mizen, and let her lie
at Hull, untill fortune appear within our Horizon; up aloft to the Top-
mast-head, and look abroad, Young men, look well to the Westward,
if you can see any Ship that hath been Nipt with the last Easterly Winds:
A Sail, a Sail, where? *Fair by us;* how stands she? *To the Eastward, and*
is two Points upon her Weather-bow, and hath her Larboard Tacks aboard:
O then she lies close by a Wind; we see her upon the Decks plainly;
a good man to Helm; up Young men, and loose the Fore-sail, Main-sail,
and Mizen; get the Larboard-tacks aboard; heave out the Main-top-
sail, and Fore-top-sail, and loose the Sprit-sail; keep her as near as she
will lie; hawl aft the sheets, and hawl up your Bowlines taught; do
you see your chase? *Yea;* how Wind you? *ENE*, then the Wind is
at *N*, hoise up your Top-sails as high as you can; heave out Sprit-sail
Top-sail, and Mizen-top-sail; hawl home the sheets, and hoise them up; a
Young man loose the Main-top-gallant-sail, and Fore-top-gallant-sail; hawl
home the sheets, and hoise them up; hoise up Main-stay-sail, and Mizen-
stay-sail, and loose the Main-top-sail, and Fore-top-sail Stay-sails, and
set them; it blows a brave chasing Gale; the Ship makes brave way
through the Sea, we raise her apace; if she keep her Course, we shall
be up with her in three Glasses; no near, keep the chase open with the
Litch of the Fore-sail: So, thus, keep her thus; come aft all hands, the
Ship will Steer the better, when you sit all quiet, by her small Sails;
for she is too much by the head, the Chase is a lusty brave Ship: So much
the better, she hath the more goods in her hold; the Ship hath a great
many Guns, it may be she is a Privateer.

Port, the Chafe is about, come fetch her Wake, and we will be about after her; we Sail far better than fhe; we have her Wake, a-lee the Helm, vere out Fore-fheet; every man ftand handfomly to his bufinefs, and mind the Bowlines and Braces, Tacks and Sheets; hawl Main-fail, hawl; let go Main-bowline, Top-bowline, Top-gallant-bowline; hawl of all, hawl, fhift the Helm; bring her to, hawl the Main-fheet and Fore-fheet, clofe aft; fet in the Lee-braces, hawl taught the Bow-lines; the Chafe keeps clofe upon a Wind; keep her open under our Lee. Gunner, fee that you have all things in readinefs, and that the Guns be clear; and that nothing pefter our Decks: —— Down with all Hammocks and Cabins that may hinder and hurt us. Gunner is all all our Geer ready? Is there ftore of Cartrages ready fill'd, all manner of Shot at the Main-maft? Is there Rammers, Sponges, Ladles, Priming-irons, and Horns, Linftocks, Wads, and Water, at their feveral quarters fufficient for them? Be fure that none of our Guns be cloy'd; and when we are in Fight, be fure to Load our Guns with Crofs-bar and Langrel; always obferve to give Fire when the word is given. See that there be Half-pikes and Javelins in readinefs, and all our Small-fhot well furnifhed, and all their Bandaleers fill'd with Powder, and Shot in their Pouches; fee that our Murtherers and Stock-fowlers have their Chambers fill'd with good Powder, and Bags of Small-fhot to Load them, that if we fhould be laid aboard, we might clear our Decks, Starboard, the Chafe pays away more room, Starboard hard; vere out fome of the Main-fheet and Fore-fheet; caft off the Larboard-braces, fteddy, keep her thus: Well Steer'd, the Chafe goes away room, her fheets are both aft, fhe is right before the Wind: Starboard hard; let rife Main-tack, let rife Fore-tack; hawl aft Main-fheet, hawl aft Fore-fheet; we have a Stern-chafe, but we fhall be up with her prefently, for we fetch upon her Hand-going; the Chafe hawls up his Main-fail and furls it; fhe puts abroad, her Waft-cloaths; fhe will Fight us; come up Young men, and furl our Main-fail, Sling our Main-yard, with the Chains in the Main-top; fling our Fore-yard, put abroad our Waft-cloaths; he will Fight us before the Wind; I fee fhe is full of men; it is a hot Ship, but deep and foul; come cheerly my hearts, it is a prize worth Fighting for; the Chafe takes in her Small-fails; up a loft and take in our Top-gallant-fails, Sprit-fail, Top-fail, Mizen-top-fail, and furl the Sprit-fail, and get the Yard alongft under the Bow-fprit; fhe puts abroad her Colours, they are Red, White, and Blew, *Dutch* Colours; no force; Boy, up and put abroad St. *George*'s Colours in our Main-top; ftep aft a-hand, and put abroad our Ancient; call all hands aloft. *Come up a loft all hands: They are all up Captain.*

Gentlemen, We are here employed and maintained by His Majefty King *J A M E S* and our Country, to do our endeavours to keep this Coaft from Pyracy and Robbers, and His Majefties Enemies; and it is our fortune to meet this Ship at this time: Therefore I defire you in His Majefties Name, and for the fake of our Country, and the honour

of

of our *English* Nation, and our selves, for every man to behave himself Couragious like *Englishmen*; and not to have the least shew of a Coward; but to observe the Words of Command, and do his utmost endeavour: Into Gods hands we commit our Cause, and our Selves. So every man to his Quarter, and shew his Courage, and God be with you.

She settles her Top-sails, we are within shot; let all our Guns be loose in the Tackles, and the Ports all knockt open, that we may be ready to run out our Guns when the Word is given: Up noise of Trumpets and hail our Prize; she answereth again with her Trumpet: Hold fast Gunner, do not fire 'till we hail them with our Voices; Port, Edge towards him, he fires his broad-side upon us: What chear my hearts? Is all well betwixt Decks? Yea, yea, only he rak'd us through and through; no force, it is his turn next; but give not Fire untill we are within Pistol-shot; Port, Edge towards him; he plies his Small-shot; hold fast Gunner; Port, right your Helm; we will run up his side; Starboard a little; give Fire, Gunner; that was well done; this Broad-side hath made their Deck thin, but the Small-shot at first did gaul us; clap in some Case-shot in the Guns you are now a loading; Brace to the Fore-top-sail, that we may not shoot a head; he lies broad off to the Southward, to bring his other Broadside to bear upon us; Starboard hard, get to Larboard Fore-tack; trim your Top-sails; run out your Larboard Guns; he Fires his Starboard Broad-side upon us, he pours in his Small-shot; Starboard, give not Fire untill he fall off, that the Prize may receive our full Broad-side, Steady, Port a little; give Fire, Gunner; his Fore-mast is by the board; this last Broad-side hath done great Execution; Cheerly my Mates, the day will be ours; he is Shot-a-head; he bears up before the Wind to stop his Leaks: Keep her thus; well Steer'd; Port, Port hard; bear up before the Wind, that we may give him our Starboard Broad-side; Gunner, is there great store of Case-shot and Langrel in our Guns? Yea, yea. Port, make ready to board him; have your Lashers clear, and able men with them: Edge towards him when you give Fire; bring your Guns to bear amongst his men with the Case-shot; well Steer'd, we are close on board; give Fire, Starboard, well done Gunner; they lie heads and points aboard the Chase; come, aboard him bravely; Enter, enter; are you latch'd fast? Yea, yea, we will have him before we go here hence; cut up the Decks; ply your hand Granadoes and Stink-pots; he cries out Quarter; *Quarter for our Lives, and we will yield up Ship and Goods:* Good quarter is granted, provided you will lay down all your Arms, open the hatches, hawl down all your Sails and furl them; loose the latchings, we will sheer off our Ship, and hoise out our Shallop; if you offer to make any Sail, expect no quarter for your Lives; go with the Shallop, and send aboard the Captain, Lieutenant, and Master and Mates, with as many more as the Shallop will carry. So we will leave the Man of War to his prize, and to secure his Prisoners.

D d   P R O P.

PROP. XLII.

A Ship deſcrib'd, and Sea-terms explain'd.

TO make what has been before ſaid more intelligent, I ſhall here give you the Draught of a Ship, with her Rigging, that ſo you may not altogether be a Stranger to the Names of the Maſts and Tackle at your firſt going aboard a Ship; and then ſhall give you the Explanation of the moſt uſual Sea-terms.

1. The

1. The Enſign.
2. The Mizon-vane.
3. The Mizon-top-ſail.
4. The Mizon-top-ſail-yard.
5. The Croſs-jack-yard.
6. The Mizon-yard.
7. The Main-vane.
8. The Main-pendant.
9. The Main-top-gallant-ſail.
10. The Main-top-ſail.
11. The Main-ſail.
12. The Fore-vane.
13. The Fore-top-gallant-ſail.
14. The Fore-top-ſail.
15. The Fore-ſail.
16. The Jack.
17. The Sprit-ſail-top-ſail.
18. The Sprit-ſail.
19. The Fore-top-gallant-ſtay.
20. The Fore-top-gallant-bowlines.
21. The Fore-top-maſt-ſtay.
22. The Fore-top-ſail-bowlines.
23. The Crane-line.
24. The Fore-ſtay.
25. The Main-ſtay.
26. The Main-top-maſt-ſtay.
27. The Main-top-gallant-ſtay.
28. The Main-top-gallant-bowlines.
29. The Fore-top-gallant-braces.
30. The Fore-top-ſail-braces.
31. The Main-top-ſail-bowline.
32. The Galleries.
33. The Poop-lanthorns.
34. The Main-top-ſail-brace.
a. The Mizon-maſt.
b. The Main-maſt.
c. The Fore-maſt.
d. The Bow-ſprit.

An Explanation of the moſt uſual Sea-terms.

A

AFT or *Abaſt*, fromward the Fore-part of the Ship, or toward the Stern, as *The Maſt hangs aft*, that is towards the Stern.

How chear ye fore and aft, that is, how fares all your Ships Company.

Amain, a Word uſed by a Man of War to his Enemy, and ſignifies, *Yield*.

Strike Amain, that is, Lower your Top-ſails.

The Anchor is a peek, that ſignifies the Anchor is right under the Hawſe (or hole) through which the Cable belonging to the Anchor runs out.

The Anchor is a Cock-bill, that is, hangs up and down by the Ship ſide.

The Anchor is foil, that is, the Cable is got about the Fluke.

An Awning, A ſail or the like, ſupported like a Canopy over the Deck, to prevent the ſcorching heat of the Sun in hot Climats.

B

To Bale, to lade Water out of the Ships Hold with Buckets, or the like.

Trench the Ballaſt, divide or ſeperate it.

The Ballaſt ſhoots, that is, runs over from one ſide to the other.

To bear with the Land, &c. To ſail towards it.

To bear in, that is, to ſail before or with a Wind into a Harbour or Channel.

A

A Piece of Ordnance doth come to bear, that is, lies right with the Mark.

Bear up, a term used in conding the Ship, when they would have her sail more before the Wind.

Bear up round, put her right before the Wind.

To Belage, to make fast any running Rope.

To Bend a Cable, is to make it fast.

A Birth, a convenient space to moor a Ship in.

A Bight, any part of a Rope between the ends.

The Bilge, the breadth of the place the Ship rests on when she is a ground.

The Ship is bilged, that is, has struck off some of her Timber on a Rock or Anchor, and springs a Leak.

A Bittake, that whereon the Compass stands.

A Bitter, a turn of a Cable about the Bits.

The Bits, two Main-square pieces of Timber, to which the Cables are fastned when the Ship rides at Anchor.

A Bonnet, an Addition to another sail, when they fasten it on, they say, *Lace on the Bonnet*; and when they take it off, *Shake off the Bonnet*; it is very rarely fasten'd to any other than the Mizon, Main, Fore-sail, and Sprit-sail, and those sails are called *Courses*, as Main-course and Bonnet, not Main-sail and Bonnet.

A Boom, a long Pole used to spread out the Clew of the Studding-sail, &c.

Board and Board, a term used when two Ships come so near as to touch one another.

To go aboard, to go into a Ship.

To make aboard, or *board it up*, is to turn to Windward.

To break Bulk, to open the Hold, and take out goods thence.

C

Careening, is bringing a Ship to lye down on one side while they trim and caulk the other.

Caulking, is driving of Ockham, Span-hair, and the like into all the seams of the Ship, to keep out Water.

To Chase, is to pursue another Ship, and the Ship so pursued is called the *Chase*.

To Cond or *Cun*, is to direct or guide, and *to cun a Ship* is to direct the Person at Helm how to Steer her: If the Ship go *before the Wind*, then he who cuns the Ship uses these terms to him at Helm, *Starboard*, *Larboard*, *Port*, *Helm a Midships*. Starboard, is to put the Helm to the Starbord (or right) side, to make the Ship go to the Larboard (or left;) for the Ship always sails contrary to the Helm. In keeping the Ship *near the Wind*, these terms are used, *Loof*, *Keep your Loof*, *Fall not off*, *Veer no more*, *keep her to*, *touch the Wind*, *have a care of the Lee-latch*. To make her go *more large*, they say, *Ease the Helm*, *no near*, *bear up*. To keep her upon the *same Point*, they use, *Steddy*, or *As you go*, and the like.

The

The Ship goes Lasking, Quartering, Veering, or Large; are terms of the same signification, *viz.* that she neither goes by a Wind nor before a Wind, but betwixt both.

The Course, is that Point of the Compass on which the Ship sails: Also. the sails are called Courses.

Cut the sail, that is, unfurl it, and let it fall down. *A sail is well cut,.* that is, well fashioned.

D

Dead-water, the Eddy-water at the Stern of the Ship.

To disembogue, is to go out of the Mouth or Strait of a Gulph.

To dispart, is to find out the Difference of Diameters of Metals betwixt the breech and mouth of a Piece of Ordnance.

The Deck is flush fore and aft, that is, is laid from stem to stern without any falls or risings.

E

End for End, a Term used when a Rope runs all out of the block, so. that it is unreeved; as when a Cable (or Hawser) runs all out at the Hawse, we say, *the Cable at the Hawse is run out End for End.*

F

A Fathom, a Measure containing six Feet.

A Fack,, is one Circle of any Rope or Cable quoil'd up round.

To faribel (or *furl*) *a Sail,* is to wrap it up close together, and bind. it with little Strings called *Caskets* fast to the Yard.

To fish a Mast, or Yard, is to fasten a piece of Timber or Plank to the Mast or Yard to strengthen it., which Plank is called a *Fish.*

To lower or strike the Flag,, is to pull it down upon the Cap, and in Fight is a token of yielding; but otherwise of great respect.

To heave out the Flag,, is to wrap it about the Staff:

Free the Boat, or *Ship,* is to bale or pump the water out.

G

The Ships Gage, is so many Foot as she sinks in the Water; or (to speak now like a Sea-man) so many Foot of Water as she draws.

Weather-Gage, is when one Ship has the Wind (or is to Weather) of another.

A loom Gale, a little Wind. '

One Ship gales away from another. In fair weather when there is but little Wind, that Ship which hath most Wind and sails fastest is said, to gale away from the other.

To greave a Ship, is to bring her to lye dry aground, to burn off her old filth.

The Ship gripes, that is, turns her Head to the Wind more than she should.

H

To Hale, is the same as to pull.

To over Hale, is when a Rope is haled too stiff, to hale it the contrary. way, thereby to make it more slack.

To

To hail a Ship, is to call to her Company to know whether they are bound, &c. and is done after this manner, *Hoa the Ship!* or only *Hoa!* To which they answer *Hâe.* Also to salute another Ship with Trumpets or the like, is called *Hailing.*

Fresh the Hawse, a term used when that part of the Cable that lies in the Hawse is fretted or chafed, and they would have more Cable veered out, that another part of it may rest in the Hawse. When two Cables that come through two several Hawses are twisted, the untwisting them is called *clearing the Hawse. Thwart the Hawse*, and *rides upon the Hawse*, are terms used when a Ship lies *thwart* or *cross*, or with her Stern just before, another Ships Hawse. *Note,* That the Hawses are the great Holes under the Head of the Ship, through which the Cables run when she lies at Anchor.

The Ship heels, that is, inclines more to one side than the other, as *she heels to Starboard*, that is, turns up her Larboard-side to lie down on the Starboard.

To Hitch, is to catch hold.

The Hold of a Ship, is that part betwixt the Keelson and the lower Deck, where all Goods, Stores, and Victuals do lye. *Rummidge the Hold*, is used for removing or clearing the Goods and things in the Hold. *Stowing the Hold*, is when they take goods into the Hold.

To Hoise, is to hale or lift up, as *Hoyse the water in*, *Hoyse up the Yards.*

Hulling, when a Ship is at Sea, and takes in all her sails, she is said to *Hull.*

L.

The Ship Labours, that is, rowls and tumbles much.

Land-fall, is a term used, when we expect to see Land; as we had a good *Land-fall*, that is made Land (or saw Land,) according to our Reckoning.

Land-locked, is when the Land lies round about us, so that no point is open to the Sea.

Land-to, A Ship is said to lye *Land-To*, when she is at so great a distance as only just to discern the Land.

To Lash, is to bind, as *lash the Fish on to the Mast*, that is bind it to the Mast.

Launch, is to put out, as to *Launch a Ship*, is to put her forth of the Dock into the Water, but it is sometimes likewise used in a Negative sense, as when a Yard is hoisted high enough, they usually call aloud *Launch-hoe*, that is hoise no more.

To lay the Land, is to lose sight of it.

The Lee-shore, is that shore against which the Wind blows.

Have a care of the Lee-latch, that is take heed the Ship go not too much to Lee-ward.

A Ship lies by the Lee, that is, has all her sails lying flat against the Masts and shrowds.

Mizon

M

Mizon Sail, hath several words peculiar to it, as *Set the Mizon*, that is, fit the Mizon fail; *Change the Mizon*, that is, bring the Yard to the other side of the Maft; *Speek the Mizon*, that is, put the Yard right up and down by the Maft; *Spell the Mizon*, that is, let go the Sheet and peek it up.

To Moor a Ship, is to lay out her Anchors in such a manner as is moft convenient for her to ride by fafely.

N

Neap-tides, are the Tides when the Moon is in the second and laft Quarter, and they are neither fo high, nor fo low, nor fo fwift as the Spring-tides.

A Ship is beneaped, a term ufed, when the water does not flow high enough to bring a Ship from off the ground, or out of a Dock, or over a Bar.

O

The Offing, that is, fromward the fhore, or out into the Sea; as *The Ship ftands for the Offing*, that is, fails from the fhore into the Sea. When a Ship keeps the middle of the Channel, and comes not near the fhore, fhe is faid to *keep in the Offing*.

Off-ward, is contrary to the fhore; as the Stern of a Ship lies to the Offward, and her head to the fhore-ward, that is, her Stern lies toward the Sea, and her head to the fhore.

Overfet, is turning over, but if a Ship turn over on a fide, when fhe is trimming a-ground, it is called *overthrown*.

P

To Parcel a feam, is (after the Seam is caulked) to lay over it a narrow piece of Canvafs, and pour thereon hot Pitch and Tar.

To Pay a feam, is to lay hot Pitch and Tar on (after Caulking) without Canvafs.

To Ride a peek, is when the Yards are fo ordered, that they feem to make the Figure of *St. Andrews Crofs*.

To Purchase, in a Ship bears the fame fenfe as down many times, as *the Capftain purchafes apace*, that is, draws in the Cable apace.

Q

Quarter Winds, are when the Wind comes in abaft the Main-maft fhrouds even with the Quarter.

A Quoil, is a Rope or Cable laid up round one Fack over another, and the laying the Fack, is called *quoiling*.

R

A Reach, is the Diftance between any two points of Land, that lie in a Right-line one from another.

To Reeve, is to put a Rope through a Block; and to pull a Rope out of a Block is called *unreeving the Rope*.

To Ride: When a Ship's Anchors hold her faft, fo that fhe does not drive with Wind or Tide, fhe is faid to *ride at Anchor*.

To

To Ride athwart, is to ride with the Ships side to the Tide.

To Ride betwixt Wind and Tide, is when the Wind and Tide are contrary and have equal strength.

To Ride Hawse-fall, is when in a rough Sea the Water breaks into the Hawses.

A Road, is any place near the Land where Ships may ride at Anchor, and a Ship riding there is called a *Roader*.

Rowse-in, (that is, Hale-in) proper only to the Cable or Hawser, and is used when the Cable or Hawser is slack to make it taught or strait.

S.

A Sail. Besides its proper signification (as belonging to the several Yards, from which it takes its various Names, as Main-sail, &c.) It signifies also a Ship, as when at Sea we descry a Ship, we cry out, *A sail! A sail!* Likewise if we speak of a Fleet (or a number of Ships together) we say the Fleet consisted of 40 or 50 sail, and not 40 or 50 Ships.

To Sarve a Rope, is to wind something about it, to keep it from fretting out.

To seaze, is to make fast, or bind.

The Ship feels, that is, when on a sudden she lies down on her side, and tumbles from one side to the other.

The Ship sends, that is, her head or stern falls deep in the trough or hollow of the Sea.

To Settle a Deck, is to lay it lower.

The Ship is sewed, that is, the Water is gone from her.

The Ship shears, that is, goes in and out, and not right forward.

To Sound, is to try with a Line or other thing how deep the Water is.

The Ship hath spent her Masts, that is, her Masts have been broke by foul Weather; but if a Ship lose her Masts in Fight, we say, *her Masts were shot by the Board*.

To Splice Ropes, is to untwist two ends of Ropes, and then twist them both together, and fasten them with binding a string about them.

The Sail is split, that is, blown to pieces.

The Ship spooms, that is, goes right before the Wind without any sail.

Spring-tides, are the Tides at New and Full-moon, which flow highest and ebb lowest, and run strongest.

The Bow-sprit Steeves, that is, stands too upright. *Steeving* is likewise used by Merchants when they stow Cotton or Wool, which being forced in with skrews, they call *Steeving* their Cotton or Wool.

T

Tack about, that is, bring the Ships head about to lye the other way.

Tallee aft the sheats, a term used for haling aft the sheats of the Main or Fore-sail.

A Windward Tide, when the Tide runs against Wind.

A Leeward Tide, when the Wind and Tide go both one way.

A Tide-gate, where the Tide runs strong.

To

To Tide it up, is to go with Tide againſt the Wind, and when the Tide alters to lye at Anchor 'till it ſerve again.

It flows Tide and half Tide, that is, it will be High-water ſooner by three hours at the ſhore than in the *Offing*.

To Tow, is to drag any thing after the Ship.

The Traverſe, is the Ships way.

V

To Veer, is to let out ; as *veer more Rope*, *veer more ſheat*.

W

The Ship is Walt, that is, wants ballaſt.

To Weather a Ship, is to go to Windward of her.

To Wind a Ship, is to bring her head about.

How Winds the Ship? that is, upon what point of the Compaſs does ſhe lie with her head.

To Would, is to bind Ropes about a Maſt or the like, to keep on a Fiſh to ſtrengthen it.

Y

The Ship Yaws, that is, goes in and out, and does not ſteer ſteddy.

The End of the Third Book.

E e   **THE**

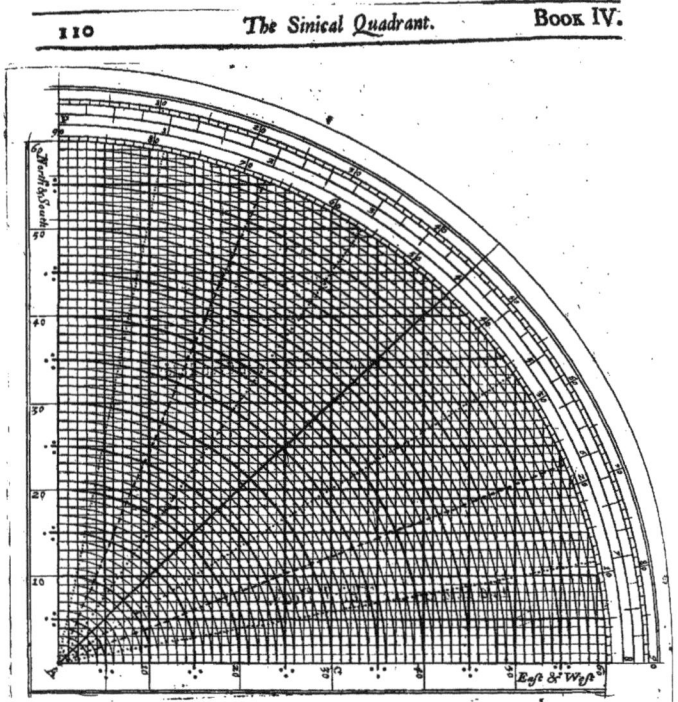

THE
Compleat ART
O F
NAVIGATION.

THE FOURTH BOOK.

*The Description and Use of such Instruments
as are proper for Navigation.*

I. *The Description and Use of the Sinical Quadrant.*

S. **W**HAT is a Sinical Quadrant?

T. It is an Instrument of Geometry, and Trigonometry; the most exact and easie that hath been yet invented, to work a Traverse at Sea by. It is Composed of *Concentrick Arches,* (or rather quarters of Circles) and of three sorts of *Right-lines.* 1. Those drawn Parallel to the Line A P represent the *Sines* and *Meridians,* and are Parallel to the North and South, upon which we count the Leagues of *Latitude.* 2. Those drawn Parallel to the Line A C represent the *Sines Complements,* and *Parallels;* and are Parallels to the East and West, upon which we count the Leagues of *Longitude* or *Departure* from the *Meridian.* And 3. Those drawn from the Center of the Quadrant A to the Limb, represent the *Rumbs* of your Compass, or several Courses; and it is upon them that we Count the *Distance,* or *Leagues*

E e 2 *Sailed;*

Sailed. Which Leagues are diftinguifhed or marked by the feveral Arches or quarters of Circles.

Note , That when you will find how many Leagues, one or many Degrees of Longitude contain by a propofed Latitude; or on the contrary, how many Degrees of Longitude a ceitain number of Leagues (of Eaft or Weft) contain alfo by a propofed Latitude : You muft count the Degrees of Latitude given, upon the neareft graduation to the Center of your Inftrument, that is to fay, upon that which begins at the fide of Eaft and Weft, and not upon the Limb or outmoft graduated Arch.

Note alfo, That when the diftances are great, you cannot then count every Arch or Line for a League, as you can do in little diftances, and therefore in that cafe you may count them for what you pleafe according as your occafion requireth, that is to fay, either 2 Leagues, or 3 Leagues, or 4 Leagues or more, if need be; you muft alfo count the Arches, for as much as you have counted the Lines of Latitude, and of Longitude. For Example, if you count an Arch for 5 Leagues, each Line of Latitude, and of Longitude, muft be alfo accounted for 5 Leagues, and fo of any other.

This Sinical Quadrant (of which the Figure before-going is a contracted Copy) is Printed in a large fheet of Royal Paper, ready to be put on Paftboard, and fitted for Practice, and is to be fold with the Book, either in a fheet or on a Board. And becaufe this Inftrument is of extraordinary and general ufe to work a Traverfe by, I fhall be the more particular in my Examples about it.

PROPOSITION I.

How to reduce Leagues of Eaft and Weft, into Degrees of Longitude.

S. HOW fhall I find (by the Quadrant) how many Degrees anfwer to a number of Leagues, by a propofed Latitude (or Parallel ?)

T. It is to be found thus: Lay the Thread upon the Degree of the propofed Latitude on the firft graduated Arch of your Quadrant, and holding it faft, count upon the fide A C, (which is taken for Eaft and Weft) the Leagues Sailed, Eaft or Weft; and obferve on what Line it falls; then follow that Line, or Perpendicular, untill the Point where it cuts the Thread, and there prick a Pin, and then count along the Thread how many Arches there are, from the Center of the Quadrant to the Pin; for that will fhew you how many Leagues (of a great Circle) there are; which Leagues muft be reduced into Degrees, and

and Minutes, by allowing 20 Leagues for a Degree; and a League for 3 Minutes.

Example 1.

Admit a Ship fail 72 Leagues Weſt, under the Parallel of 50 Degrees of Latitude; and it is demanded how many Degrees of Longitude are altered?

Firſt lay the Thread upon 50 Degrees upon the neareſt graduated Arch of your Quadrant; then count 72 Leagues, upon the ſide proper for the Leagues of Longitude (to wit from A to C) then follow that Line or Perpendicular C, untill the Point where it cuts the Thread, and there pricking a Pin, count upon the Arches how many Leagues there are from the Center A, to the ſaid Point, or Pin; and you will find 112 Leagues, which you muſt reduce into Degrees, allowing 20 Leagues for a Degree, and a League for 3 Minutes, and there will come 5 Degrees 36 Minutes of Longitude; which is what the propoſed Ship hath altered, as was required.

Example 2.

Admit a Ship fail 64 Leagues Eaſt, under the Parallel of 47 Degrees: I demand how many Degrees of Longitude are altered?

Lay the Thread upon 47 Degrees, on the neareſt graduated Arch of your Quadrant, then count on the ſide A C, 64 Leagues, and follow that Line as before, untill the Point where it cuts the Thread; and then counting the Leagues to that Point upon the Arches, you will find 93 Leagues, which comes to 4 Degrees 39 Minutes of Longitude, which is what the Ship hath altered.

P R O P. II.

How to change Degrees and Minutes of any Parallel into Leagues.

S. **H**OW ſhall I find how many Leagues a Degree of Longitude containeth, under a propoſed Parallel or Latitude?
T You muſt lay the Thread upon the Degree of the propoſed Latitude, then count upon the Arches along the Thread 20 Leagues, (which a Degree of a great Circle containeth) and look how many Leagues the Perpendicular gives upon the ſide of Eaſt and Weſt, for that will be the Leagues that a Degree of Longitude containeth by the propoſed Latitude, or Parallel.

Example.

Example.

Admit it was required to tell how many Leagues a Degree of Longitude containeth under the Parallel of 44 Degrees of Latitude.

You muſt firſt lay the Thread upon 44 Degrees, on the neareſt graduated Arch of your Quadrant, then count upon the Arches (along the Thread) 20 Leagues; and you will find that the Arch of 20 Leagues will cut 14 Leagues and a half of Longitude, by which you may conclude that a Degree of Longitude containeth 14 Leagues ½ under the Parallel of 44 Degrees of Latitude.

S. What muſt I do to know how many Leagues ſeveral Degrees of Longitude contain under a propoſed Parallel?

T. You muſt reduce the Degrees and Minutes (of Longitude) into Leagues, allowing 20 Leagues to a Degree, and a League for 3 Minutes; then laying the Thread upon the Degree of the propoſed Latitude (or Parallel) count upon the Arches (along the Thread) the Leagues proceeding from the propoſed Degrees of Longitude; then look how many Leagues of Longitude the ſaid Arch cuts, for it will be the Leagues that the propoſed Degrees of Longitude containeth.

Example.

Admit you have altered 5 Degrees 45 Minutes of Longitude, under the Parallel of 50 Degrees of Latitude, and it be required to find the Leagues and Parts anſwering thereto.

You muſt reduce the 5 Degrees 45 Minutes into Leagues, and there will come 115 Leagues; then laying the Thread upon 50 Degrees, count upon the Arches (along the Thread) 115 Leagues; then count upon the ſide A C, or of Eaſt and Weſt, how many Leagues of Longitude the ſaid Arch cuts, and you will find it to be 74 Leagues; by which you may conclude, that 5 Degrees 45 Minutes of Longitude under the Parallel of 50 Degrees containeth 74 Leagues, which was required.

PROP.

P R O P. III.

How to turn Leagues and Miles of Eaſting or Weſting, into Degrees and Minutes of Longitude.

S. WHAT muſt I do to turn Leagues of Eaſting or Weſting into Degrees and Minutes of Longitude?

T. If the difference in Latitude or diſtance between two Parallels is not above 3 Degrees, take the middle Parallel in diſtance between the two Latitudes; for that being known you may eaſily with your Sinical Quadrant, turn or reduce any Number of Leagues into Degrees and Minutes of Longitude, as ſhall be ſhown in the following Propoſitions.

S. How do you find the middle Parallel in diſtance?

T. The way to find the middle Parallel in diſtance, is to add the two Latitudes together, and half of the Product or Sum, will be the middle Parallel required.

Example.

Admit you depart from the Parallel of 38 Degrees 30 Minutes of Latitude North, and are arrived in the Parallel of 35 Degrees 40 Minutes alſo of Latitude North, and then would know the middle Parallel in diſtance.

Firſt add the two Latitudes together, to wit, 38 Degrees 30 Minutes, and 35 Degrees 50 Minutes, and there will come 74 Degrees 20 Minutes, whoſe half 37 Degrees 10 Minutes is the middle Parallel required.

S. What muſt I do when the diſtance between two Parallels, (or difference in Latitude) is greater then 3 Degrees?

T. When the difference in Latitude is greater then 3 Degrees, you muſt then make uſe of the Table in the Fifth Book, for reducing Leagues or Miles of Eaſting or Weſting into Degrees of Longitude, as its uſe ſheweth; and not of any middle Parallel, becauſe it is defective in great diſtances.

Note, The Leagues accounted on the Meridians or Sines are Leagues of Latitude, and thoſe accounted on the ſide of Eaſt or Weſt or Parallels, are Leagues of Longitude.

P R O P.

PROP. IV.

The Courfe and Diftance being given how to find the difference of Latitude and of Longitude.

Example 1.

The Courfe N E b N, (or the Third Rumb) and diftance 50 Leagues: I demand the difference of Latitude, and of Longitude?

FIRST, count 50 Leagues upon the Third Rumb, (from the North) to wit, upon the Arches; and from that Point (where you are to prick a Pin) you will find upon the Meridian 41 Leagues ⅔ which is the difference in Latitude, and upon the Parallels 27 Leagues ¼ which is the difference in Longitude (or Departure from the Meridian.)

Example 2.

The Courfe is S W b W, (or the fifth Rumb) and the diftance 54 Leagues; the difference of Latitude, and of Longitude is required.

Count 54 Leagues upon the fifth Rumb (from the South) and from the Point where the 54th. Arch cuts the faid Rumb, you will find upon the Meridians 30 Leagues, which is the difference in Latitude; and upon the Parallels 45 Leagues which is the difference in Longitude (or departure from the Meridian.)

S. Doth this Quadrant ferve for the 32 Points of the Compafs or 4 quarters of the Chard or Fly ?

T. Yes, and therefore every Rumb you fee on it reprefents four of thofe Two and Thirty Points; for the fame Rumb which ferveth for the *S W b W,* muft be taken alfo for the *S E b E,* the *N W b W,* and the *N E b E,* this being underftood: I come now to

Example 3.

Admit a Ship depart from 46 Degrees 30 Minutes of Latitude North; and 10 Degrees 20 Minutes of Longitude and fail 53 Leagues upon the S S E, (or fecond Rumb) the Latitude and Longitude of the faid Ship is required: I demand what Latitude, and Longitude fhe is in ?

Firft, count 53 Leagues upon the fecond Rumb, and from the Point where the 53 Arch cuts the faid Rumb, you will find upon the Meridians

49

49 Leagues of Latitude towards the South, and upon the Parallels 20 ⅓ of Longitude Easterly; then turn or reduce the 49 Leagues of Latitude in Degrees, allowing 20 Leagues to a Degree, and a League for 3 Minutes, and there will come 2 Degrees 27 Minutes for the difference of Latitude South, which you must subtract from 46 Degrees 30 Minutes, the Latitude of the place from whence you parted, (since it is North, and the Ship did fail Southerly) and the remainer 44 Degrees 3 Minutes North, will be the Latitude the Ship is arrived at.

Now to find the difference in Longitude, you must first take the middle Parallel (as before taught) between 46 Degrees 30 Minutes, and 44 Degrees 3 Minutes, and you will find it to be 45 Degrees 16 Minutes, therefore lay the Thread upon 45 Degrees 16 Minutes on the nearest graduation of your Instrument; then count upon the side of East or West 20 Leagues ½, and observe the Point where the imagined Perpendicular, or Meridian cuts the Thread, and how many Leagues there are from the Center of the Quadrant to the said Point, (counting upon the Arches) and you will find 29 Leagues, which being turned or reduced into Degrees by allowing 20 Leagues to a Degree, and a League for 3 Minutes, there will come 1 Degree 27 Minutes, which is the difference of Longitude, which must be added to 10 Degrees 20 Minutes, the Longitude of the place from whence the Ship parted (since her Course was Easterly) and the Sum 11 Degrees 47 Minutes will be the Longitude the Ship is arrived at.

Example 4.

Admit a Ship depart from 40 Degrees 30 Minutes of Latitude North, and 8 Degrees 45 Minutes of Longitude, and sail 72 Leagues upon the W N W, (or sixth Rumb) the Latitude and Longitude of the said Ship is required: I demand what Latitude and Longitude the said Ship is arrived at?

First, count 72 Leagues upon the 6th. Rumb, (making each Arch worth 2 Leagues) and you will find upon the Meridians 27 Leagues ½ of Latitude, (towards the North) and upon the Parallels 66 Leagues ½ of Longitude, (towards the West) then reduce the 27 Leagues ½ of Latitude into Degrees, allowing 20 Leagues to a Degree, and a League for 3 Minutes, and there will come 1 Degree 22 Minutes for the difference of Latitude (Northerly) which you must add to 40 Degrees 30 Minutes, the Latitude of the place from whence you parted, (since it is North and the Ship did fail Northerly) and the Sum 41 Degrees 52 Minutes North will be the Latitude the Ship is arrived at.

To find the difference in Longitude, take the middle Parallel between 40 Degr. 30 Min. and 41 Degr. 52 Min. viz. 41 Deg. 11 Min. then lay the Thread upon 41 Deg. 11 Min. on the nearest graduation of your Instrument, and count upon the side of East or West 66 Leagues ½, and observe (as before) the Point where the imagined Meridian cuts the Thread,

F f   and

and how many Leagues there are from the Center of your Inftrument to the faid Point, and you will find 72 Leagues (by counting each Arch for two Leagues) which being reduced into Degrees, there will come 3 Degrees 36 Minutes for the difference of Longitude, (Wefterly) which muft be Subtracted from 8 Degrees 45 Minutes, the Longitude of the place from whence the Ship parted (fince fhe failed towards the Weft, or Wefterly) and the remainer 5 Degrees 9 Minutes, will be the Longitude the Ship is arrived at.

S. This is eafie enough, but what muft I do to work feveral fhort Traverfes of 24 hours Run; as it happens when the Wind is contrary and often fhifts about?

T. In that cafe you muft firft make a Traverfe Table of 7 Column's; on the firft of which write (at the head or top of it) *the Courfe*; on the fecond *the Points* of the Compafs; on the third *the Diftance*; and in the four others the Cardinal Rumbs, that is to fail *North, South, Eaft,* (and) *Weft*; then find by your Inftrument how many Miles (or Leagues) each Courfe gives to the *North* or *South, Eaft* or *Weft,* to write it down in its proper Column; as you will better underftand by this

Example 5.

Admit a Ship parts from the Parallel of 50 Degrees 30 Minutes of Latitude North, and 11 Degrees 20 Minutes of Longitude, and being Bound to the Southward is forced by fhifting Winds to fail upon thefe feveral Courfes, (taken from the Log-board:)

Courfe.	Points.	Diftance	North.	South.	Eaft.	Weft.
		Miles.	Miles.	Miles.	Miles.	Miles.
S S W	2	20		18 $\frac{1}{2}$		7 $\frac{3}{5}$
S W $\frac{1}{2}$ W ly	4 $\frac{1}{2}$	24		15 $\frac{3}{4}$		18 $\frac{1}{4}$
W b N	7	32	6 $\frac{1}{4}$			31 $\frac{1}{2}$
S E b S	3	28		23 $\frac{1}{4}$	15 $\frac{1}{2}$	
S b E $\frac{1}{4}$ Ely	1 $\frac{1}{2}$	36		34 $\frac{1}{4}$	10 $\frac{3}{4}$	
			6 $\frac{1}{4}$	91 $\frac{1}{4}$	26 $\frac{2}{4}$	57 $\frac{1}{4}$
				6 $\frac{1}{4}$		26
				85 $\frac{1}{4}$		31 $\frac{1}{4}$

I demand the Latitude and Longitude of the faid Ship, and the iffue or direct Courfe and diftance from the firft place of departure?

Firft, count upon the *S S W,* or fecond Rumb, 20 Miles, (to wit, upon the Arches) and you will find 18 $\frac{1}{2}$ Miles upon the Meridians (or Sines,) which.

which is the Southing of the firſt Courſe, and is to be placed in the South Column; and 7¼ upon the Parallels, which is the Weſting, to be placed in the Weſt Column.

The ſecond Courſe is *S W ½ Wby*, or 4 Rumbs ¼, 24 Miles: Therefore lay the Thread upon the fourth Point ½ (from the South) and count 24 Miles upon it, and from that point you will find upon the Meridians 15¼ Miles which is the Southing, and 18¼ Miles in the Parallels which is the Weſting of this Courſe.

And what hath been ſaid of theſe two Courſes, the ſame is to be underſtood of the reſt in the Table. All the Courſes being reduced and laid down in the ſame order or manner, you muſt ſum up the Miles in each Column, and ſo you will find under the North Column 6¼ Miles in the South 91½, in the Eaſt 26, and in the Weſt Column 57¾ Miles.

Then compare the North and South Column's together, and Subtract the leſſer out of the greater, and there will remain 85¼ Miles for the difference in Latitude Southward; do as much for the Eaſt and Weſt Columns, and the remainer 31½ Miles, is the departure to Weſtward of the Meridian.

Next, then reduce or turn the 85¼ Miles (under the South Column) into Degrees of Latitude; and there will come 1 Degree 25 Min (the quarter is neglected) for the difference in Latitude, South; which muſt be Subtracted from 50 Degrees the Latitude of the firſt place of departure (ſince the Ship did ſail to the Southward) and the remainer 48 Degrees 35 Minutes North, is the Latitude the Ship is in.

To find the difference of Longitude, take the middle Parallel between 50 Degrees, and 48 Degrees 35 Minutes, and you will find it to be 49 Degrees 17 Minutes, therefore lay the Thread upon 49 Degrees 17 Minutes of the neareſt graduated Arch, then count upon the ſide of Eaſt and Weſt 31¼ Miles, and obſerve the Point where the ſaid 31¼ Miles cuts the Thread, and how many Lines there are from the Center of the Quadrant to the ſaid Point, (counting upon the Arches) and you will find 48 Miles, which is 48 Minutes for the difference of Longitude Weſt, which being Subtracted from 11 Degrees 20 Minutes, the Longitude of the place from whence the Ship parted (ſince ſhe ſailed Weſterly) the remainer 10 Degrees 32 Minutes will be the Longitude the Ship is arrived at.

To find the direct Courſe and neareſt diſtance from the place where the Ship began this Traverſe, to that where ſhe now is ſuppoſed to be; find where the 85¼ Miles of Latitude interſect the 31¼ Miles of Longitude, and bring the Thread to that Point, and you will find that the direct Courſe hath been *S S W* a Degree Southerly, (ſince the Ship hath ſailed between the South and the Weſt) and the diſtance run (counted upon the Arches from the Center of your Inſtrument, to the ſaid Point of interſection) will be Miles.

Note, If your Diſtance be very great, your beſt way will be to reduce your Miles into Leagues.

Ex

Example 6.

Admit a Ship depart from the Parallel of 30 Degrees of Latitude North, and 1 Degree 30 Minuts of Longitude; and being bound to the Southward, is forced by shifting Winds to sail upon these several Courses (taken from the Log-board) viz.

West South West	. . . .	40 Miles.
South West	. . . .	50
South by West	. . . .	35
South West by South	. . . . . .	42
West by South	. . . .	54
South	. . . .	28 Miles.

I demand the Latitude and Longitude she is in, and the direct Course and Distance from the first place of Departure?

Course.	Points.	Distance Miles.	North. Miles.	South. Miles.	East. Miles.	West. Miles.
W S W	6	40		15 ¼		37
S W	4	50		35 ¼		35 ¼
S b W	1	35		34 ½		6 ¼
S ₰ S	3	42		35		23 ¼
W b S	7	54		10 ¼		53
S	0	28		28		09
				158 ¾		155 ¼

Having reduced these several Courses as before taught, there comes 158 ¼ Miles Southing, and 155 ½ Miles Westing. Then reduce the 158 Miles into Degrees, and there will come 2 Degrees, 38 Minutes, which is the difference in Latitude to be Subtracted from 30 Degr. (since the Ship sailed to the Southward) and the remainder 27 Degrees 22 Minutes North is the Latitude she is in.

To find the difference in Longitude, take the middle Parallel between 30 Deg. and 27 Deg. 22 Min. and you will find it to be 28 Deg. 41. Min. by which the 155 Miles Westing give 2 Deg. 57 Min. for the difference of Longitude West; which must be Subtracted from 1 Deg. 30 Min. (since the Ship sailed Westward) but because it cannot be done; I add 360 Deg. to 1 Deg. 30 Min. and from the sum 361 Deg. 30 Min. I Subtract the 2 Deg. 57 Min. of difference, and the remainder 358 Deg. 33 Min. is the Longitude she is in, which was required.

To

To find the direct Course and nearest distance, find where the $158\frac{1}{2}$ Miles of Latitude intersects the $155\frac{1}{3}$ Miles of Longitude, and lay the Thread to that Point, and it will shew you the direct Course, and the distance is to be counted upon the Arches, as before directed.

Example 7.

Admit a Ship departs from the Parallel of 2 Degrees 40 Minutes of Latitude North, and 357 Degrees 30 Minutes of Longitude, and being bound to the Southward is forced by contrary or shifting Winds to sail upon these several Courses, viz.

South South East	26 Leagues.
South East	30
East by North	16
South East by South	38
South by East	45
South East by East	52 Leagues.

I demand the Latitude and Longitude she is in, and the direct Course and Distance from the first place of Departure?

Course.	Points.	Distance. Leagues.	North. Leagues.	South. Leagues.	East. Leagues.	West. Leagues.
S. S. E	2	26		24	10	
S. E	4	30		$21\frac{1}{4}$	$21\frac{1}{4}$	
E. b N	7	16	$3\frac{1}{8}$		$15\frac{2}{3}$	
S E b S	3	38		$31\frac{1}{2}$	$21\frac{1}{8}$	
S b E	1	45		$44\frac{1}{8}$	$8\frac{3}{4}$	
S. E b E	5	52		$28\frac{7}{8}$	$43\frac{1}{4}$	
			$3\frac{1}{8}$	$149\frac{1}{4}$ $3\frac{1}{8}$	120	
				$146\frac{1}{8}$		

146 Leagues $\frac{1}{8}$ Southing (or of Latitude) being reduced into Degrees and Minutes, there will come 7 Deg. 20 Min. which is the difference of Latitude towards the South, to be Subtracted from 2 Deg. 40 Min. of Latitude North, (since the Ship did sail towards the South) but it cannot be done, and so the Ship hath passed the Equinoctial; therefore I Subtract the 2 Deg 40 Min. of Latitude North, from 7 Deg. 20 Min. which is the difference South of Latitude, and the remainder 4 Deg. 40 Min. is the Latitude she is in.

Now

Now to find the difference of Longitude by the forementioned Table, (turning first the Leagues of Easting into Miles, or the Miles of the Table into Leagues) and you will find that the 120 Leagues (or 360 Miles) Easting, answer to 6 Degrees, which is the difference of Longitude towards the East, to be added to 357 Degrees 30 Minutes (since she did sail Easterly) and the sum will be 363 Degrees 30 Minutes, from which you must Subtract 360 Degrees, and the remainder 3 Degrees 30 Minutes is the Longitude she is in. For the direct Course, and nearest Distance, find where the 146 Leagues ⅜ of Latitude, intersect the 120 Leagues of Longitude, and lay the Thread to that Point, and you will find the direct Course, (since the Ship did sail between the South and the East) and the nearest Distance is to be counted upon the Arches to that Point, as aforesaid.

Example 8.

Admit a Ship depart from the Parallel of 42 *Degrees* 20 *Minutes of Latitude North, and from the first Meridian; and being bound to the Northward is forced by contrary Winds to sail upon these several Courses*, viz.

North by East	25 Leagues.
North East by North	38
East by South	20
North North East ½ Ely	45
North East by East	50
North	32 Leagues.

I demand the Latitude and Longitude she is in, and the direct Course and Distance from the first place of Departure?

Course.	Points.	Distance. Leagues.	North. Leagues.	South. Leagues.	East. Leagues.	West. Leagues.
N b E	1	25	24½		4⅞	
NE b N	3	38	31½		21⅛	
E b S	7	20		3⅞	19⅝	
NNE ½ Ely	2½	45	39¾		21¼	
NE b E	5	50	27¾		41½	
N	0	32	32			
			155½ 3⅞	3⅞	108⅜	
			151⅝			

151 ¾ Leagues Northing, (or of Latitude) being reduced into Degrees and Minutes as before taught, there comes 7 Degrees 35 Minutes, which is the difference in Latitude, to be added to 42 Degrees 20 Minutes of Latitude North, (since the Ship hath failed towards the North) and the sum 49 Degrees 55 Minutes is the Latitude she is in.

Next, find by the Table as before directed how many Degrees and Minutes of Longitude answer to 108 ⅞ Leagues Easting, or rather 326 Miles, and you will find it to be 7 Degrees 51 Minutes, which is the difference of Longitude; and Longitude she is in (since she hath failed from the first Meridian Eastward.)

For the Course and distance, lay the Thread upon the Point where the 151 ¼ Leagues of Latitude, intersect the 108 ⅞ Leagues of Longitude; and it will shew that the direct Course is *N E b N* 1 Degree 50 Minutes Easterly, (since she hath failed between the North and the East) and the nearest distance (counted upon the Arches, to that Point) 186 Leagues, which was required.

I have been long upon this Proposition, because it is the most useful in the practice of Navigation.

PROP. VI.

The Distance Run, and difference of Latitude being given, how to find the Course, and difference of Longitude.

FIRST find the difference of Latitude, by Subtracting the Latitude of the place from whence you departed from the observed Latitude, if they be both of the same Denomination: That is to say, both Latitude North, or both Latitude South; but if they be of different sides, that one be Latitude North, and the other Latitude South, you must add them together, and the sum will be their difference of Latitude, to be reduced into Leagues.

S. This is easie enough, but what must I do next?

T. The next thing you are to do, is to count the said Leagues (or Miles) of Latitude, upon its proper side, to wit, upon that taken for North, and South, (begining at the Center of your Instrument,) you must also count upon the Arches the Distance or Leagues (or Miles) failed, and observe the Point where the said Arch intersects the said Leagues (or Miles of Latitude;) for the Thread being laid upon that Point, will shew the Course, and the Leagues (or Miles) of Longitude, which being reduced into Degrees and Minutes, as before taught, will be the difference of Longitude.

Example.

Example 1.

Admit a Ship sail from the Parallel of 48 Degrees 30 Minutes of Latitude North, on some Point between the South and the West 56 Leagues; and then finds her self in 45 Degrees 54 Minutes Latitude of North: I demand the Course, and Departure from the Meridian?

First find the Difference between 48 Degrees 30 Minutes, and 45 Degrees 54 Minutes, (by Subtracting the lesser Latitude out of the greater) and there will remain 2 Degrees 36 Minutes for the difference of Latitude; which being reduced into Leagues (or Miles) there comes 52 Leagues (or 156 Miles) which must be counted upon the side of North and South; and the distance 56 Leagues (or 168 Miles) upon the Arches, then observe where that Arch and Line of Latitude intersect one another, and lay the Thread thereon, and it will shew the Course to be South south West 30 Minutes South, (since she did sail between the South and the West) and upon the Parallels you will find 21 Leagues (or 63 Miles) of Longitude Westward, which is the Departure from the Meridian.

Example 2.

Admit a Ship sail from the Parallel of 46 Degrees 10 Minutes of Latitude North, and 329 Degrees 30 Minutes of Longitude on some Point between the North, and the East 108 Leagues, and then finds her self in 49 Degrees 40 Minutes of Latitude also North: I demand the Course and Longitude she is in?

I Subtract the lesser Latitude out of the greater (since they are both on the same side of the Equinoctial) and there remaineth 3 Degrees 30 Minutes, which is the difference North of Latitude; which being reduced into Leagues, there comes 70 Leagues to be counted upon the side of North and South; count also upon the Arches 108 Leagues, then lay the Thread on the Point of Intersection of the Arch of 108 Leagues, and the 70 Leagues of Latitude, and it will shew the Course to be North East 4 Degrees 45 Minutes Easterly (since she did sail between the North and the East) and upon the Parallels you will find 82 ½ Leagues of Longitude Eastward; or departure from the Meridian) which is to be reduced into Degrees and Minutes of Longitude by the precedent Table, (after it is reduced into Miles) by which, and a Rule of Three, as before directed, you will find it answer to 6 Degrees 6 Minutes ¹⁴⁸⁄₃₀, which being more than the half of a Minute, may be taken for a Min: and so the sum will be 6 Degrees 7 Minutes, which is the difference of Longitude, to be added to 329 Degrees 30 Minutes (since she did sail Eastward) there comes 335 Degrees 37 Minutes, which is the Longitude she is in as was required.

Example 3.

Example 3.

Admit a Ship departs from the Parallel of 2 Degrees 20 Minutes of Latitude South, and 3 Degrees 30 Minutes of Longitude, and sail on some Point between the North and the West 250 Leagues; and then finds her self in 5 Degrees 30 Minutes of Latitude North: I demand the Course and Longitude she is in?

Latitude South departed	2 Deg.	20 Min.
Add the Latitude North arrived at . .	5	30
The Sum . .	7	50

Which being reduced into Leagues, there comes 156⅔ Leagues, which being counted upon the side of North and South, and the 250 Leagues upon the Arches, and the Thread placed or laid upon the Point of Interfection, sheweth the Course to be *N W* 6 Degrees 40 Minutes Westerly, (since she did sail between the North and the West) and upon the Parallels you will find 196 Leagues of Longitude or departure from the Meridian Westward, which by the Table answers to 9 Degrees 48 Minutes, which is the difference of Longitude to be Subtracted from 3 Degrees 30 Minutes, (since she did sail to the Westward) but because it cannot be, I add 360 Degrees to 3 Degrees 30 Minutes, and from the sum 363 Degrees 30 Minutes, I Subtract the 9 Degrees 48 Minutes of difference, and the remainder 353 Degrees 42 Minutes, is the Longitude she is in, as was required.

Example 4.

Admit a Ship departs from the Equinoctial and the first Meridian, and sail on some Point between the South and the East 188 Leagues, and then finds her self in 6 Degrees 10 Minutes of Latitude South: I demand the Course and Longitude she is in?

Answ. The 6 Degrees 10 Minutes of Latitude South arrived at, is the difference of Latitude, which being reduced or turned into Leagues, there comes 123⅓ Leagues to be counted upon the side of North and South, and the 188 Leagues upon the Arches, and the Thread being laid upon the Point of their Interfection, will shew the Course to be *S E* 4 Degrees 20 Minutes Easterly, (since she did sail between the South and the East) and upon the Parallels you will find 142⅔ Leagues of Longitude Eastward; which by the Table (being reduced into Miles) Answers to 7 Degrees 8 Minutes, which is the difference of Longitude towards the East, and the Longitude she is in, as was required, (since she departed from the first Meridian.)

G g   P R O P.

PROP. VI.

The Course and difference of Latitude being given, to find the distance and difference of Longitude.

FIRST, find the difference of Latitude by Subtracting one from another, if they be both Latitude North, or both Latitude South; but if they be of contrary side, that one be South and the other North, you must add them together, and the sum will be the Degrees and Minutes of their difference, which must be reduced into Leagues; then count the Leagues proceeding from the difference of Latitude upon the side of North and South, and observe the Point where it intersect the Course or Rumb given for it will mark the required distance, which is to be counted upon the Arches from the Center of your Instrument to the said Point; count also upon the Parallels how many Leagues of Longitude there are, and it will be the departure from the Meridian; which Leagues being turned into Degrees and Minutes of Longitude, (as before taught) and then added or Subtracted from the Longitude given, will shew you the Longitude your Ship is in.

Example 1.

Admit a Ship sail from the Parallel of 46 Degrees 40 Minutes of Latitude North SW b S, 'till she be in Latitude 44 Degrees 30 Minutes North: I demand the distance sailed, and the departure from the Meridian?

First, Subtract 44 Degrees 30 Minutes from 46 Degrees 40 Minutes, and the remainder 2 Degrees 10 Minutes is the difference of Longitude to be reduced into Leagues, as before directed; and there comes 43 $\frac{1}{3}$ Leagues of Latitude, which I count upon the side of North and South; then I count upon the Arches along the *SW b S*, (or third Rumb) how many Leagues there are from the Center of the Quadrant, to the Line of 43 $\frac{1}{3}$ Leagues; and I find 52 Leagues, which is the distance Run; and upon the Parallels I find 29 Leagues, which is the departure from the Meridian Westward.

Example 2.

Admit a Ship sail from the Parallel of 42 Degrees 10 Minutes of Latitude North, and 358 Degrees 20 Minutes of Longitude N E b E, 'till she be in Latitude 45 Degrees 28 Minutes North: I demand the distance sailed and Longitude she is in?

First,

First, Subtract the two Latitudes one from another, (since they are both on the same side of the Equinoctial) and there remaineth 3 Degrees 18 Minutes for their difference to the Southward, which being reduced into Leagues, there comes 66 Leagues of Latitude to be counted upon the side of North and South; then count upon the Arches along the *N E b E*, how many Leagues there are to that Line of 66 Leagues, and you will find 118 Leagues, which is the distance, and upon the Parallels you will find 98 Leagues of Longitude Eastward, which being reduced into Miles, you will find by the Table that it answers to 6 Degrees 45 Minutes, which is the difference of Longitude to be added to 358 Degrees 20 Minutes, (since she did sail to the Eastward) and there will come 365 Degrees 5 Minutes, from which I Subtract 360 Degrees, and the remainder 5 Degrees 5 Minutes is the Longitude she is in, as was required.

Example 3.

Admit a Ship sail from the Parallel of 3 Degrees 16 Minutes South, and 2 Degrees 45 Minutes of Longitude N N W, 'till she be in Latitude 2 Degrees 14 Minutes North: I demand the distance sailed, and the Longitude she is in?

	Deg.	Min.
Latitude departed	3	16
Add the Latitude arrived at . .	2	14
Difference in Latitude Northward	5	30

5 Degrees 30 Minutes reduced into Leagues, there comes 110 Leagues of Latitude, by the intersection of which, with the given Rumb, I find the distance run to be almost 118¾ Leagues; and the departure from the Meridian 45¼ Leagues; which Leagues of Departure being reduced into Degrees and Minutes of Longitude, (as before taught) there will come 2 Degrees 16 Minutes, which is the difference of Longitude to be Subtracted from 2 Degrees 45 Minutes (since she did sail to the Westward) and the remainder 29 Minutes is the Longitude she is in, as was required.

PROP. VII.

The Alteration of the Latitude and Longitude being given,
how to find the Course and distance.

S. WHAT is the first thing I must take notice of in this Proposition?
T. The first thing you are to take notice of (to find the
distance) is, that when the Latitudes are of the same side of
the Equinoctial, and that there is no difference in Longitude (both places
being under the same Meridian) you must only Subtract the two Latitudes one from another, and the remainder will be their difference, which
being reduced into Leagues, will shew you the distance Run.

Example.

Admit a Ship sail from the Parallel of 40 Degrees 20 Minutes of Latitude
North, and 10 Degrees 30 Minutes of Longitude, and would sail to 37
Degrees 10 Minutes of Latitude North, and 10 Degrees 30 Minutes of
Longitude: I demand what Course she must Steer, and the distance?

First, consider that both places are in the same Meridian or Longitude,
therefore Subtract only the two Latitudes one from another, and there
will remain 3 Degrees 10 Minutes for the difference of Latitude Southward, which being reduced into Leagues, there will come 63 ¼ Leagues,
which is the distance she is to Run South (since that Latitude she would
go to, is less than that she is in.)
S. What must I do when both Latitudes are of different sides of the
Equinoctial?
T. If one Latitude is North and the other South, (or of different side
as you call it) and both places be also under the same Meridian, you must
add the two Latitudes together, and the sum will be their difference.

Example.

Admit a Ship sail from 2 Degrees 15 Minutes of Latitude North, and 4
Degrees 20 Minutes of Longitude, and would sail to Latitude 5 Degrees
30 Minutes South, and 4 Degrees 30 Minutes of Longitude: I demand what
Course she must Steer, and the distance?

	Deg.	Min.
Latitude North departed	2	15
Latitude South arrived	5	30
Difference South	7	45

The

The two Latitudes being added together, their sum 7 Degrees 45 Minutes is the difference in Latitude, which being reduced into Leagues, there comes 155 Leagues, which is the distance she is to Run South, (since the Latitude of her departure is North, and that she will go to is South.)

S. What must I do to find the difference in Latitude, if one place was under the Equinoctial, and the other distant from it, but both in the same Longitude?

T. In that case, the Latitude of the place which is out of the Equinoctial will be the difference in Latitude, which being reduced into Leagues, will shew the distance she is to Run, either North or South; this is so plain that it needs no Example.

S. What is there to be done if both places were under the Equinoctial, and should differ in Longitude?

T. You must Subtract the two Longitudes one from another, and the remainder will be their difference, which being reduced into Leagues, will shew the distance between both places; but when both places are under the same Parallel but of (or distant from) the Equinoctial, you must first find the difference in Longitude, or departure from the Meridian, then lay the Thread upon the given Latitude, and count upon the Arches along the Thread the departure from the Meridian, and from that Point you will find upon the Parallels the Leagues of Longitude that answer to it.

Example.

Admit a Ship sail from the Parallel of 48 Degrees 50 Minutes of Latitude North, and 341 Degrees 50 Minutes of Longitude, and would sail to a place whose Latitude is also 48 Degrees 50 Minutes North, and 336 Degrees 30 Minutes of Longitude : I demand what Course she must Steer, and the distance?

Longitude departed	341 Deg.	50 Min.
Longitude arrived	336	30
Difference West	5	20

The two Longitudes being Subtracted one from another; there remaineth 5 Degrees 20 Minutes, which is the difference of Longitude Westward, (since the Longitude she is in, is lesser then that from whence she departed) this being known, lay the Thread upon 48 Degrees 50 Minutes of the graduation, which begins at the side of East and West; then count upon the Arches along the Thread 106½ Leagues, which is the value of 5 Degrees 20 Minutes turned into Leagues, and from that Point you will find upon the Parallels, or side of East and West, 70 Leagues, which is the distance she is to Run West.

S. What

S. What muſt I do to find the diſtance and Courſe, when both places differ both in Latitude and Longitude?

T. In that caſe, find firſt the difference in Latitude and Longitude, as in the precedent Examples; then reduce the difference of Latitude into Leagues: Do as much for the difference of Longitude, and then ſee by the before-mentioned Table of reducing Leagues or Miles of Eaſting or Weſting into Degrees of Longitude, how many Degrees and Minutes anſwers to the Leagues proceeding from the difference of Longitude, and reduce as before thoſe Degrees into Leagues; then count them upon the ſide of Eaſt and Weſt, and the Leagues proceeding from the difference of Longitude upon the ſide of North and South, obſerving where theſe two Lines interſect one another, for the Thread being laid on the Point of their interſection will ſhew the Courſe, then count upon the Arches along the Thread how many Leagues there are from the Center of your Inſtrument to the ſaid Point, and it will be the diſtance required.

Example.

Admit a Ship departs from the Latitude 43 Degrees 30 Minutes North, and 11 Degrees 10 Minutes of Longitude, and will go into 45 Degrees 30 Minutes of Latitude North, and 8 Degrees 45 Minutes of Longitude: I demand the Courſe and Diſtance?

Latitude North arrived . . 45 Deg. 30 Min.
Latitude North departed . . 43   30

Difference North 2   00
Multiplyed by 20 Leag. that a Deg. con. 20

       There comes . . 40 Leagues of Latitude.

Longitude departed . . . 11 Deg. 10 Min.
Longitude arrived 8   45

Difference Weſt 2   25
       Multiplyed by . . 20

               40
               8 $\frac{1}{3}$

       There comes . . . 48 $\frac{1}{3}$ Leagues of Longitude,

Or 145 Miles, which is the departure from the Meridian, and anſwers by the precedent Table to 3 Degrees (near) 22 Minutes, which being reduced into Leagues (by allowing alſo 20 Leagues to a Degree, and a League to 3 Minutes,) there comes 67 $\frac{1}{3}$ Leagues of Longitude, which you muſt count upon the ſide of Eaſt and Weſt: Count alſo the Forty Leagues of Latitude upon the ſide of North and South; then lay

                                     the

the Thread on the Point of interfection of thofe two Lines, and it will
fhew you that the Courfe is *N W b W* 3 Degrees 10 Minutes Wefterly;
and upon the Arches you will find 78¼ Leagues, which is the diftance
required.

Example 2.

*A Ship from the Latitude of 38 Degrees 40 Minutes North fail South Weft-
ward, 'till fhe hath altered her Latitude 2 Degrees 30 Minutes, or 50 Leagues,
and is departed from the Meridian 36 Leagues: I demand the Courfe and
Diftance?*

First, fee what Latitude fhe is in, by Subtracting the 2 Degrees 30
Minutes, (difference of Latitude) from 38 Degrees 40 Minutes, (fince
fhe did fail Southerly) and there will remain 36 Degrees 10 Minutes,
which is the Latitude fhe is in. Then reduce the 36 Leagues (departure
from the Meridian) into Miles, (as before) and there will come 108
Miles, which by the precedent Table anfwers to 2 Degrees (almoft) 17
Minutes, which Degrees being reduced into Leagues, there comes 45¾
Leagues to be counted upon the fide of Eaft and Weft; Count alfo upon
the fide of North and South 50 Leagues, and obferve where thefe two
Lines interfect on another, for the Thread being laid on it will fhew
you that the Courfe hath been *S W* 2 Degrees 40 Minutes South; then
count upon the Arches along the Thread, how many Leagues there is
to that Point, and you will find almoft 68 Leagues which is the diftance
Run, as was required.

PROP. VIII.

*The Courfe and Departure from the Meridian given,
to find the diftance and difference in Latitude.*

THIS Propofition being of no ufe in the Practice of Navigation,
fince there is not (yet) found a way (at Sea) to obferve the
Longitude, I fhall fay but little concerning it, being fufficient
to give you an

Example.

*Admit a Ship Sail N E b N (or third Rumb) until fhe be departed 32 Leagues
from the Meridian: I demand the diftance and difference in Latitude?*

First,

Firſt, count upon the ſide of Eaſt and Weſt 32 Leagues (departed from the Meridian) and obſerve where that Line (of 32 Leagues) interſect the *N E b N*, then count upon the Arches along the given Rumb, how many Leagues there are from the Center of your Inſtrument to the ſaid Point of Interſection, and you will find 57 ½ Leagues, which is the diſtance, and upon the Meridians you will find 47 ¼ Leagues, which is the difference of Latitude required.

PROP. IX.

To know how many Leagues you muſt ſail upon any Point of the Compaſs, to riſe a Degree of Latitude.

YOU muſt count upon the ſide of North and South 20 Leagues, (that a Degree of the Equinoctial containeth) and obſerve the Point where it interſects the given Rumb, then count upon the Arches how many Leagues there is to that Point, and it ſhall be the Leagues you muſt ſail, to raiſe or depreſs the Pole of a Degree upon any Point of the Compaſs.

Example.

Admit you would know how many Leagues you muſt ſail upon the N E b E or fifth Rumb, to raiſe a Degree of Longitude.

Firſt, count upon the ſide of North and South 20 Leagues (which is a Degree of the Equinoctial) and obſerve the Point where that Line (of 20 Leagues) interſect the *N E b E*, or fifth Rumb; then count upon the Arches from the Center of your Inſtrument to the ſaid Point, and you will find 36 Leagues, which you muſt ſail on that Point to raiſe a Degree Latitude: and on the Parallels you will find 30 Leagues, which is the departure from the Meridian.

PROP.

PROP. X.

How to Correct your Dead Reckoning by the Sinical Quadrant.

S. IS all kind of Reckoning Corrected the same way by the Sinical Quadrant?

T. No, for it differs according to what the Course hath been.

S. How many sorts of Corrections is there then?

T. There is of 3 sorts, the first of which serveth only for the first and second Rumb, (from the North or South) which Rumbs are never Corrected, except there be Variation.

The second Correction serveth for the sixth and seventh Rumb, (or thereabout) and then the Longitude is not Corrected, but the Course and Distance is Corrected.

The third Correction serveth for the third, fourth, and fifth Rumb; as you will easily understand by the following Examples.

S. When do you make use of the first Correction?

T. We make use of the first Correction, when a Ship hath sailed upon one only Point of the Compass, thus: We count upon the Arches along the Course, the Leagues sailed by Dead Reckoning, and then we count to that Point upon the Meridians, how many Leagues it gives in Latitude; and upon the Parallels, how many Leagues it gives in Longitude: Next, we find the difference between the Latitude departed, and the Latitude observed, and reduce it into Leagues, then we see if the Leagues by Dead Reckoning, agree with the Leagues of Latitude by Observation; and if they agree, the Dead Reckoning is right, is good; but if they do not agree, there is some errour in the Dead Reckoning; and therefore it must be Corrected by counting upon the side of North and South, the Leagues of the difference in Latitude proceeding from the Observation; count also upon the Arches along the Rumb, how many Leagues there are from the Center of your Instrument to the said Lines of Leagues, and it will be the Distance Corrected; and upon the Parallels you will find the Leagues of Departure Corrected, which being reduced into Degrees of Longitude (as before taught) will be the difference of Longitude Corrected.

Example.

Admit your Ship sail (by Dead Reckoning) S b E 51 Leagues, and by your Observation you find to have altered in Latitude 2 Degrees 54 Minutes: I demand the Distance Corrected, and the Departure Corrected?

it is of to write it down, and so forth; observing the same Rules in every thing else as you did with the first, in the three precedent Examples.

And *Note*, that those Observations are most to be trusted to, which are made when the Sun is farthest from the Meridian, provided he be not so near to the Horizon as to suffer any very considerable refraction.

PROP. IX.

To find the Variation by the Scale and Compass.

S. WHAT must I do to find the Variation by the Scale and Compass?

T. You must first observe (with your Azimuth Compass) the Magnetical Amplitude of the Sun, and by the Tables of Amplitude you must find the Sun's true Amplitude for that Day: For these two things being known, you may with ease find the Variation (and which way it is) thus:

With the Chord of 60 Degrees describe a Circle, and quarter it with two Diameters at Right-angles, at the Extremities of which write N for *North*, S for *South*; E for *East*, and W for *West*; as in this following Figure: Then take with your Compass the true Amplitude, and prick it from the East or West Point as the occasion requireth, prick off also the Magnetical Amplitude, then take the Distance between the Magnetical Amplitude and the North, and set it from the true Amplitude Northward, and the Foot of the Compass will fall in a certain Point, from which, if you draw a Line through the Center that Line will represent the North and South Line of the Compass or Magnetical Meridian; and that way, that this Point lyeth from the North, that way is the Variation, which is just so much as is the Arch Comprehended between the said Point and the North.

Example. 1.

Admit that in Latitude 46 Degrees North, the Sun having 20 Degrees North Declination: I observe the Magnetical Amplitude at Sun Rising to be 42 Degrees from the East, Northward; and then would know the Variation of the Compass.

First, with the Chord of 60 Degrees I describe a Circle, and having quarter'd it as before directed, with the 20 Degrees Declination, and 46 Degrees Latitude; I enter the Table of Amplitude, and find the true Amplitude to be 29 Degrees 30 Minutes, which I prick from E, the East Point towards the North to G; and taking the Magnetical Amplitude 42 Degrees, I prick that also from E, Northward to H; then taking the Distance between H and N, the North Point, I set one Foot of my Compass in G, the Point of the true Amplitude, and the other being

being turned toward the North reaches to L, the Point to which the North Point of the Compass is directed, so that L, N, 12 Degrees 30 Minutes, is the quantity of Variation, which is *Easterly;* because L falls towards E, that is, between the North and the East.

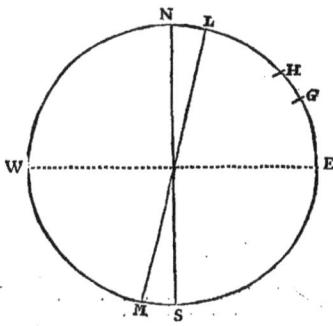

	$D.$	$M.$
Magnetical Amplitude	42	00
True Amplitude	29	30
Variation	12	30

Example. 2.

Admit that in Latitude 38 Degrees North, the Sun having 11 Degrees Declination South, the Magnetical Amplitude at Sun Set, is observed to be 35 Degrees from the West Southward: *I demand what is the Variation, and which way it is?*

First enter the Table of Amplitudes with the Latitude and Declination, as before directed; and you will find the true Amplitude to be 14 Degrees 1 Minute; which being pricked from W, to G; and also the Magnetical Amplitude to H, as before directed, the Distance between L, N, shews the Variation to be 20 Degrees 59 Minutes to the Eastward, from the North.

Mag-

Courſe.	Points.	Diſtance. Leagues.	North. Leagues.	South. Leagues.	Eaſt. Leagues.	Weſt. Leagues.
N N E	2	28	$25\frac{2}{8}$		$10\frac{3}{4}$	
N b W	1	16	$15\frac{1}{8}$			$3\frac{1}{8}$
N	0	24	24			
N b E	1	32	$31\frac{1}{3}$		$6\frac{2}{4}$	
N E	4	21	$14\frac{7}{8}$		$14\frac{7}{8}$	
N b W	1	43	$42\frac{1}{8}$			$8\frac{3}{8}$
			$153\frac{1}{3}$		$31\frac{7}{8}$ $11\frac{4}{8}$	$11\frac{4}{8}$
					$20\frac{1}{8}$	

Latitude North arrived	42 Deg.	48 Min.
Latitude North departed	36	30
Difference North	6	18

Having found the Northing, Eaſting and Weſting of the ſeveral Courſes and Diſtance given, ſum up the ſeveral Columns and compare the Eaſt and Weſt Column together, there will come $153\frac{1}{3}$ Leagues Northing, and $20\frac{1}{8}$ Eaſting; then count upon the ſide of North and South $153\frac{1}{3}$ Leagues; and upon the ſide of Eaſt and Weſt $20\frac{1}{3}$ Leagues, lay the Thread on the Point of interſection of theſe two Lines, and it will ſhew you the Courſe which is North 7 Degrees 30 Minutes Eaſterly; then Subtract the two Latitudes one from another, and there will remain 6 Degrees 18 Minutes, which is the difference in Latitude Northward, to be reduced into Leagues, and there will come 126 Leagues, which being counted upon the ſide of North and South, obſerve where the ſaid Line of 126 Leagues interſect the Courſe, (which the Thread ſheweth) and then count upon the Arches how many Leagues there is, from the Center of your Inſtrument to the Point of interſection, and you will find 127 Leagues, which is the diſtance Corrected; and upon the Parallels you will find $16\frac{1}{2}$ Leagues of Longitude or Departure Corrected, which being turned into Miles, you will find by the precedent Table that it anſwers to 1 Degree almoſt 4 Minutes, which is the difference of Longitude Eaſtward, to be added to the Longitude departed, *viz.* 345 Degrees 45 Minutes, (ſince ſhe did ſail Eaſtward) and there will come 346 Degrees 49 Minutes, which is the Longitude (Corrected) ſhe is in.

As

As to the second Correction.

Having already told you , that it ferveth only for the 6 and 7 Rumb : I have only to add , that when your Ship fail but upon one Point of the Compafs, you muft only count upon the Arches along the Rumb or Courfe, the Leagues failed by Dead Reckoning , then find (or look). how many Leagues of Latitude and of Longitude it giveth ; that done, Subtract the two Latitudes one from another, if they be both of the fame fide, (but if they be of different fides, add them together) and the remainder (or fum) fhall be the difference of Latitude, which being reduced into Leagues , muft be counted upon the fide of North and South ; count alfo the Leagues of Longitude or Departure (found by your Dead Reckoning) upon the fide of Eaft and Weft , and obferve the Point where they interfect one another , for the Thread being laid on it , will fhew you the Courfe Corrected , and the Diftance Corrected.

But if your Ship hath failed upon feveral Points of the Compafs; firft reduce the feveral Courfes under the four chief Rumbs of your Traverfe Table; to wit, under the North, South, Eaft and Weft , then Subtract the Leagues Northing and Southing one from another, (as before taught, directed) and the remainder will be the difference of Latitude ; Subtract alfo the Leagues of Eafting and Wefting one from another , and the remainder will be the departure from the Meridian , or difference in Longitude by Dead Reckoning ; that done, count upon the fide of North and South , the Leagues of the difference of Latitude proceeding from your Obfervation ; count alfo upon the fide of Eaft and Weft , the Leagues of Longitude by Dead Reckoning , and lay the Thread on the Point where they interfect one another , and it will fhew you the Courfe Corrected , and the Diftance Corrected.

Example 1.

Admit a Ship fail (by Dead Reckoning) E S E 50 Leagues , and by Obfervation find to have altered the Latitude 39 Minutes : I demand the Courfe Corrected, and the Diftance Corrected ?

Firft, count 50 Leagues upon the Arches along the *E S E*, or 6 Rumb, and from that Point you will find upon the Meridians 19 Leagues of Latitude , and upon the Parallels you will find 46 ¼ Leagues of Longitude , or departure from the Meridian ; but becaufe the Obfervation is more certain than the Dead Reckoning , reduce the 39 Minutes difference of Latitude (found by Obfervation) into Leagues, and there will come 13 Leagues which be counted upon the fide of North and South , count alfo the 46 ¼ Leagues of Longitude (by Dead Reckoning) upon the fide of Eaft and Weft , lay the Thread on the Point where they interfect one another , and it fhall fhew you the Courfe Corrected , which is *E S E* almoft almoſt 7 Degrees Southerly, and upon the Arches you will find 48 Leagues, which is the Diſtance Correſted.

Example 2.

Admit a Ship departs from the Parallel of 46 Degrees 30 Minutes of Latitude North, and 7 Degrees 40 Minutes of Longitude, and ſail by Dead Reckoning Weſt by South 96 Leagues, and then find by Obſervation to be in Latitude 45 Degrees 10 Minutes North: I demand the Courſe Correſted, and the Diſtance Correſted?

Latitude North departed	46 Deg.	30 Min.
Latitude North arrived, Subtracted .	45	10
Difference South	1	20

Count upon the Arches along the *W b S*, or 7 Rumb, the 96 Leagues ſailed by Dead Reckoning, and from that Point you will find upon the Meridians 19 Leagues of Latitude, and upon the Parallels you will find 94 ¼ Leagues of Longitude, which by the Table (being firſt reduced into Miles) anſwers to 6 Degrees 50 Minutes, which is the difference of Longitude Weſtward, and is to be Subtracted from the 7 Degrees 40 Minutes of Longitude departed, (ſince ſhe did ſail Weſtward) and there will remain 0 Degrees 50 Minutes, which is the Longitude ſhe is in.

Longitude departed	7 Deg.	40 Min.
Difference Weſtward, Subtracted . .	6	50
Longitude arrived	0	50

This done, Subtract the two Latitudes one from another, (ſince they are both of the ſame ſide of the Equinoctial) and there will remain 1 Degree 20 Minutes for their difference, which being reduced into Leagues, there will come 26 ⅔ Leagues, to be counted upon the ſide of North and South; count alſo the 94 ¼ Leagues of Longitude or Departure (by Dead Reckoning) upon the ſide of Eaſt and Weſt, then lay the Thread on the Point of their interſection, and it will ſhew you the Courſe Corrected, which is *W b S* 4 Degrees 40 Minutes Southerly; and upon the Arches you will find 97 ½ Leagues, which is the Diſtance Corrected.

Example 3.

Admit a Ship departs from 40 Degrees 45 Minutes of Latitude North, and 347 Degrees 30 Minutes of Longitude, and ſail (by Dead Reckoning)

Eaſt

Eaſt North Eaſt . . . . 32 Leagues.
North Eaſt 20
North Eaſt by Eaſt 16
Eaſt by North 38
Eaſt . . . . 42 Leagues.

and then find by Obſervation to be in Latitude 43 Degrees 15 Minutes North.
I demand the Courſe Corrected, and the Diſtance Corrected ?

Courſe.	Points.	Diſtance	North.	South.	Eaſt.	Weſt.
		Leagues.	Leagues	Leagues.	Leagues.	Leagues.
E N E	6	32	12 $\frac{1}{4}$		29 $\frac{1}{2}$	
N E	4	20	14 $\frac{1}{8}$		14 $\frac{1}{8}$	
N E b E	5	16	8 $\frac{1}{3}$		13 $\frac{1}{2}$	
E b N	7	38	7 $\frac{1}{2}$		37 $\frac{1}{4}$	
E	8	42			42	
			42 $\frac{1}{4}$		136 $\frac{1}{4}$	

Latitude North arrived . . . . . 43 Deg. 15 Min.
Latitude North departed 40   45
Difference North 02   30

Having reduced the ſeveral Courſes as before directed, (or taught)
there comes 42 $\frac{1}{4}$ Leagues Northing, and 136 $\frac{1}{4}$ Leagues Eaſting, which
by the Table (or middle Parallel) anſwers to 9 Degrees 10 Minutes,
which is the difference of Longitude Eaſtward, to be added to 347 Degrees 30 Minutes of Longitude departed, (ſince ſhe did ſail Eaſtward)
and there will come 356 Degrees 40 Minutes, which is the Latitude ſhe
is in.

Longitude departed . . . . . . 347 Deg. 30 Min.
Difference Eaſt, add 9   10
Longitude arrived 356   40

Now Subtract the two Latitudes one from another (ſince they are both
of the ſame ſide) and there will remain 2 Degrees 30 Minutes for their
difference Northward, which are worth or anſwers 50 Leagues, and,
muſt be counted upon the ſide of North and South ; count alſo upon the
ſide of Eaſt and Weſt, the 136 $\frac{1}{4}$ Leagues of Longitude or Departure by
Dead

Dead Reckoning, and lay the Thread on the Point of their interſection, and it will ſhew you the Courſe Corrected, which is *E N E* 2 Degrees 30 Minutes Eaſterly ; and upon the Arches you will find 145 Leagues, which is the Diſtance Corrected.

Example 4.

Admit a Ship departs from 1 *Degree* 10 *Minutes of Latitude North, and* 3 *Degrees* 30 *Minutes of Longitude, and ſail by Dead Reckoning,*

Weſt by South	45 Leagues.
South Weſt by Weſt	28
South Weſt	19
Weſt	41
Weſt South Weſt	50 Leagues.

and then finds by Obſervation to be in Latitude 2 *Degrees* 8 *Minutes South : I demand the Courſe Corrected, and the Diſtance Corrected ?*

Courſe.	Points.	Diſtance.	North.	South.	Eaſt.	Weſt.
		Leagues.	Leagues.	Leagues.	Leagues.	Leagues.
W. b S	7	45		8¼		44⅞
S W b W	5	28		15½		23¼
S W	4	19		13⅜		13⅜
W	8	41				41
W S W	6	50		19¼		46¾
				56¼		167⅞

Latitude North departed	1 Degr.	10 Min.
Latitude South arrived, add	2	8
Difference South	3	18

Having reduced all the Courſes, there comes 56¼ Leagues of Latitude Southing, and 167⅞ Leagues of Longitude Weſting, which by the Table (being reduced into Miles) anſwers or gives 8 Degrees 24 Minutes, which is the difference of Longitude Weſtward, and is to be Subtracted from 3 Degrees 30 Minutes of Longitude, (ſince ſhe did ſail Weſtward) but becauſe it cannot be done, add 360 Degrees to the 3 Degrees 30 Minutes of Longitude departed, and from the ſum, *viz.* 363 Degrees 30 Minutes, Subtract the 8 Degrees 24 Minutes of difference, and there will remain 355 Degrees 6 Minutes, which is the Longitude ſhe is in.

Lon-

Longitude departed . . . .   3 Deg. 30 Min.
              Add   360
                                   _____
                                   363   30
Difference Weſt Subtracted . .   8   24
                                   _____
Longitude arrived   355   6

Now to find the Courſe and Diſtance Corrected, add the two Latitouds together, (ſince they are of different ſides) and there will come 3 Degrees 18 Minutes, which is the difference of Latitude Southward, to be reduced into Leagues, and ſo there comes 66 Leagues, which muſt be counted upon the ſide of North and South; count alſo upon the ſide of Eaſt and Weſt the 167¾, or rather 168 Leagues Weſt, and the Thread being laid on the Point of their interſection, will ſhew you the Corrected Courſe, which is *W S W* 1 Degree Weſterly, and upon the Arches you will find 180 Leagues, which is the Diſtance Corrected.

As to the Third Correction,

Serving for the Third, Fourth, and Fifth Rumb; if the Ship hath ſailed but upon one Point of the Compaſs, count only upon the Arches along the Rumb or Courſe, the Leagues ſailed by Dead Reckoning, then ſee or look upon the Parallels, how many Leagues it gives in Longitude; that done, count upon the ſide of North and South the Leagues proceeding from the difference between the Latitude departed, and the Latitude obſerved, and by its interſection with the Rumb, (by Dead Reckoning) you will find upon the Parallels, how many Leagues of Longitude it giveth, which Leagues being added to the Leagues of Longitude or Departure by Dead Reckoning, take half of the ſum for the Leagues of Longitude Corrected; then count again upon the ſide of North and South, the Leagues proceeding from the difference of Latitude; (found by obſervation) count alſo upon the ſide of Eaſt and Weſt the Leagues of Longitude Corrected, and obſerve the Point where they interſect one another, for the Thread being laid on it, will ſhew you the Courſe Corrected, and the Diſtance Corrected.

But if ſhe had ſailed upon the ſeveral Points of the Compaſs, (before your Obſervation) reduce the ſeveral Courſes by the Traverſe Table, as in the precedent Examples; then count upon the ſide of North and South the Leagues of Latitude by Dead Reckoning; count alſo upon the ſide of Eaſt and Weſt, the Leagues of Longitude or Departure, and obſerve the Point where they interſect one another; for the Thread being laid on it, will ſhew you the Courſe and Diſtance, (by Dead Reckoning) that done, count upon the ſide of North and South, the Leagues proceeding from the difference of Latitude by Obſervation, and obſerve where it interſects the Courſe; then count upon the Parallels, how many

I i   Leagues

Leagues of Longitude there is to that Point, and add them to the Leagues of Longitude by Dead Reckoning, and the half of the sum shall be the Leagues of Longitude, or Departure Corrected; then count again upon the side of North and South, the Leagues of difference of Latitude by Observation; count also upon the side of East and West, the Leagues of Longitude Corrected, and the Thread being laid on the Point of their intersection, will shew you the Course Corrected, and the Distance Corrected.

Example 1.

Admit a Ship sail (by Dead Reckoning) S E b S, 56 Leagues, and by Observation find to have altered her Latitude 2 Degrees 15 Minutes : I demand the Course Corrected, the Departure or Longitude Corrected, and the Distance Corrected ?

First, count upon the Arches along the *S E b S*, or third Rumb, the 56 Leagues sailed by Dead Reckoning, and you will find upon the Meridians 46½ Leagues of Latitude Southing, and upon the Parallels 31 Leagues of Longitude Easting; then reduce (or turn into Leagues the 2 Degrees 15 Minutes difference of Latitude proceeding from your Observation, and there will come 45 Leagues, which being counted upon the side of North and South, observe the Point where it intersects the *S.E b S*, (or third Rumb, which is the Course by Dead Reckoning) and you will find upon the Parallels 30 Leagues of Longitude, which being added to the 31 Leagues of Longitude found by Dead Reckoning, the sum will be 61 Leagues, whose half 30½ Leagues is the Departure or Longitude Corrected, which must be counted upon the side of East and West; count also once more the 45 Leagues of Latitude (proceeding from your Observation) upon the side of North and South, and observe the Point where it intersects the Line of 30½ Leagues of Longitude, for the Thread being laid on it, will shew you the Corrected Course, which is *S E b S*, 15 Minutes Easterly, and upon the Arches along the Thread, you will find 54½ Leagues, which is the Distance Corrected.

Example 2.

Admit a Ship departs from 44 Degrees 30 Minutes of Latitude North, and 9 Degrees 20 Minutes of Longitude, and sail by Dead Reckoning S W 96 Leagues, and then by Observation find to be in Latitude 41 Degrees 42 Minutes North : I demand the Course Corrected, the Distance Corrected, and the Departure or Longitude Corrected ?

First, count upon the Arches, along the *S W*, or fourth Rumb, the 96 Leagues sailed by Dead Reckoning, and you will find upon the Meridians 68 Leagues of Latitude Southward, and upon the Parallels you will find also 68 Leagues of Longitude Westward, (by Dead Reckoning ;) this

done,

done, Subtract the two Latitudes one from another, (since they are both on the same side of the Equinoctial) and there will remain 2 Degrees 48 Minutes, which is the difference of Latitude found by Observation, to be reduced into Leagues, and there will come 56 Leagues, which must be counted upon the side of North and South, and from the Point where it intersects the Course; to wit, the *S W*, (or fourth Rumb) you will find upon the Parallels 56 Leagues of Longitude, which being added to the 68 Leagues of Longitude by Dead Reckoning, there will come 124 Leagues, whose half 62 Leagues, is the Departure or Longitude Corrected, which (being reduced into Miles) by the precedent Table, answers to almost 4 Degrees 15 Minutes, which is the difference of Longitude Westward, and therefore must be Subtracted from 9 Degrees 20 Minutes of Longitude departed, (since she did sail Westward) and there will remain 5 Degrees 5 Minutes, which is the Longitude she is in.

Longitude departed	9 Deg.	20 Min.
Difference West Subtracted	4	15
Longitude arrived	5	5

For to find the Course Corrected, and Distance Corrected, count upon the side of North and South, the 56 Leagues proceeding from the difference of Latitude by Observation; count also upon the side of East and West the 62 Leagues of Longitude or Departure Corrected, and observe the Point where they intersect (or cuts) one another, for the Thread being laid on it, will shew you the Corrected Course, which is *S W*, 3 Degrees Westerly, and upon the Arches you will find 83½ Leagues, which is the Distance Corrected.

Example 3.

Admit a Ship departs from 39 Degrees 50 Minutes of Latitude North, and 334 Degrees 30 Minutes of Longitude, and sails by Dead Reckoning,

North East	33 Leagues.
North East by East	44
North East by North	50
East North East	28
North by East	39
East	22 Leagues.

and then (by Observation) find to be in Latitude 46 Degrees 20 Minutes North: I demand the Course Corrected, the Departure or Longitude Corrected, and the Distance Corrected?

Course.

Course.	Points.	Distance Miles.	North. Leagues	South. Leagues	East. Leagues	West. Leagues
N E	4	33	23 $\frac{1}{3}$		23 $\frac{1}{3}$	
N E b E	5	44	24 $\frac{1}{2}$		36 $\frac{1}{2}$	
N E b N	3	50	41 $\frac{1}{2}$		27 $\frac{1}{2}$	
E N E	6	28	10 $\frac{1}{4}$		25 $\frac{3}{8}$	
N b E	1	39	38 $\frac{1}{4}$		7 $\frac{3}{4}$	
E	8	22			22	
			138 $\frac{4}{3}$		143 $\frac{1}{2}$	

Having reduced the several Courses, as you were taught, there will come 138 $\frac{1}{3}$ Leagues Northing, which is the difference of Latitude by Dead Reckoning; and 143 $\frac{1}{2}$ Leagues Easting, which is the Departure or difference of Longitude; (also by Dead Reckoning) then count upon the side of North and South, the 138 $\frac{1}{3}$ Leagues Northing, and upon the side of East and West, the 143 $\frac{1}{2}$ Leagues Easting, and lay the Thread on the Point of their interſection, and it will ſhew you the Course by Dead Reckoning, which is N E, 1 Degree Easterly; that done, Subtract the 2 Latitudes one from another, (since they are of the same side) and there will remain 6 Degrees 30 Minutes, which is the true difference of Latitude Northward, to be reduced into Leagues, and there will come 130 Leagues, which being counted upon the side of North and South, obſerve the Point where it interſects the Course; (by Dead Reckoning which the Thread maketh) to wit, the North-Eaſt 1 Degree Easterly, and you will find upon the Parallels 135 Leagues of Longitude, which muſt be added to the 143 Leagues of Longitude by Dead Reckoning, and there will come 278 Leagues, whose half 139 Leagues is the Departure, or Longitude Corrected, which (being reduced into Miles) by the precedent Table, Anſwers to 9 Degrees 32 Minutes, which is the difference of Longitude Eaſtward, and therefore muſt be added to 334 Degrees 30 Minutes of Longitude Departed, (since ſhe did ſail Eaſtward) and there will come 344 Degrees 2 Minutes, which is the Longitude ſhe is in.

Longitude departed . . . . . 334 Deg. 30 Min.
Difference Eaſt, add   9   32
Longitude arrived 344   02

To

To find the Courſe Corrected, and the Diſtance Corrected, you muſt count upon the ſide of North and South, the 130 Leagues proceding from the difference of Latitude by Obſervation, and the 139 Leagues of Longitude Corrected upon the ſide of Eaſt and Weſt, then obſerve where they interſect one another, and lay the Thread on that Point, and it will ſhew you the Courſe Corrected, which is *N E* 2 Degrees Eaſterly, and upon the Arches you will find 190 Leagues, which is the Diſtance Corrected.

Example 4.

Admit a Ship departs from 2 Degrees 10 Minutes of Latitude South, and 3 Degrees of Longitude, and by Dead Reckoning ſail;

North . . . . 118 Leagues.
South 36
Eaſt 27
Weſt . . . . 94 Leagues.

and then finds to be in Latitude 5 Degrees 46 Minutes North: I demand the Diſtance Corrected, the Courſe Corrected, and the Longitude Corrected?

North.	South.	Eaſt.	Weſt.
118	36	27	94
36			27
82			67

Latitude North arrived . . . . 5 Deg. 46 Min.
Latitude South departed 2 10
Difference North 7 56

Firſt, Subtract the Leagues Northing and Southing one from another, the leſſer from the greater, and there will remain 82 Leagues Northing, which muſt be counted upon the ſide of North and South; Subtract likewiſe the Leagues Eaſting and Weſting one from another, and there will remain 67 Leagues Weſting, which muſt be counted upon the ſide of Eaſt and Weſt, and then laying the Thread on the Point of their interſection, it will ſhew you the Courſe by Dead Reckoning, which is *N W* 5 Degrees 45 Minutes Northerly; that done, add the two Latitudes together, (ſince they are of different ſides) and the ſum 7 Degrees 56 Minutes is the true difference of Latitude Northward, which being reduced into Leagues, there will come 158¾ Leagues, and are to be counted upon the ſide of North and South, then obſerving where the Line of 158¾ Leagues interſect the Courſe (by Dead Reckoning) which the Thread ſheweth, you will find to that Point upon the Parallels 130
Leagues

Leagues, which muſt be added to the 67 Leagues of Longitude by Dead
Reckoning, and there will come 197 Leagues, whoſe half 98½ Leagues,
is the Departure or Longitude Corrected, to be reduced into Miles, and
then by the precedent Table you will find that it anſwers to 4 Degrees
55¼ Minutes, which is the difference of Longitude Weſtward, which
muſt be Subtracted from 3 Degrees of Longitude Departed, (ſince ſhe
did ſail Weſtward) but becauſe it cannot be, add 360 Degrees to the
3 Degrees of Longitude, and from the ſum, *viz.* 363 Degrees, Subtract the 4 Degrees 55¼ Minutes difference of Longitude, and there
will remain 358 Degrees 4¾ Minutes, which is the Longitude ſhe is in.

Longitude departed	3 Deg.	00 Min.
Add	360	00
	363	00
Difference Weſt Subtracted . .	4	55½
Longitude arrived	358	04½

Firſt, to find the Diſtance Corrected, and the Courſe Corrected,
count upon the ſide of North and South the 158¾ Leagues, proceeding
from the difference of Latitude, found by Obſervation; count alſo upon
the ſide of Eaſt and Weſt, the 98½ Leagues of Longitude Corrected,
then lay the Thread on the Point of their interſection, and it will ſhew
you the Courſe Corrected, which is *NW b N*, 1 Degree 45 Minutes
Northerly, and upon the Arches you will find 187 Leagues, which is
the Diſtance Corrected.

PROP. XI.

*How to Correct the Courſe when there is Variation, or
when your Compaſſes varies.*

TO underſtand well how to Correct the Variation of your Compaſſes by the Sinical Quadrant, obſerve well the following
Rules, ſince they are General, and will render it as eaſie as
you can deſire.

Firſt Maxim.

If the Variation is Eaſterly when your Ship ſail Eaſtward, retire
from the North, and draw nearer the South ſo many Degrees and
Minutes as there is Variation; but if your Ship ſail Weſtward when
the

the Variation is Eafterly, you muft draw nearer the North, and retire from the South fo many Degrees and Minutes as there is Variation.

Example 1.

Admit a Ship fail E N E 57 Leagues, by a Compafs that vary of a Point, or 11 Degrees 15 Minutes Eafterly: I demand the difference of Latitude, and Departure from the Meridian?

Firft, count upon the Arches along the E b N, or 7 Rumb, the 57 Leagues, (fince the Variation is Eafterly, and fhe hath failed Eaftward, I muft retire from the North, and draw near the South of 11 Degrees 15 Minutes, which is a Point of the Compafs) and you will find upon the Meridians 11 Leagues Northing, which is the difference of Latitude, and upon the Parallels you will find 56 Leagues Eafting, which is the difference of Longitude, or Departure from the Meridian required.

Example 2.

Admit a Ship fail S W b S 106 Leagues, by a Compafs that varies 9 Degrees Eafterly: I demand the difference of Latitude, and Departure from the Meridian?

Firft, lay the Thread upon the S W b S 9 Degrees Wefterly, (fince the Variation is Eafterly, and fhe hath failed Weftward, you muft draw near the North, or retire from the South of 9 Degrees, that the Compafs varies,) and count upon the Arches along the Thread 106 Leagues, and you will find upon the Meridians 78 Leagues Southing, which is the difference of Latitude, and upon the Parallels you will find 72 Leagues Wefting, which is the difference of Longitude, or Departure from the Meridian Weftward.

Example 3.

Admit a Ship departs from 49 Degrees 45 Minutes of Latitude North, and 7 Degrees 30 Minutes of Longitude, and fail,

South by Weft	51 Leagues.
South South Eaft	29
South Eaft by Eaft	34
Eaft South Eaft	47
Eaft by North	16
South Eaft	53 Leagues.

by a Compafs that varies 14 Degrees 30 Minutes Eafterly: I demand the Latitude and Longitude fhe is in?

Courfe.

Courſe.	D. M.	Diſtance. Leagues.	North. Leagues.	South. Leagues.	Eaſt. Leagues.	Weſt. Leagues.
S S W	3 15 *W*	51		45 1/4		21 1/3
S b E	3 15 *S*	29		28 2/3	4	
S E	3 15 *S*	34		25 1/4	22 2/3	
S E b E	3 15 *S*	47		28 1/4	37 2/3	
E	3 15 *S*	16		00 1/4	16	
S E b S	3 15 *S*	53		45 2/3	27	
				174 1/3	107 1/3 21 1/3	21 1/3
					86	

Having obſerved the 14 Degrees 30 Minutes of Variation upon each Courſe Weſting, there will remain 174 1/3 Leagues Southing, which being reduced into Degrees (by dividing it by 20) there will come 8 Degrees 43 Minutes, which is the difference of Latitude Southward, and muſt be Subtracted from 49 Degrees 45 Minutes of Latitude departed, and there will remain 41 Degrees 2 Minutes, which is the Latitude ſhe is in.

Latitude North departed	49 Deg.	45 Min.
Difference South, Subtracted . . .	8	43
Latitude North arrived	41	02

Subtract alſo the Leagues Eaſting and Weſting one from another, and there will remain 86 Leagues Eaſting, which by the precedent Table (being reduced into Miles) anſwers to 6 Degrees 8 Minutes, which is the difference of Longitude Eaſtward, and therefore muſt be added to the 7 Degrees 30 Minutes of Longitude departed, and there will come 13 Degrees 38 Minutes, which is the Longitude ſhe is in.

Longitude departed	7 Deg.	30 Min.
Difference Eaſtward	6	08
Longitude arrived	13	38

For to find the Courſe and direct Diſtance, count upon the ſide of North and South 174 1/3 Leagues Southing, and upon the ſide of Eaſt and Weſt the 86 Leagues Eaſting, lay the Thread on the Point of their interſection, and it will ſhew you the Courſe, which is *S S E* 3 Degrees 30 Minutes Eaſterly, and you will find upon the Arches 195 Leagues, which is the Diſtance.

Ex-

Example 4.

Admit a Ship departs from 37 *Degrees* 40 *Minutes of Latitude North*, *and* 351 *Degrees of Longitude*, *and sail*,

North North East . . . . 21 Leagues.
North West 47
North North West 33
West by South 24
North West by West 19
West North West . . . . 54 Leagues.

by a Compass that varies 8 *Degrees* 30 *Minutes Easterly: I demand the Latitude, and Longitude she is in?*

Course.	D. M.	Distance. Leagues.	North. Leagues.	South. Leagues.	East. Leagues.	West. Leagues.
NNE	8 30 *S*	21	20 $\frac{1}{3}$		5 $\frac{1}{3}$	
NW	8 30 *N*	47	37 $\frac{1}{4}$			28
NNW	8 30 *N*	33	32			8 $\frac{1}{4}$
WbS	8 30 *W*	24		1 $\frac{1}{4}$		24
NWbW	8 30 *N*	19	12 $\frac{2}{3}$			14 $\frac{1}{4}$
WNW	8 30 *N*	54	27 $\frac{1}{2}$			46 $\frac{1}{2}$
			130 $\frac{1}{4}$ 1 $\frac{1}{4}$	1 4	5 $\frac{1}{3}$	121 5 $\frac{1}{3}$
			129			115 $\frac{2}{3}$

After having observed the 8 Degrees 30 Minutes upon every Course proposed, there will come 130 $\frac{1}{4}$ Leagues Northing, 1 $\frac{1}{4}$ Leagues Southing, 5 $\frac{1}{3}$ Leagues Easting, 121 Leagues Westing; that done, Subtract the 1 $\frac{1}{4}$ Leagues Southing, from the 130 $\frac{1}{4}$ Leagues Northing (since it is the least) and there will remain 129 Leagues Northing, which being reduced into Degrees, there will come 6 Degrees 27 Minutes, which is the difference of Latitude Northward, to be added to the 37 Degrees 40 Minutes of Latitude departed, and the Sum, *viz.* 44 Degrees 7 Minutes, is the Latitude she is in.

Latitude North departed . . . . . 37 Degr. 40 Min.
Difference North, add 6   27
Latitude North arrived . . . . . 44 .   07

Subtract alfo the Leagues Eafting and Wefting one from another, the leffer out of the greater, and there will remain 115¾ Leagues Wefting, which (being reduced into Miles) by the precedent Table, anfwers to 7 Degrees 52 Minutes, which is the Difference of Longitude Weftward, to be Subtracted from 351 Degrees of Longitude departed, (fince fhe did fail Weftward) and there will remain 343 Degrees 8 Minutes, which is the Longitude fhe is in.

Longitude departed . . . . . 351 Deg. 00 Min.
Difference Weft Subtract . . .   7   52
Longitude arrived. . . . . . : 343   08

For to find the Courfe and Diftance, count upon the fide of North and South the 129 Leagues Northing, and upon the fide of Eaft and Weft the 115¾ Leagues Wefting, lay the Thread on the Point of their interfection, and it will fhew you the Courfe, which is *N W* 3 Degrees North, and you will find upon the Arches 173 Leagues, which is the Diftance.

Second Maxim.

If the Variation is Wefterly when you fail Eaftward, draw nearer the North, or retire from the South, fo many Degrees and Minutes as you find there is Variation; but if you fail Weftward, you muft retire from the North, and draw nearer the South fo many Degrees as there is Variation.

Example 1.

Admit a Ship Sail, S E b E 53 Leagues, by a Compafs that varies 7 Degrees 30 Minutes Wefterly: I demand the difference of Latitude, and Departure from the Meridian?

Firft, count upon the Arches along the *S E b E*, (or 5 Rumb) 7 Degrees 30 Minutes North, (fince the Variation is Wefterly, and fhe hath failed Eaftward, I muft draw nearer the North of 7 Degrees 30 Minutes that the Compafs varies) and you will find upon the Meridians 35 Leagues (Southing,) which is the difference of Latitude, and upon the Parallels you will find 40 Leagues (Eafting,) which is the difference of Longitude, or Departure from the Meridian required.

Ex-

Example 2.

Admit a Ship fail W N W 82 Leagues, by a Compaſs that varies 11 Degrees 15 Minutes Weſterly: I demand the difference of Latitude, and Departure from the Meridian?

Firſt, lay the Thread upon the *W b N*, (ſince the Variation is Weſterly, and ſhe hath ſailed Weſtward, you muſt draw nearer the South, or retire from the North of 11 Degrees 15 Minutes, which is a Point that the Compaſs varies) and count upon the Arches along the Thread 82 Leagues, and you will find upon the Meridians 16 Leagues (Northing,) which is the difference of Latitude, and upon the Parallels you will find 80¼ Leagues (Weſting,) which is the difference of Longitude, or Departure from the Meridian Weſtward.

Example 3.

Admit a Ship departs from 36 Degrees 30 Minutes of Latitude North, and 335 Degrees 20 Minutes of Longitude, and fail,

North . . . . 17 Leagues.
North North Weſt 25
North Eaſt by North 33
Eaſt North Eaſt 44
Eaſt by North 39
North Eaſt by Eaſt . . . . . . 50 Leagues.

by a Compaſs that varies 5 Degrees 30 Minutes Weſterly: I demand the Latitude, and Longitude ſhe is in?

Courſe.	D. M.	Diſtance. Leagues.	North. Leagues.	South. Leagues.	Eaſt. Leagues.	Weſt. Leagues.
N	5 30 W	17	17			1 2/3
N N W	5 30 W	25	22¼			11 1/3
N E b N	5 30 N	33	29		15¼	
E N E	5 30 N	44	20¼		38¾	
E b N	5 30 N	39	11¼		37¾	
N E b E	5 30 N	50	31 1/3		38 1/3	
			132		130¼ 13¼	13 1/3
					117	

After, having obferved the 5 Degrees 30 Minutes upon every Courfe propofed, there will come 132 Leagues Northing, 130 ¼ Leagues Eafting, and 13 ¼ Leagues Wefting; that done, reduce the 132 Leagues Northing into Degrees, and there will come 6 Degrees 36 Minutes, which is the differenco of Latitude Northward, to be added to the 36 Degrees 30 Minutes of Latitude departed, and the fum, *viz.*

Eatitude North departed	36 Deg.	30 Min.
Difference North, add	06	36
Latitude. North arrived.	43.	06.

Subtract the Leagues Eafting and Wefting one from another, the leffer out of the greater, and there will remain 117 Leagues Eafting, which (being reduced into Miles) by the precedent Table anfwers to 7 Degrees 37 Minutes, which is the difference of Longitude Eaftward, to be added to the 335 Degrees 20 Minutes of Longitude, (fince fhe did fail Eaftward) and there will come 342 Degrees 57 Minutes, which is the Longitude fhe is in.

Longitude departed	335 Deg.	20 Min.
Difference Eaft, add.	7	37
Longitude arrived	342	57

For to find the Courfe and Diftance, count upon the fide of North and South the 132 Leagues Northing, and upon the fide of Eaft and Weft the 117 Leagues Eafting, lay the Thread on the Point of their interfection, and it will fhew you the Courfe, which is *N E* 3 Degrees 30 Minutes North, and you will find upon the Arches 176 Leagues, which is the Diftance.

Example 4.

Admit a Ship departs from 46 Degrees 35 Minutes of Latitude North, and 10 Degrees 30 Minutes of Longitude, and fail,

Weft by South	37 Leagues.
South Weft by Weft	49
Weft North Weft	58
South South Eaft	41
South Weft by South	27
South by Weft	19 Leagues.

by a Compafs that varies 13 Degrees 45 Minutes Wefterly: I demand the Latitude, and Longitude fhe is in?

Courfe.

Courſe.	D. M.	Diſtance.	North.	South.	Eaſt.	Weſt.
		Leagues.	Leagues.	Leagues.	Leagues.	Leagues.
W S W	2 30 S	37		15 $\frac{1}{3}$		33 $\frac{1}{3}$
S W	2 30 S	49		36 $\frac{1}{4}$		33 $\frac{1}{4}$
W b N	2 30 W	53	9			57 $\frac{1}{4}$
S E b S	2 30 E	41		33 $\frac{1}{4}$	24 $\frac{1}{4}$	
S S W	2 30 S	27		25 $\frac{1}{3}$		9 $\frac{1}{3}$
S	2 30 E	19		19	$\frac{1}{2}$	
			9	129 $\frac{1}{2}$	25	133 $\frac{1}{2}$
				9		25
				120 $\frac{1}{2}$		108 $\frac{1}{2}$

After, having obſerved the 13 Degrees 45 Minutes upon every Courſe propoſed, there will come 9 Leagues Northing, 129 $\frac{1}{2}$ Leagues Southing, 25 Leagues Eaſting, and 133 $\frac{1}{2}$ Leagues Weſting; that done, Subtract the 9 Leagues Northing from the 129 $\frac{1}{2}$ Leagues Southing, (ſince it is the leaſt) and there will remain 120 $\frac{1}{2}$ Leagues South, which being reduced into Degrees, there will come 6 Degrees 1 $\frac{1}{2}$ Minutes, which is the difference of Latitude Southward, to be Subtracted from 46 Degrees 35 Minutes of Latitude departed, and there will remain 40 Degrees 34 Minutes, which is the Latitude ſhe is in.

Latitude North departed . . . . 46 Deg. 35 Min.
Difference South, Subtract 6 01

Latitude North arrived 40 34

Subtract alſo the Leagues Eaſting and Weſting one from another, and there will remain 108 $\frac{1}{2}$ Leagues Weſting, which by the precedent Table (being reduced into Miles) anſwers to 7 Degrees 30 Minutes, which is the difference of Longitude Weſtward, and therefore muſt be Subtracted from 10 Degrees 30 Minutes of Longitude departed, and there will remain 3 Degrees, which is the Longitude ſhe is in.

Longitude departed . . . . . . 10 Deg. 30 Min.
Difference Weſtward, Subtract . . 7 30

Longitude arrived 3 00

Ferr

For to find the Courſe and direct Diſtance, count upon the ſide of North and South the 120¼ Leagues Southing, and upon the ſide of Eaſt and Weſt the 108¼ Leagues Weſting, lay the Thread on the Point of their interſection, and it will ſhew you the Courſe, which is *S W* 3 De-grees *S*, and you will find upon the Arches 162 Leagues, which is the Diſtance.

P R O P. XII.

The Deſcription and Uſe of ſuch Inſtruments as are proper for Navigation.

S. BEFORE you ſhew me the Uſe of theſe Inſtruments, pray in-form me upon what Principles Celeſtial Obſervations by Inſtru-ments are found?

T. Upon theſe two, 1. That when many Circles are drawn from one and the ſame Center, but of different Diameters, Lines drawn from their Center to their ſeveral Circumferences, divide each Circle in the ſame manner and proportion, as the Circles A B C D, E F G H, being deſcribed from the ſame Center I, and the Lines I E A, I K L drawn, the Arches E K or K H ſhall have the ſame proportion to the Circumference E F G H, as the Arches A L or L B have to the Circum-ference A B L D, and contain the like number of Degrees.

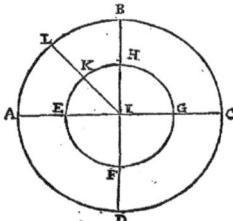

2. That the Globe whereon we live is of no ſenſſible quantity compared with the large Sphere of the Sun, and therefore any part of the Earth's Surface may be taken for the Center of the Sun or Heavens, and by con-ſequence the Center of every Inſtrument may be accounted alſo the Center of the Sun or Heavens.

S. This

S. This is very plain, and I apprehend very well, that this Globe of Earth and Water compared with the vaſt Orbs that ſurrounded it may be reckoned only as a Point or Center; therefore pray proceed to the Deſcription and Uſe of Inſtruments.

I. Of the *Aſtrolabe.*

S. WHAT do you ſay of the *Aſtrolabe?*
    T. I ſay, before you make uſe of it, you ought to examin if it be good, to prevent thoſe errors, which elſe might cauſe the loſs of your Ship, and it may be of your life.
    S. How ſhall I find if my *Aſtrolabe* be well made?
    T. Firſt, holding it as in Time of Obſervation, let fall a Thread with ſome Lead at the end, from the Point A, (which repreſents the Zenith) and if the ſaid Thread paſs over the Center of your *Aſtrolabe* and Point B, or cover the Line A B, it is a ſign that it is good; for it muſt hang ſo when ever you obſerve the Suns Altitude; therefore it would not be amiſs to faſten 2 or 3 Pound weights of Lead on the lower part of it, at B, (which repreſents the *Nadir*) that it may the better keep its Equilibrium at Sea.
    The next thing required for a good *Aſtrolabe*, is, that the Horizontal Line C D be Parallel to the Horizon, wherefore placing your Label exactly upon the ſaid Horizontal Line, look through the two Vanes (or ſights) on any ſenſible Point on the Horizon, then turning your Inſtrument ſo, that the ſight which was next to your Eye be now fartheſt from you (without altering, or moving in the leaſt, the Label from the Horizontal Line;) look again, and if you can ſee the ſame Point as before through the two ſights, you may conclude that the Horizontal Line of your *Aſtrolabe* C D is Parallel to the Horizon, which is a thing ſo neceſſary to a good *Aſtrolabe*, that without it one can truſt to his Obſervations, in which there muſt need be ſome errors when that faileth.
    S. If the Point A of the *Aſtrolabe* repreſents the *Zenith*, B the *Nadir*, and K L the *Horizon*, pray what doth the Center N of my *Aſtrolabe* repreſent?
    T. It repreſents the Center of the Earth, and of the Heavens too, for if the Earth is but a Point, being compared with the Firmament, and Orbe of the Sun, (as all Mathematicians agree) then whereſoever we be upon its Superficies may be taken for its Center, and alſo for that of the Heavens, ſince the Semidiameter of the Earth is ſo little and inconſiderable being compared to that of the Firmament, that the error it can cauſe is ſaid to be inſenſible.
    S. It is plain enough, that the Center of the *Aſtrolabe*, repreſents the Center of the Earth and Heavens too; but what conſequence do you draw from that to any advantage to me, or for my Inſtruction?

T. I

T. I draw this Confequence, that the Lines drawn from the Center of the *Aftrolabe* divide the Circumference of the Heavens, in the fame manner as it divides the Circumference of the *Aftrolabe*, for all Circles that have the fame Center, are divided in the fame manner, by the Lines drawn from the Common Center.

For Example.

If the Circle L M K of the *Aftrolabe*, C V D the *Sphere* the Sun is in, and F T O the Firmament, have the fame Center N; if you draw the Lines N L C F and N M E G, the Arches F G, C E and L M, will be the fame parts of each Circle, and will contain the fame number of Degrees.

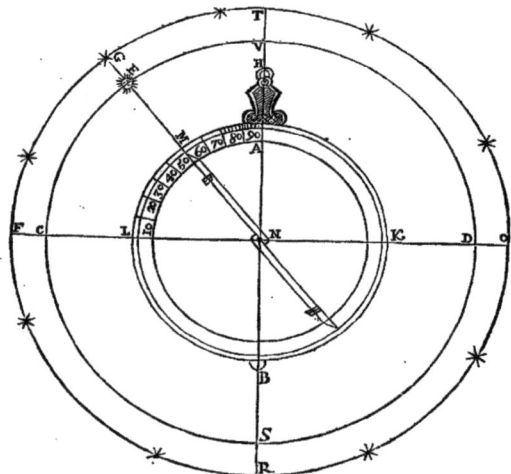

This being the Foundation of *Celeftial Obfervations*, you may conclude, that the *Aftrolabe* is the moft Natural of all Inftruments, fince it is divided into as many equal parts as the Heavens; to wit, 36 Degrees, (however, a quarter of it is fufficient, as the precedent Figu

S. Is the Ufe of the *Aftrolabe* very hard?.

T. No, it is rather eafie, there being no more in it than to hold it fo, (with your left Thumb thorow the Ring H) that it may have a free Motion, to the end, that the Horizontal Line L K may be Parallel to the Horizon; then turn your right fide, and the graduated part of your Inftrument fo to the Sun, that the Sun fhine neither on one fide, nor on the other; then lift up the Label with your Right-hand, 'till the Beams of the Sun entring through the hole of the uppermoft Vane or Sight, doth alfo pafs through the hole in the lowermoft Vane, and the Label will fhew you the Altitude or height of the Sun above the Horizon, (if you count from the Horizontal Line to the Label, from the Zenith to the Label, it will be the Diftance of the Sun from the Zenith.)

For Example.

Admit (in the precedent Figure) that the Sun is at the Point E, the Horizon Line is C D, and the Zenith of the obferver is V, if you hold your *Aftrolabe* as before directed, and raife the Label until the Sun's Beams pafs through the holes on the Vanes or Sights, the Degrees contain'd by the Arch L M, to wit, 50 Degrees, will be the height of the Sun (at F) above the Horizon, fince the Arches L M, C E and F G are alike, and contain the fame number of Degrees.

S. Doth not the agitation or motion of the Sea, render the ufe of this Inftrument difficult?

T. Yes, for when it bloweth hard, and the Sea is any thing ruff or agitated, it is not eafie to make the Sun's Beams pafs through the Sights or holes on both the Vanes, as before directed, becaufe of the fmallnefs thereof; and therefore I fhall fhew you here, what may be added to the faid *Aftrolabe* to render it more eafie and ferviceable at Sea: Make the Label fo long that it pafs the Limb of your *Aftrolabe* about Three Fingers breadth on both fides, then place the Vanes (or Sights) on the ends of it; but let the lowermoft reach or extend it felf 2 or 3 Degrees on each fide the hole, by defcribing the Arch of a Circle A B, from the foremoft Vane or Sight C, to the lowermoft D E, and having marked fome Degrees draw *fecret* Lines from C, and divide the Vane E D in Degrees, and becaufe they will be very great, you may fubdivide them by Tranfverfal (or Crofs) Lines, and then it is ready for ufe, and this is the way to obferve with it. Before you begin to obferve, place the Label exactly upon the Degree, which near hand (you judge) will be the height of the Sun, then obferve upon what part of the lowermoft Vane the Beam falls, for if it falls lower than the mark O, you may conclude that the Sun is higher than you thought, therefore you muft add the Degrees and Minutes, which the Beam marks upon the lowermoft Vane, to the Degrees upon which the Label was placed, and the fum will be the true height of the Sun; but if the Beam is higher than the mark O, you muft Subtract what is upon the lowermoft Vane from the Degrees upon

L l   which

which the Label was placed, and the remainer will be the Sun's height above the Horizon, which being Subtracted from 90 Degrees, the remainer will be the Distance of the Sun from your Zenith.

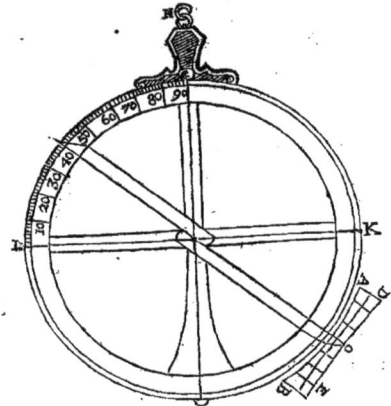

II. *Of the Cross-Staff.*

T. The Cross-staff consists of a strait square Graduated Staff, and four Crosses or Vanes of different Lengths, *viz.* 1. The *Ten-Cross*, which belongs to that side of the Staff called the *Ten-side*, where the Graduations begin at about 3 Degrees, and proceeding towards the Center or Eye-end encreafe (by 10 Minutes) to 10 Degrees. 2. The *Thirty-Cross*, which belongs to the *Thirty-side* of the Staff, where the Divisions or Graduations begin at 10 Degrees, and end at 30 Degrees. 3. The *Sixty-Cross*, which belongs to the *Sixty-side*, where the Divisions begin at 20 Degrees, and end at 60 Degrees. And 4. The *Ninety-Cross*, which belongs to the *Ninety-side*, where the Divisions begin at 30 Degrees, and end at 90 Degrees. Sometimes the several sides of the Staff are numbred likewife with their Complements to 90, as against 10 stands 80, &c..

S. How

S. How fhall I know, if the Staff and Croffes are well made?

T. If your Staff and Croffes are exact, they will have this proportion, *viz.* 1. Half the *Ten-Crofs* being laid on the *Ten-fide* fhall reach from 10 Degrees to 9 Degrees 12 Minutes, and the whole *Ten-Crofs* from 10 Degrees to 8 Degrees 31 Minutes. 2. Half the *Thirty-Crofs* meafured on the *Thirty-fide*, fhall reach from 30 Degrees to 23 Degrees 52 Minutes, and the whole *Thirty-Crofs* from 30 Degrees to 19 Degrees 47 Minutes. 3. Half the *Sixty-Crofs* meafured on the *Sixty-fide*, will reach from 60 Deg. to 40 Deg. 13 Minutes, and the whole Sixty-Crofs from 60 Degrees to 30 Degrees. And 4. Half the *Ninety-Crofs* meafured on the *Ninety-fide*, will reach from 90 Degrees to 53 Degrees 7 Minutes, and the whole Ninety-Crofs from 90 Degrees to 36 Degrees 52 Minutes. You may alfo examin the Truth of your *Staff* by obferving the Latitude on fhore with your *Aftronomical Ring*, and (if the place permits) fee if you can find the fame with your Crofs-ftaff, whofe Crofs is to make right Angles with it.

S. How muft I hold my Staff in time of Obfervation?

T. To know this, you muft find the Center of your Eye, thus: Put on the Sixty and Ninety Croffes on their proper fides, and place them exactly upon 45 Degrees of each proper graduation, then bring the End of your Staff to reft upon the out-corner of your right Eye, and if you can fee each End of the two Croffes in a right Line one with another (as in this Figure) your Staff ftands parallel with the Center of your Eye, and fo you muft hold it every time you obferve the Sun or Stars Height.

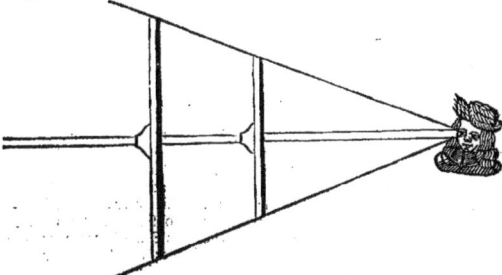

S. What is the Ufe of this Inftrument?

T. Its chief Ufe is to take the *Altitude* of the Sun or Stars, and this may be done either by a forward or backward Obfervation. It likewife fhews the *Diftance* between two Stars.

S. Since

S. Since the ſtaff has four Croſſes, pray tell me, whether they are to be uſed altogether or ſeverally?

T. The Croſſes are to be uſed ſeverally, according as the Altitude or Diſtance you are to obſerve is greater or leſſer; for if it be leſs than 10 Degrees, the *Ten-Croſs* muſt be uſed; if more than 10 Degrees, but leſs than 30 Degrees, the *Thirty-Croſs*; if above 30 Degrees, but leſs than 60 Degrees, the *Sixty-Croſs*; and if above 60 Degrees, the *Ninety-Croſs*; but this laſt is not ſo exact as other Inſtruments.

S. How muſt I take a forward Obſervation of the Sun or Stars Meridian Altitude at Sea, thereby to find what Latitude I am in?

T. Conſider firſt, whereabouts the greateſt Meridian Altitude of the Sun or Star may be at that time, thereby to know which Croſs is moſt proper for your uſe; then put that Croſs on with the flat ſide towards you, remembring to place it on the ſide of the Staff to which it belongs. This being done, Place the Eye-end of your Staff parallel to the Center of your Eye as before taught, and holding it there ſteddy, move the Croſs 'till you ſee the Center of the Sun or Sar by the upper End of the Croſs, and the Horizon by the lower End; and the Degrees and Minutes cut by the inner Edge of the Croſs, upon the ſide of the ſtaff peculiar to the Croſs you uſe is the preſent Altitude of the Sun or Star; but becauſe it is the greateſt Altitude you deſire to know, you muſt continue your Obſervation as long as you find the Altitude to increaſe, ſtill drawing the Croſs nearer to your Eye, 'till you find the Altitude to decreaſe, (which you will ſoon perceive, for then the Sea will appear by the lower End of the Staff inſtead of the Horizon) and then forbear any farther Obſervation, counting the Degrees and Minutes cut by the inner Edge of the Croſs on the proper ſide, which gives the Meridian Altitude required, and that ſubtracted from 90 Degrees, gives the Zenith-diſtance, if it be not graduated on the ſtaff. Note, That your beſt time to begin your Obſervation, to find the Meridian Altitude, will be about half an hour before Noon.

S. Is it not difficult to cut exactly the Center of the Sun by the upper End of the Croſs?

T. Yes, for it is only to be done in hazy or thick weather, except you make uſe of a coloured, fixed on the top of the Croſs, to defend the Eye from the brightneſs of the Sun. But if the Sun ſhine out, you may obſerve either the upper Limb of the Sun, and afterwards ſubtract 15 Minutes from the Altitude for the Suns Semidiameter, and add 15 Minutes to the Zenith-diſtance; or you may obſerve the Suns lower Limb, and afterwards add 15 Minutes the Suns Semidiameter to the Altitude, and ſubtract it from the Zenith-diſtance; for by obſerving the upper Limb, your Croſs will be nearer your Eye by the Semidiameter of the Sun, and ſo make the Altitude appear more than really it is, and by obſerving the lower Limb, your Croſs will be ſo much too far from your Eye, and ſo make the Altitude appear leſs than it is.

S. Do

S. Do Aftronomers agree, that the apparent Semidiameter of the Sun is 15 Minutes?

T. Yes, When he is fartheft from the Earth, according to the Obfervations of *Ticho-Brahe*, but when he is neareft, they reckon 16 Minutes, which is fo inconfiderable a Difference in Navigation, that moft Pilots negleft it, generally reckoning 15 Minutes for the Suns Semidiameter all the Year.

S. How muft I obferve the Diftance between two Stars?

T. To obferve the Diftance between two Stars, or the Diftance of the Moon from a Star, is much after the fame manner as to find the Suns Altitude, for having placed the Staff to you Eye as before directed, look to both Ends of you Crofs, moving it farther or drawing it nearer, 'till you can fee the two Stars (or Moon and Star) one at one end, and the other at the other End of the Crofs, and the Degrees and Minutes cut by the Crofs (as aforefaid) is the true Diftance fought.

S. Is the backward Obfervation as fure and eafie as the forward?

T. Yes, and eafier, and therefore I recommend it to you as better than the firft; as to the ordinary way to make ufe of the Crofs-ftaff for back Obfervations, place that Crofs which you think beft for your ufe upon the Eye-end of your Staff, fo that the flat fide of it be exaftly even with the End of the Staff, then flide on the Horizon Vane, and looking at the lower End of the Crofs, remove the Horizon Vane 'till you can fee the Horizon, and in the fame time the fhadow made by the upper End of your Crofs, exaftly upon the Line drawn upon the Horizon Vane, and fo continue obferving 'till the Sun be on your Meridian, at which time the Horizon Vane will fhew you (on that fide proper to the Crofs you ufe) the Degrees and Minutes of the Suns Altitude, and its Complement; but this way of obferving without a Brafs fhooe fitted on the end of the Crofs you ufe, is very defeftive and fubjeft to errors, as many have found it by experience. There is befides fome defefts in this praftice, becaufe of the *Penumbra*, (that is to fay, almoft fhadow) and therefore take notice of it, if you will not omit an error of 15 Minutes, by taking the Altitude of the Suns Limb inftead of the Center thereof.

S. If this way of obferving is fo defeftive, pray tell me of a better that may Correft what is amifs in this?

T. The

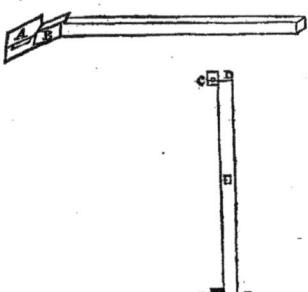

T. The beſt way for a back Obſervation with the Croſs-ſtaff, will be to have an Horizon Vane covered with White paper, and ſet on upon the Eye-end of your Staff, as A B, then to fix on the upper end of the Croſs a little Plate of Braſs, with a little hole in it, as C D, and a Vane at the lower End of it as F G; the Croſs being thus fitted, put it on the Staff, and turning your back to the Sun, look through the ſight F G, (at the lower end of the Croſs) and move your Croſs 'till the Beams of the Sun, which paſs through the Sun which paſs through the hole at C D, falls ſo upon the Line A, as to be equally divided by it, and you perceive the Horizon exactly at the ſame time, then have you the Suns preſent Altitude, and ſo you muſt continue to obſerve as often as your judgment ſhall direct you, untill you find that the Sun is exactly in your Meridian.

III. *The Deſcription of the Quadrant, commonly called* Davis's *Quadrant, but by the* French, *the* Engliſh *Quadrant, the Inventer being an* Engliſh-man.

S. WHAT are the chief parts of this Inſtrument?
 T. The chief parts of it are two Arches, (a leſſer and a greater) and three Vanes.
 S. How do you call the leſſer Arch H L?
 T. It is called the *Sixty-Arch*, becauſe it containeth 60 Degrees.
 S. Why is this Arch made leſſer than that of N O, which contains but half ſo many Degrees?
                 T. It

T. It is to the end that the shadow of the Vane R, which is placed on it, (or rather the Beams of the Sun which pass through the Hole and *Glass* of the said Vane at S) may appear the better upon the Vane T.

S. How do you call the greatest Arch N O?

T. It is called the *Thirty-Arch* because it containeth 30 Degrees.

S. Why is this Arch made upon a large Radius than H L?

T. It is to the end that it may be the better divided into Degrees and Minutes, and that those Degrees being bigger, the Observation may be the more exact.

S. How do you call the Vane T?

T. The *Horizon Vane*, and that next to it at R is called the *Shadow Vane*; and the third Vane V, is named the *Sight Vane*, because it is to be placed to your Eye in time of Observation.

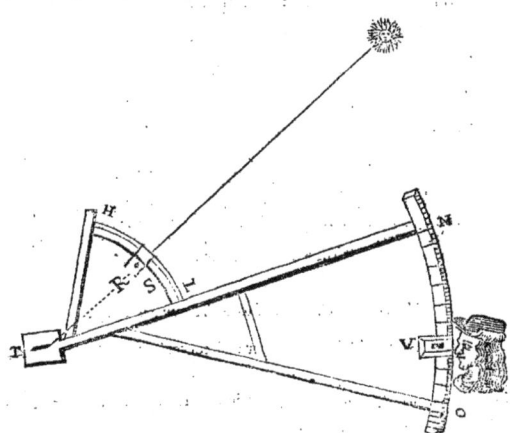

S. What is the use of this Quadrant?

T. To observe the Suns Meridian Altitude by his shadow.

S. How is it to be used?

T. First, put on the Horizon Vane on the Center or End of the Quadrant as T; then consider what will be near hand the Complement of the Suns Meridian Altitude that day, and set the Vane R 10, 15 or 20

Degrees.

Degrees lefs than the Complement of the Suns Altitude, making it Parallel to the Horizon Vane ; (if you can) then turn your back to the Sun, and looking through the fight at V, bring the upper Edge of the fhade Vane, to fall upon the upper Edge of the flit in the Horizon Vane, (but if there is a little hole with a Glafs in the middle of the fhade Vane of your Quadrant , bring the Beams of the Sun which pafs through the faid hole , to fall fo upon the upper Edge of the flit in the Horizon Vane, as to be equally divided by it.) and at the fame time look through the faid flit for the Horizon, and if you fee only Sea, then flide your Eye-vane a little lower towards B , but if on the contrary you fee all Sky , then remove your fight-vane a little higher towards C , then obferve again as before, continuing to move your fight-vane higher or lower until you fee the fhadow upon the upper Edge of the flit , and at the fame time the Horizon, through the fame flit on the Horizon-vane , then have you the Suns prefent Altitude ; and when he is rifen a little higher, obferve again, and fo continue to do from time to time , and very often when you perceive he is almoft upon your Meridian; and when you find that the Sun is at his higheft, do not alter your Vanes, but continue obferving as they ftand , until you perceive the Sun defcends.

S. How fhall I know when the Sun is defcending or leffening his Altitude ?

T. As the Sun defcends from the Meridian , you will lofe fight of the Horizon , and when you fee nothing but Sky , you have done obferving for that day, for your Vanes muft not be removed any more , until you have worked your Obfervation , Thus :

Look how many Degrees and Minutes are cut by the infide of the fight-vane , and by the upper Edge of the fhadow-vane , for thofe two fums being added together, will be the Suns Diftance from your Zenith , to which , add or fubtract the Suns Declination as you fhall be directed in the following difcourfe.

IV. *Of the Graduated or Aftronomical Ring.*

S. IS this Graduated Ring as proper and convenient for the Sea as the Aftrolabe ?

T. Yes, and more eafie, and it muft needs be more exact , becaufe the Degrees are as big again as they are in an Aftrolabe of the fame Diameter.

S. How is it to be made ?

T. The

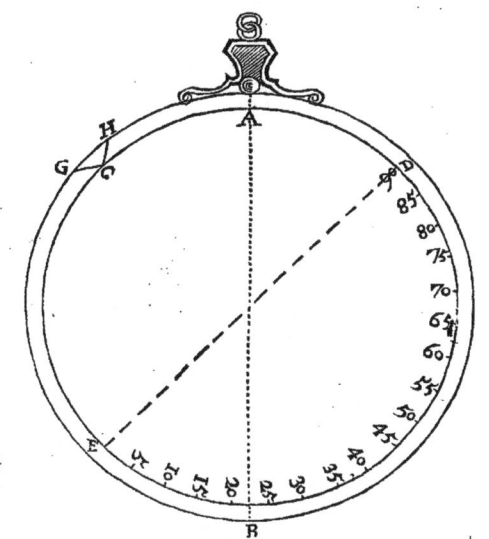

T. The way to make it, is to get a Brafs Circle as big as a common *Aftrolabe*, equally heavy every where, or in all its parts, and of a full Inch in Breadth and Thicknefs; draw through the Center the Diameter A B, and fet on the Point A a Ring to hold it by, as in the *Aftrolabe*; from the Point A, take 45 Degrees on each fide of it, which end in C and D; from the Point D, draw through the Center the Diameter D E; divide the infide of the half Circle D, B, E, into 90 equal parts, beginning at E; and at the Point C, (within-fide) make a hole as little as poffible you can, but let it be wider and wider outward, 'till it be as a Tundifh, and becaufe it cannot be done without taking from your Circle the part G C H, the half Circle A C B by that means will become lighter then the other, by which your Inftrument will lofe its Equilibrium, and the Suns Beams cannot fall in its due place: Therefore do not forget to take from the half
Circle

Circle A D B, as much in weight as is necessary to set it on the same Equilibrium, as it was before you made the hole G C H.

S. How shall I know if this Instrument is in its Equilibrium, or is equally poised?

T. You may know it by fastening a Thread (with a Lead at the End) exactly upon the Perpendicular at A, and if the said Thread (being Perpendicular) falls exactly upon the Line A B, marked at A and B, or upon 22 Degrees 30 Minutes from E, you may assure your self that your Instrument is in its Equilibrium, in which it will keep best, if you add two Pound weights to the Point B.

S. How do you use it?

T. Its use is thus; first hold the Instrument with your right Thumb, (which is to be put on the Ring, set on it for that purpose) so, as to give it a free motion, then turn the hole C to the Sun, and mind on what Degree its Beams fall, for that will be the present Distance of the Sun from your Zenith: That is to say, if you count from E, but if from D, it is the Suns present Altitude.

V. *The Use of the Quadrant to observe the Stars Altitude without Horizon.*

S. IS it better to observe the Stars Altitude by this Quadrant than with the Cross-staff?

T. Yes, for with the Cross-staff one must see the Horizon (as well as the Star) which renders the work very uncertain, and so much the more as vapours in the Night-time darken the Horizon; and therefore I recommend to you this Quadrant, since with it you may observe exactly without the Horizon, provided it be made as followeth.

First, see that it be of well season'd Box or Peartree, and of 15 or 16 Inches Semidiameter; its Limb is to be divided into 90 Degrees, and each Degree into 12 equal parts of 5 Minutes.

S. How is a Degree to be divided into 12 equal parts of 5 Minutes each?

T. To understand how it is to be done, admit that B F is a Degree: First, describe six (Concentrick) Circles, which divide the Line B C into 6 equal parts, then divide the Degrees B F into two equal parts, and make an Isoscele Triangle upon it, as the Example sheweth, and so you shall divide a Degree into 12 parts of 5 Minutes, for each Segment of those Lines Comprehended between two Concentrick Circles is of 5 Minutes: This done, place two moveable Vanes or Sights upon the side A G, through which you may see the Star, and instead of a Thread put on a pretty wide Ruler as H, which must have a free motion upon the Center A of the Quadrant, not forgetting that the said Ruler must have three or four Pound weights of Lead at the lower end, as you see in K, and must be open in length that you may see the Degrees and Minutes of

the

the Stars Altitude, which the Thread M K will shew; and although I said that it must have a free motion upon the Center of the Quadrant, it must not be understood more then is necessary to be drawn by its weights towards the Center of the Earth; and then it will not be subject to be tossed to and fro, as a loose and light one would.

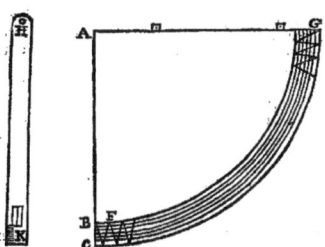

S. How must I observe with this Quadrant the Altitude of a Star?

T. You must turn the Center A to the Star, and holding the Limb to your Cheek-bone, place your Left-hand near your Ruler to stop it, as soon as you can see the Star through the holes on the Sights, for then the Thread K M, will shew you on the Limb the Degrees and Minutes of the Stars Altitude, which being subtracted from 90, you shall have its Distance from your Zenith.

Do not forget to wink with your left Eye, while you are looking through the Sights for the Stars.

P R O P. XIII.

How to know on a Year given, which of the four Tables
of the Suns Declination you are to make use of.

S. SINCE you give me four different Tables of the Suns Declination, I wish you would give me also some directions concerning it, that I may not mistake when I shall have occasion of them?

T. The way not to mistake, is to cut off from the proposed years the Thousands, Hundreds, Scores, and Fours, and the remainder will shew you in which of the Columns you are to look for the Suns Declination,

Mm 2   clination,

clination, and if there remains nothing, you must look for it in the Leap-year.

Example.

Admit you would know what Table to make use of in the year 1685.

First, cut off from the proposed year the Thousand-and Hundreds, and there will remain 85 , from which taking all the Scores, there will remain 5, from which taking also the 4, there will remain but 1, which sheweth, that 1685 is the first year after Leap-year; from whence it followeth that you are to make use of the first Table after Leap-year; that is to say, that you are to look for the Suns Declination in the Table or Column, on the Top of which is written *The first Year.*

Another Example.

Admit you would know which of these four Tables you are to make use of in the Year 1688.

Cut off as before from the proposed year the Thousand and Hundreds, and there will remain 88 ; from which, take off the Scores and Fours, and there will remain nothing ; by which you know that 1688 is a Leap-year; and that you must make use of that Table, on the Top of which is written *Leap-Year.*

S. Would it not be better to give me a Table of it, for fear I should mistake?

T. It is not necessary after what hath been said ; however to satisfie you, to prevent any mistake , I shall give you the following Table, which needs no other Directions for its Use , than what is exprest on the head of it.

Leap-Year.	First-Year.	Second-Year.	Third-Year.
1684	1685	1686	1687
1688	1689	1690	1691
1692	1693	1694	1695
1696	1697	1698	1699

Advertisement.

WHEN you look for the Suns Declination in its proper Table, take great care that you do not mistake in the Month , or Day of the Month , and that you do not take a South Declination for a North Declination , or on the contrary, a North Declination for a South Declination ;

clination; for this is of confequence, and you muft be very fure of what you do, when you will know what Latitude your Ship is in; take alfo great care before you go to Sea, to get the beft or laft Calculated Tables of the Suns Declination that are to be had, for how exact foever you be in your Obfervation, there will be fome error if your Tables be not good; befides, you muft needs know for what Meridian the faid Tables have been Calculated, that by the difference in Longitude you may Correct the errors which infallibly happen, when we fail under a Meridian far diftant from that Meridian for which they were Calculated, as you will better underftand by the following Propofition.

P R O P. XIV.

How to proportion the Suns Declination to any other Meridians Diftance from that, for which thefe Tables were Calculated.

S. IS this of any ufe to Pilots, or any that Commands a Ship?

T. Yes, of very great ufe, chiefly when the Sun is near the Equinoctial, and therefore if you be under a Meridian far diftant from that for which your Tables were Calculated (as for Example: 50, 60, 70 Degrees or more) you muft Correct your Tables, (of the Suns Declination) and you are to confider if you are Eafterly, or Wefterly from the Meridian, for which your Table is Calculated; alfo if the Declination of the Sun increafes or decreafes, that you may not miftake and make an Adition for a Subtraction, or a Subtraction for an Adition.

S. When is it that the Suns Declination increafes?

T. The Suns Declination increafes whilft the Sun is a going from the Equinoctial to the Tropicks, (that is to fay, from the 10th. of March to the 11th. of June, and from the 12th. of September to the 11th. of December,) but it decreafes whilft the Sun is a going back from the Tropicks to the Equinoctial (that is, from the 11th. of December to the 10th. of March, and from the 11th. of June to the 12th. of September.) This being underftood, it will be eafie to Correct your Tables after the following Examples.

Example 1.

Admit you would know the Suns Declination on the 18th. of March, 1685. Your Ship being then in 12 Degrees of Latitude North, and 300 Degrees of Longitude (Weft.)

Firft, you muft confider that your Ship is Wefterly from the Meridian of the Lizard, for which your Tables were Calculated, and therefore the

the Sun will come to it later than to the Meridian of the Lizard, for if the Sun rife Eafterly, and continue his ordinary Courfe Wefterly, he muft needs come fome hours fooner to the Meridian of the Lizard, than to that your Ship is at, which will be the caufe that the Suns Declination will not be fo great at the Lizard as where your Ship is, becaufe that all the time that the Sun hath been a going from the Meridian of the Lizard, to your prefent Meridian, he hath moved (in the Ecliptick) more Northerly, and fo being further off from the Equinoctial, his Declination is increafed; from whence it followeth, that you muft add fome Minutes proportionably to the difference between the Declination for the day of Obfervation, and that for the day following, (that is to fay, between the Declination of the 18th. and 19th. of *March*,) which is eafily done with a Rule of Three.

But when the Suns Declination is decreafing, you muft fubtract proportionably to the fame difference: This being underftood, we will now come to the Practice of what I have propofed, Thus:

Firft, find the difference of Longitude between the Meridian of the place, for which your Tables were Calculated, and that where you fuppofe your Ship to be.

Longitude of the Lizard	12 Deg.	37 Min.
	360	00
	372	37
Longitude my Ship is in, Subtract . .	300.	00
Difference in Longitude	72	37

Find out alfo the difference between the Suns Declination on the 18th. and 19th. of *March*, 1685.

Declination on the 18th. of *March*, . .	3 Deg.	24 Min.
and on the 19th. of *March*	3	47
Difference in Declination	0	23

Then fay, If 360 Degrees (that the Sun runs through in a day) gives 23 Minutes (differ.) What will 73 Degrees (differ. in Longit.) give?

If 360 Deg. give 23: What 73?

$$
\begin{array}{r}
23 \\
\hline
219 \\
146 \\
\hline
1679
\end{array}
$$

$$360)\ 1679\ (4$$
$$239$$

As

As you see, there will come almoſt 5 Minutes, which being added to 3 Degrees 24 Minutes, the Suns Declination (on the 18th of *March*, 1685,) comes 3 Degrees 29 Minutes for the Declination of the Sun at Noon, where my Ship is that day by Reckoning; that is to ſay, in 300 Degrees of Longitude (Weſt.)

S. Why do you add 360 Degrees to the Longitude of the Lizard?

T. It is, becauſe the Longitude that my Ship is in is greater than the Longitude of the Lizard, from which it cannot be ſubtracted without adding to it the 360 Degrees of the Circumference of the World.

Example 2.

The 28 *of* Auguſt, 1686. *Sailing along the Coaſt of* Florida, *and finding by my Reckoning to be arrived in* 282 *Degrees* 30 *Minutes of Longitude (Weſt:) I deſire to know the Suns Declination (at Noon) in the ſaid place?*

Now that the Declination decreaſes, remember you muſt ſubtract from the Declination of the day given, what the difference of the two Meridians give, for the Suns Declination will now be leſs, when the Sun comes to your (preſent) Meridian, then your Table ſheweth you.

	Deg.	Min.
Longitude of the Lizard	12	37
	360	00
	372	37
Longitude of my Ship, Subtr.	282	30
Difference in Longitude	090	07
Declination of the 28th. of *Auguſt*, 1686.	5	56
Declination of the 29 of *Auguſt*	5	33
Difference in Declination	0	23

If 360 Degrees give 23 Minutes: What will 90 Degrees (of difference in Longitude) give?

If 360 give 23: What 90?

$$90$$

360) 2070 (5   2070

270

	Deg.	Min.
Subtract	0	6
In all	5	50 Declination.

You

You see that 90 Degrees (difference in Longitude) give 5 Minutes 45 Seconds, or almost 6 Minutes, which being subtracted from 5 Degrees 56 Minutes (because the Suns Declination decreases) remaineth 5 Degrees 50 Minutes for the Declination of the Sun at Noon, the 28th. of *August*, 1686, where I reckon my Ship to be; that is, in 282 Degrees 30 Minutes of Longitude (West.)

Example 3.

The 30th. *of* August, 1685. *Sailing in Latitude North, and finding by my Reckoning to be arrived in 298 Degrees 20 Minutes of Longitude (West:) I desire to know the Suns Declination at Noon on the said place?*

	Deg.	Min.
Longitude of the Lizard	12	37
	360	00
	372	37
Longitude of my Ship, Subtract . . .	298	20
Difference of Longitude	074	17
Declination of the 30th. of *August*, 1685. .	5 Deg.	05 Min.
Declination of the 31th. of *August*	4	42
Difference in Declination	0	23

If 360 Degrees give 23 Minutes: What will 74 Degrees give?

$$
\begin{array}{r}
23 \\
\hline
222 \\
148 \\
\hline
1702
\end{array}
$$

$$
36c)\ 1702\ (4 \\
\underline{262}
$$

And there will come almost 4 Minutes 45 Seconds, and so we may count or reckon 5 Minutes; which being subtracted from 5 Degrees 5 Minutes, remaineth 5 Degrees for the Suns Declination at Noon, as was required.

Example 4.

The 18th. *of September*, 1684: *my Ship being then Southerly from the Equinoctial, (or in South Latitude) and by Reckoning in 112 Degrees 40 Minutes of Longitude East: I desire to know the Suns Declination at Noon in the said place?*

Now that my Ship is Easterly from the Meridian of the Lizard, it will be sooner Noon to me, than to those that dwell under the Meridian of the Lizard; from whence it followeth, that the Sun is not declined

clined so much as my Table sheweth (on the propofed day) and therefore I look for the Declination of the 18th. and 17th. of *September*, (in the Leap-year) which I find to be 2 Degrees 24 Minutes, and 2 Degrees 1 Minute, which being subtracted one from another, there remaineth 23 Minutes, for the difference in Declination, and since my difference in Longitude, is of 100 Degrees 3 Minutes; I say as before:

If 360 Degrees give 23 Minutes: What will 100 Degrees give?

$$\begin{array}{cc} 100 & 360)\ 2300\ (6 \\ \hline 2300 & 140 \end{array}$$

And it will give 6 Minutes, which being subtracted from 2 Degrees 24 Minutes, the Declination of the 18th. of *September*, 1684. remaineth only 2 Degrees 18 Minutes for the Declination which was required.

Example 5.

The 22th. of August, 1687. my Ship being then in South Latitude, and by Reckoning in 98 Degrees 45 Minutes of Longitude (East:) I defire to know the Suns Declination at Noon, at the said place?

Longitude of my Ship	98 Deg.	45 Min.
Longitude of the Lizard	12	37
Difference in Longitude	86	08

Declination of the 22th. of *August*, 1687 .	8 Deg.	15 Min.
Declination of the 21th. of *August* . . .	8	36
Difference in Declination	0	21

If 360 Deg. gives 21 Minutes difference: What will 86 Deg. give?

$$\begin{array}{cc} & 21 \\ .\ .\ .\ .\ 360)\ 1806\ (5 & 86 \\ \hline 6 & 172 \\ & \overline{1806} \end{array}$$

And it will give 5 Minutes, which I add to the 8 Degrees 15 Minutes Declination, the 22th. of *August* 1687. For as I am Easterly of the Meridian, for which my Table was Calculated, and the Suns Declination decreases, it is 5 Minutes more where my Ship is, than my Table sheweth, (because the Sun is not yet come to that Meridian) and so there will come 8 Degrees 20 Minutes for the Declination of that day, in the proposed place.

N n

PROP.

PROP. XV.

*How to work your Observation from the 10th. of March,
to the 12th. of September, when you sail in Latitude
North.*

S. IS there not certain Rules for to know when I must add the Suns
Declination to my Observation?

    T. Yes, and to know it, remember that from the 10th. of *March,*
to the 12th. of *September,* the Sun hath North Declination, because
that all that time he is on the North side, or Northerly of the Equinoctial, and therefore if in that time you are sailing in Latitude North,
you are to add the Suns Declination to the Complement of his Altitude;
that is to say, to the Degrees and Minutes that you find the Sun distant
from your Zenith, when he is on your Meridian Southerly from your
Zenith, for if you had sailed so near the Equinoctial, that the Sun should
be on the North of your Zenith, (and his shadow should fall Southerly)
then instead of adding the Suns Declination to your Observation, you
must subtract the Complement of the Suns Altitude from his Declination,
and the remainder will be your Latitude, or distance of your Ship from
the Equinoctial.

Example 1.

   *Admit my Ship at Sea in Latitude North, and the Sun 35 Degrees 30
Minutes distant from my Zenith, whose Declination is at that time 11
Degrees 30 Minutes North: I demand the Latitude of the place, the Sun
being then South from my Zenith?*

The Complement of the Suns Meridian Altitude,
  or Zenith distance is . . . . 35 Deg. 30 Min.
The Suns Declination North, add 11   30
                          ———————
      The Latitude of the place is 47   00 North.

Example 2.

   *My Ship being in Latitude North, and the Sun 27 Degrees 20 Minutes from
my Zenith, whose Declination is 4 Degrees 27 Minutes North, the Latitude
is required, the Sun being then South from my Zenith?*

Distance of the Sun from my Zenith . . . 27 Deg. 20 Min.
Declination North, add 04   27
                          ———————
    The Latitude I am in 31   49 North.

Ex-

Example 3.

The 15th. of August, 1685. *my Ship being in Latitude North, and the Sun 44 Degrees 45 Minutes distant from my Zenith, the Latitude is required, the Sun being then South from my Zenith ?*

Distance from my Zenith	44 Deg. 45 Min.
Declination North, add	10 34
The Latitude my Ship is in .	55 19 North.

Example 4.

The 26th. of May, 1686. *my Ship being then in Latitude North, and the Sun 10 Degrees 20 Minutes distant from my Zenith, the Latitude of the place is required, the Sun being then North from my Zenith ?*

Advertisement.

R Emember here what was told you (page 270.) that when you are in Latitude North, and the Sun is North from your Zenith, you must subtract the distance of the Sun from your Zenith (or the Complement of his Meridian Altitude) from his Declination, and the remainder will be the Latitude you are in.

Declination North of the Sun	22 Deg. 41 Min.
Distance of the Sun from my Zenith, Subtract	10 20
The Latitude I am in	12 21 North.

Example 5.

The 18th. of June, 1684. *my Ship being then in Latitude North, and the Sun 14 Degrees 30 Minutes distant from my Zenith, the Latitude of the place is required, the Sun being then North from my Zenith ?*

Declination North of the Sun	23 Deg. 19 Min.
The Distance of the Sun from my Zenith, Subtr.	14 30
The Latitude required . . .	08 49

Example,

Example 6.

The 19th. of April, 1685. my Ship being in Latitude North, and the Sun on my Zenith, the Latitude I am in is required?

Advertisement.

SINCE the Sun is on my Zenith, his Declination must needs be my Latitude: Therefore, I look in my Table for the Suns Declination (on the proposed day) which I find to be 14 Deg. 48 Min. North, the Latitude my Ship is in, the Suns Declination is the requir'd Latitude 14 Deg. 48 Min. North.

S. How shall I know the Suns distance from my Zenith, by the Suns Meridian Altitude?

T. You shall know it by subtracting the Suns Meridian Altitude from 90 Degrees, and the remainder will be the Suns distance from your Zenith.

PROP. XVI.

How to work your Observations in North Latitude, from the 12th. of September, to the 10th. of March.

S. WHAT Rules have you for this?

T. This is the Rule; you are to remember that from the 12th. of *September* to the 10th. of *March*, the Suns Declination is South; and therefore if in that time you are sailing in Latitude North, you are to subtract the Suns Declination, from the Complement of the Suns Meridian Altitude, which is the distance of the Sun from your Zenith.

Example 1.

The 3d. of October, 1687. my Ship being then in Latitude North, and the Sun 52 Degrees 20 Minutes distant from my Zenith, the Latitude of the place is required; the Sun being then South from my Zenith?

Distance from my Zenith	52 Deg.	20 Min.
Declination South, Subtract . . .	07	54
The Latitude	44	26 North.

Ex-

Example 2.

The 12th. of February, 1689. my Ship being in Latitude North, and the Sun 45 Degrees 20 Minutes distant from my Zenith, the Latitude is required, the Sun being then Southerly from my Zenith?

Distance of the Sun from my Zenith . .	45 Deg.	20 Min.
Declination South, Subtract	10	06
The Latitude I am in . .	35	14 North.

Take notice here, that if the Sun have no Declination, his distance from your Zenith is your Latitude, which is North, if the Sun is Southerly from your Zenith, and South if the Sun is Northerly from you; this is so plain that it needs no Example.

P.R.O.P. XVII.

How to work your Observations in Latitude South.

S. **W**HAT Rules must I observe, being in Latitude South?

T. You are to observe the same Rules in Latitude South, as you have done in Latitude North; that is to say, that if your Ship is in Latitude South, and the Sun Northerly from your Zenith, and his Declination be South, you are to add the Suns Declination to the Complement of his Altitude or distance from your Zenith; but if the Suns Declination is North, and you be in Latitude South, then you are to subtract from it the Suns Declination: Besides, also if the Sun is South from your Zenith, and his Declination be South, (whilst you are in Latitude South) you are to subtract the Complement of the Suns Altitude from his Declination, and the remainder will be your Latitude; as the following Examples will make it plain to you.

Example 1.

The 16th. of February, 1685. my Ship being in Latitude South, and the Sun 20 Degrees 15 Minutes distant from my Zenith, the Latitude is required, Sun being then Northerly from my Zenith?

Distance from my Zenith	20 Deg.	15 Min.
Declination South, add	08	20
The Latitude I am in	28	35 South.

Exe-

Example 2.

The 4th. of November, 1684. my Ship being in Latitude South, and the Sun 18 Degrees 10 Minutes distant from my Zenith, the Latitude is required, the Sun being then Northerly from my Zenith?

Distance from my Zenith	18 Deg.	10 Min.
Declination South, add	18	35
The required Latitude	36	45 South.

Example 3.

The 2d. of December, 1686. my Ship being in Latitude South, and the Sun 12 Degrees 4 Minutes distant from my Zenith, the Latitude is required, the Sun being then Southerly from my Zenith?

Advertisement.

Remember here that when you are in Latitude South, and the Sun is Southerly from your Zenith (or that you are between the Sun and the Equinoctial) you must subtract the Complement of the Suns Altitude (that is to say, his distance from your Zenith) from his Declination, and the remainder shall be the Latitude of the place, as you see by this Example.

Declination South of the Sun . . .	23 Deg.	12 Min.
Distance from my Zenith, Subtract .	12	04
The Latitude	11	08 South.

Example 4.

The 15th. of November, 1684. my Ship being in Latitude South, and the Sun 8 Degrees distant from my Zenith Southerly: The Latitude is required?

Declination South	21 Deg.	03 Min.
Distance from my Zenith, Subtract .	08	00
The Latitude I am in . . .	13	03 South.

Example 5.

The 8th. of February, 1687. the Sun being on my Zenith: The Latitude I am in is required?

(Remember

(Remember here what was faid (or told you) pag. 272. that when the Sun is on your Zenith, his Declination is your Latitude.)

Anfwer. Since the Suns Declination is (on the propofed day) 11 Degrees 27 Minutes South: The Latitude I am in is South 11 Deg. 27 Min.

Example 6.

Admit the Sun is 18 Degrees 15 Minutes diftant from my Zenith Northerly, and that he hath no Declination (that day:) The Latitude is required?

Anfwer. Since the Sun hath no Declination, his diftance from my Zenith is my Latitude; and therefore I fay, that the Latitude I am in is 18 Degrees 15 Minutes South, fince I am to the Southermoft of the Sun, who then is on the Equinoctial.

Example 7.

The 2d. of May, 1685. being at Sea in Latitude South, and the Sun 52 Degrees 30 Minutes diftant from my Zenith, the Latitude is required, the Sun being then Northerly from my Zenith?

Advertifement.

REmember now that the Declination is North, and you are in Latitude South : Therefore fubtract the Declination from the Suns diftance from your Zenith, and the remainder will be your diftance from the Equinoctial, which is your Latitude.

	Deg.	Min.
Diftance from my Zenith	52	30
Declination North, Subtract	18	24
Latitude South	34	06

Example 8.

The 24th. of Auguft, 1684. being at Sea in Latitude South, and the Sun 36 Degrees 20 Minutes diftant from my Zenith Northerly: The Latitude is required?

	Deg.	Min.
Diftance from my Zenith	36	20
Declination North, Subtract	07	14
Latitude South	29	06

Adver-

Advertisement.

TAKE notice, that if the Sun is on your Zenith and have no Declination, you are on the Equinoctial.

S. How shall I know if the Sun is exactly on my Zenith, or what side of it he is of, being near it?

T. The best way to know it then, is with the Sea Compass and Thread fastened over the Glass, (already Treated of in the 6th. Proposition of the Third Book), which for this, is to be used thus: When you perceive by your Observation that it is Noon, or the Sun is on your Meridian, dispose your Compass so that the Thread be East and West, and if its shadow falls exactly upon the Point of the socket, and upon the East and West Point of your Compass, you may conclude, that the Sun is on your Zenith, but if it do not, mind on what side the shadow falls in respect of the East and West Point of your Compass; (that is to say, if it falls Northerly, or Southerly from it) for that will shew you if the Sun is South or North from your Zenith, since the shadow falls always on the contrary side the Sun is of.

Advertisement.

TAKE also notice, that if you be where the Sun doth not set (as within the Artick or Antartick Circle) and there would observe the height of the Pole by the lowest Altitude of the Sun; (that is to say, when the Sun is on your Meridian under the Pole) add the Complement of his Declination (or Pole distance) to the Suns lowest Altitude, and you shall have the height of the Pole, (which is always equal to your distance from the Equinoctial.)

Example.

The 20th. of June, 1684. being at Sea where the Sun doth not set, I observe (at Midnight) the Suns lowest Meridian Altitude, and find it to be 8 Degrees 40 Minutes: The height of the Pole is required?

	Deg.	Min.
The distance of the Pole from the Equinoctial	90	00
The Suns Declination, Subtract	23	12
The Compl. of the Suns Declina. (or Pole distance)	66	48 North.
To which add the Suns Meridian Alt. under the Pole	08	40
The height of the Pole Artick is the Latitude I am in	75	28

Adver-

Advertisement.

TAKE notice, that when you obferve the Suns loweft Altitude (under the Pole) it may be Midnight to thofe who dwells under the Meridian, for which your Table of Declination is Calculated; and therefore it is neceffary to Correct the Declination fet down in the faid Table, becaufe of the 12 hours difference from Noon to Midnight, for it is certain that the Declination of the Sun at Midnight cannot be the fame, being it was Calculated but for Noon; fince then, the Sun hath increafed or decreafed his Declination half a day more than your Table fhews, you muft needs Correct it; but firft, you are to confider how far you are to the Eaft or Weft of the Meridian, for which your Table is Calculated, and if the diftance is confiderable, (which you may know by the difference in Longitude) you ought to Correct the Declination according to the directions in page 265. and that not only for the propofed day, but alfo for the day next before it.

Then for the 12 hours (from Noon to Midnight) take the difference between the Declination of the propofed day, and that of the day before, (by fubtracting one from another) and if the Declination increafes, add half of that difference to the Declination of the day before the propofed day; but if it decreafes, fubtract it from the fame from the Declination thereof, and the remainder will be Corrected Declination, as by the following Example.

S. Is this always to be obferved?

T. Yes, when you will know the height of the Pole or Latitude, by the loweft Meridian Altitude of the Sun; except it be about a Fortnight after the *Solftice* or 11th. of *June*, becaufe then the Sun moves fo flowly in the Ecliptick, that his Declination changes but little in a day, and fo in 12 hours cannot caufe any errors worth taking notice of, no more than when you are only 20 or 30 Degrees to the Eaft or Weft, from the Meridian, for which the fame Table was Calculated; for fo little difference requires no Correction, chiefly when the Sun is near the *Solftice*, and therefore will take no notice of it in this Example, fuppofing we are but 25 Degrees diftant from the Meridian of the Lizard.

Example 2.

The 19th. of July*, 1684. I obferve the Suns loweft Meridian Altitude, and find it to be 10 Degrees 15 Minutes, the height of the Pole is required?*

Since the Suns Declination is Calculated and fet down in my Table but for Noon, which in thefe Parts happen, but when the Sun is South from my Zenith, and that notwithftanding for all that, he is North from it at the time of my loweft Obfervation, which is the true time of Midnight

Oo

to

to those who dwells in a Paralel Sphere, or where the Sun rises and sets, it followeth, that for to find the Suns Declination at that time, which is 12 hours before what is call'd Noon, I must look for the Declination of the 18th. and 19th. of *July* which is 18 Degrees 51 Minutes, and 18 Degrees 37 Minutes, that I subtract one from another, and there comes for the difference 14 Minutes, whose half (7 Minutes) I subtract from the Declination of the 18th. of *July*, 18 Degrees 51 Minutes, (since the Suns Declination decreases) and the remainder 18 Degrees 4 Minutes, is the Declination at that time.

See the Practice.

Declination of the 18th. of *July*, 1684. .	18 Deg.	51. Min.
Declination of the 19th.	18	37
Difference in Declination . . .	00	14
Half of that Difference . . .	00	07 Min.
Declination of the 18th. of *July*, 1684. .	18 Deg.	51 Min.
Half of the difference, Subtract . . .	00	07
Declination of the 18th. of *July*, at Midnight	18	44 North.

Now that I know the Suns Declination, and his lowest Altitude, it is easie to know the height of the Pole, there being no more to do than in the precedent Example; that is, to subtract this Declination from 90 Degrees, and to add the remainder (which is the Complement of the Suns Declination) to the Suns lowest Altitude, for these two sums being added together, will shew you the height of the Pole required.

Distance of the Equinoctial from the Pole .	90 Deg.	00 Min.
The Declination, Subtract	18	44
Complement of the Suns Declination . .	71	16
The Suns lowest Meridian Altitude, add .	10	15
The height of the Pole	81	31

Or Latitude I am in, out of which you must allow for the refraction as before directed.

P R O P.

P R O P. XVIII.

How to work your Observation made at any Stars
that rise and set.

S. WHAT Rule do you give me for the Stars?
T. The very same that I gave you for the Sun, there being no difference (but in the Declination) if the Star which
you observe at rising : Therefore observe this General Rule.
The Declination which is of contrary Denomination, as for Example,
which is South when you are in Latitude North, or North when you are
in Latitude South, it to be subtracted from the Complement of the
Meridian Altitude, or distance of the Star from your Zenith; but if the
Declination is North, when you are in Latitude North, or South when
you are in Latitude South, you are to add it to the Complement of the
Star Altitude, except you be between the Equinoctial and the Star, or
the Stars Declination is greater then its distance from your Zenith, for
then you must subtract the Complement of the Stars Altitude, or Zenith
distance from the Declination of the same Star, and the remainder will
be the distance of your Zenith from the Equinoctial, which is your Latitude.

Example 1.

Being at Sea in Latitude North, I observe the Great Dog Rising on the Meri-
dian, and find it 48 Degrees 30 Minutes distant from my Zenith Southerly,
and his Declination is 16 Degrees 15 Minutes Southerly : The Latitude
is required ?

Distance of the *Great Dog* from my Zenith	48 Deg.	30 Min.
Declination South, Subtract	16	15
Latitude North	32	15

Example 2.

I observe at Sea the Eagles Heart on the Meridian, and find it 40 Degrees
25 Minutes distant from my Zenith Southerly, and his Declination is 8
Degrees 5 Minutes Northerly: The Latitude is required ?

Distance of the *Eagles Heart* from my Zenith	40 Deg.	25 Min.
Declination North, add	08	05
The Latitude I am in	48	30 North.

O o 2   Ex-

Example 3.

Being at Sea in Latitude North, I observe the Star Capella, *(on the Meridian) and find it* 26 *Degrees* 20 *Minutes distant from my Zenith Northerly, and its Declination is* 45 *Degrees* 38 *Minutes Northerly: The Latitude is required?*

Declination North . . . . 45 Deg. 38 Min.
Distance from my Zenith Northerly, Subtr. 26   20

The Latitude 19   18 North.

The same may be understood of the Stars you observe at in Latitude South, for as you see it is the very same as to the Sun, except this, that you cannot observe their Altitude by their shadow as to the Sun; but in all the rest it is the same, therefore look over for further instruction what hath been said of the Sun, for I know no difference but in the Name.

P R O P. XIX.

How to find the height of the Pole by the highest and lowest Meridian Altitude of the Stars, (that doth not set) whose Declination or Polar distance is set down in your Tables.

WHAT must I do for to know the height of the Pole, by the highest Meridian Altitude of a Star, (near it?)

T. If you observe the Meridian Altitude of a Star, when it is above the Pole, whose height you would know, you must only subtract the Complement of that Stars Declination from its (highest) Meridian Altitude, and the remainder is the height of the Pole.

Example 1.

Admit that you observe the brightest Star of the Guards, whose Declination is 75 *Degrees* 32 *Minutes North (when it is on the Meridian above the Pole) and find its Altitude to be* 62 *Degrees* 12 *Minutes: The Latitude is required?*

The Distance from the Equinoctial to the Pole . 90 Deg. 00 Min.
Declination North of the Star, subtract 75   32

Complement of the Declination, or distance of the 14   28
Star from the Pole.   The

The Meridian Altitude of the Star . . . 62 Deg. 12 Min.
The Complement of its Declination, Subtract 14   28

The height of the Pole or Latitude I am in .   47   44 North.

Example 2.

Admit that I observe the Cocks Foot (in the Crosiers) when it is on the Me-
ridian above the South Pole, and find its Altitude to be 59 Degrees 50
Minutes, his Declination being then 61 Degrees 17 Minutes South: The
Latitude is required?

	90 Deg.	00 Min.
	61	17
Distance of the Star from the Pole	28	43
The Meridian Altitude of the Star	59	50
Distance of the Star from the Pole, Subtract .	28	43
The height of the Pole or Latitude	31	07 South.

You are to take notice, that the Stars which do not set, comes also
to the Meridian under the Pole, and then their Altitude is the least of all,
for to know when a Star draws near the Meridian under the Pole, take
a Plummet (or Thread with Lead at the end) and hold it so, that it cuts
the North Star in the middle, and if the Star you will observe begins to
draw near the Thread, it will not be far from the Meridian; observe
then its Altitude several times until you find it ascend or rise, for the least
of all will be the Stars Meridian Altitude, to which if you add the Complement of its Declination or Pole distance, you shall have the height of
the Pole.

As for Example.

Admit that I observe the upper of the two foremost Stars of the Square in the
Great Bear, when it is on the Meridian under the Pole, and find its Al-
titude to be 9 Degrees 10 Minutes, its Declination being then 63 Degrees
28 Minutes North: The Latitude or height of the Pole is required?

Distance of the Equinoctial from the Pole,	90 Deg.	00 Min.
Declination of the Star, Subtract	63	28
Distance of the Star from the Pole	26	32
The Altitude of the Star	09 Deg.	10 Min.
Distance from the Pole, add	26	32
The height of the Pole, or Latitude I am in 35		42 North.

Example.

Example 2.

The Cocks Foot, *whose Declination is* 61 *Degrees* 17 *Minutes Southerly, being on the Meridian under the Pole* 10 *Degrees* 40 *Minutes above the Horizon, the height of the Pole is required?*

	90 Deg. 00 Min.	
	61	17
Diſtance of the Star from the Pole . .	28	43
The Altitude of the Star, add . . .	10	40
Height of the South Pole or Latitude I am in	39	23

Suppoſing you have hollowed for the Refraction, as you were directed page 87 of the Second Book.

PROP. XX.

How to know to find the height of the Pole by the Stars (that doth not ſet) without their Declination or Polar Diſtance.

S. IS there no way to find the height of the Pole by the Stars, without their Declination or Pole diſtance?

    T. Yes, for if you obſerve the higheſt and loweſt Meridian Altitude of a Star that doth not ſet, and add them together, the half of the Sum will be the height of the Pole.

Example.

Admit that by ſome accident of a Sea-fire, Rats, or any other, you have loſt your Tables of Declination, and yet would find now (by the Stars) the height of the Pole; you muſt firſt obſerve the higheſt Meridian Altitude of a Star that doth not ſet, to which we'll ſuppoſe thee clear of the Guards, whoſe Meridian Altitude above the Pole I find to be 62 Degrees 12 Minutes, then 12 hours after I obſerve again when the ſame Star is on the Meridian under the Pole, and find its loweſt Altitude to be 33 Degrees 16 Minutes, which being added to its higheſt Altitude 62 Degrees 12 Minutes, comes 95 Degrees 28 Minutes, whoſe half 47 Degrees 44 Minutes is the height of the North Pole, or Latitude my Ship is in.

                                                          The

The fame is to be underftood of any other Star that doth not fet, as well in Latitude South, as in Latitude North.

But take notice, that for to fucceed in this Practice, you muft not alter your Paralel 'till both Obfervations are made, for if you fhould fail by any other Rumb than Eaft or Weft, it muft needs caufe fome error, which will be fo much the greater the further you are (from the Paralel of your firft Obfervation) when you obferve the loweft Altitude of the Star.

PROP. XXI.

Of the North (or Pole) Star, with a Table of its Diftance from the Pole Artick, for the four chief Points of the Compafs the Guards are upon, with the way to know when the Guards are on each Point of the Compafs, named on the faid Table.

S. WHY do you call it North (or Pole) Star?

T. Becaufe it is the neareft Star to the North Pole.

S. Is it not on the very Pole it felf?

T. No, for it is 2 Degrees 22 Minutes diftant from it, and hath been at a greater diftance, however it will be nearer and nearer 'till the Year 2100. when it will be lefs than 30 Minutes diftant from the Pole, but then after that will move further and further from it, fo that in the Year 12700. if the World laft fo long, it will be 48 Degrees diftant from the faid Pole, by which you may underftand that it is the Pole Star, but by chance, fince it is but by its own proper motion from Weft to Eaft (as the Zodiack moves) that it comes to be fo near it only for a time.

S. Hath the Pole Star any motion from Eaft to Weft, as other Stars have?

T. Yes, but not fo fwift, for as the Circle that it defcribes about the Pole is very little, it moves more flowly to finifh it in 24 hours, (as all the other Stars do theirs) from whence it followeth however, that it will be fome time above the Pole, and fome time under it.

S. By what do they commonly know, when the North Star is above or under the Pole, or even with it?

T. It is commonly known by the Point of the Compafs, that the brighteft of the Guards is upon (in refpect of the North Star) but I cannot recommend it to you, becaufe of the Excentricite of the North Star, which although Corrected, is not without errors, and therefore I fhall give you only a Table for the four chief Rumbs, not only becaufe then you cannot miftake in judging which of them the Guards are upon, but becaufe you may truft to it, being more juft or exact than for any other Rumb the Table could be Calculated for.

A Table shewing the Declination or distance of the North Star above or beneath the Pole, upon the four chief (or Cardinal) Points of the Compass the Guards are upon, being what must be added or subtracted from its Altitude for to have the height of the Pole.

When the brightest of the Guards is		D.	M.	
	North Subtract	2	4	From the Pole Star Altitude.
	East add	1	10	To the Pole Star Altitude.
	South add	2	4	To your Observation at the North Star.
	West Subtract	1	10	From the Pole Star Altitude, and the remainder will be the height of the Pole, or Latitude you are in.

S. How shall I know what Point of the Compass the Guards are upon?

T. For to know which of these four Points of the Compass the Guards are upon (from the North Star) imagin a Line drawn from your Zenith through the Pole Star to the Horizon, and that shall be a Line of North and South, by which you shall know when the Guards are North or South, for when you shall perceive the brightest of the Guards exactly on that Line under the Pole Star, (that is to say, Perpendicularly under it) you may conclude that the Guards are North, but when the brightest of the Guards is on the same Line Perpendicularly above the Pole Star, then are the Guards South; as for the other two Points (set down in the Table) imagin likewise a Line of East and West drawn Parallel to the Horizon through the Pole Star (that is, Cross or at Right Angles upon the Line of North and South) and when you see the brightest of the Guards on that Line with the Pole Star, and to the Right-hand, you may be sure that the Guards are East or else West, if it be on the other side of the Pole Star to your Left-hand, and then is the best time to make your Observation at the Pole Star, that is, when the Guard are either East or West.

PROP.

P R O P. XXII.

How to find the height of the Pole (Artick) by the
North Star , and the Guards.

S. **W** HAT muſt I do to find the height of the Pole by the
North Star?
T. You muſt ſtay 'till the brighteſt of the Guards be
exactly upon one of the Points of the Compaſs ſet down
in your Table, (that is to ſay, either Eaſt, Weſt, North or South, from
the Pole Star) for that is the beſt time to obſerve the Altitude of the
North Star, to which if you add or ſubtract what your Table ſheweth,
you ſhall have the height of the Pole, as you will better underſtand
by the following Example.

Example 1.

Admit that I obſerve the Altitude of the North Star , and find it to be 42
Degrees 20 Minutes above the Horizon, the Guards being then Eaſt from
it : The height of the Pole is required?

Firſt, I ſet down the obſerved Altitude of the North Star 42 Degrees
20 Minutes, to which I add what my Table ſheweth, that the North
Star is under the Pole 1 Degree 10 Minutes, and the ſum 43 Degrees 30
Minutes, is the height of the Pole.

	Deg.	Min.
Altitude of the North Star	42	20
The brighteſt of the Guards is Eaſt, add .	1	10
Height of the North Pole or Latitude I am in	43	30

Example 2.

Admit that I obſerve the North Star , and find its Altitude to be 47 De-
grees 14 Minutes , when the brighteſt of the Guards is North from it :
The height of the Pole is required?

Altitude

Altitude of the North Star . . . . . 47 Deg. 14 Min.
The brightest of the Guards is North, Subtr.   02   04

Height of the North Pole . .   45   10

Take notice, That the Guards are East or West, when you find that the brightest of them is neither higher nor lower than the North Star, which you may easily discover with your Cross-staff or Quadrant, as soon as you have observed the Altitude of the North Star.

THE

THE
LOG-BOOK,
FOR

A Voyage intended by Gods affiftance from the Lizard in the Latitude 50 Deg. 00 Min. North, and Longitude (from the Pike of Teneriff*) 12 Deg. 37 Min. to the Ifland of* Barbadoes, *in the Lat. 13 Deg. 12 Min North, Longitude 319 Deg. 40 Min. the difference of Longitude between the* Lizard *and the* Barbadoes *is 52 Deg. 57 Min.*

Anno 1684.

H.	K.	HK	F.	Courfes.	Winds.	
						March the 26th. being *Wednefday* at Noon, we faw the *Lizard* Point
2	5	1	0	S W b S	N	bear *N b E* about 5 Leagues of us, we
4	5	1	0			had a frefh Gale and fair weather,
6	6	0	0			and were in Company of a *Dutch*
8	6	0	0			Ship bound for *Genova*, the
10	6	0	0			Courfe I make to be *S W b S*, the
12	6	0	0			diftance 142 Miles, the difference of
2	6	0	0			Latitude 118 Min. and the difference
4	6	0	0			of Longitude or Meridian diftance
6	6	1	0			79 Min. *W.*
8	6	1	0			
10	6	0	0			Latitude departed . 49° 45'
12	5	0	0			Difference of Latitude 1 58
12	71	0	0	*Whichbeing doubl'd coms 142 Mil. run in 24 hours*		Lat. by Dead Reckoning 47 47 N. the 27 of *March*

H.	K.	HK	F.	Courses.	Winds.
2	5	0	0	SWbS	NNW
4	4	0	0		
6	3	1	0		
8	4	1	0		
10	4	1	0		NW
12	5	0	0		
2	5	0	0	SSW ½ Sly	
4	5	1	0	SWbS	NWbW
6	5	1	0	SWbS ½ Sly	WbN
8	5	1	0		
10	5	0	0		
12	4	0	0		
12	56	1	0	Doubled, comes 113 Miles run.	

March the 27th. being *Thursday* was fair weather, and smooth water, the wind decreasing by degrees 'till fix a Clock; we faw a fail on our Larboard quarter. The true Courfe I make to be *SWbS* 3° 15' *Sly*, the diftance 112 Miles, the difference of Latitude 97 Min. and the Meridian diftance 57 Min. *W.*

Zenith diftance . 38° 43'
Declination . .   7 17

Lat. by Obfervation 46 00 N.
the 28th of *March.*

H.	K.	HK	F.	Courses.	Winds.
2	3	0	0	SSW ½ Sly	W
4	3	1	0		
6	3	0	0		
8	2	1	0	SbE	WSW
10	2	1	0		
12	3	0	0	SSE ½ Ely	SW
2	2	1	0		
4	2	1	0		
6	3	0	0		
8	3	0	0	SE ½ Ely	SSW
10	2	1	0		
12	2	0	0		
12	33	0	0	Doubled, comes 66 Miles run.	

March the 28th. on *Friday* the Wind very fcant and bad weather, a great Sea coming from the South Weft. The true Courfe I judge to be S b E 4° *Ely*, the diftance 58 Miles, the difference of Latitude 56 Min. and the Meridian diftance 15 Min. *E.*

Latitude . . . 46° 00'
Difference in Lat. . 00 56

Lat. by Reckoning . 45 04 N.
the 29th of *March.*

H.	K.	HK	F.	Courses.	Winds.	
2	4	0	0	S W ¼ Wly	S S E	*March* 29th. being *Saturday* thick foggy weather, and a fresh Gale with a head Sea. The true Course I make to be *S W* 2° *Wly*, the distance 93 Miles, the difference of Latitude 63 Min. and the Meridian distance 68 Min.
4	4	0	0			
6	4	1	0			
8	4	1	0			
10	4	0	0			
12	3	1	0			
2	3	1	0			Latitude . . 45° 04'
4	3	1	0	S W b S ½ Wly	S E b S	Difference in Lat. 01 03
6	4	1	0			Lat. by Reckoning 44 01 N.
8	4	0	0			the 30th. of *March.*
10	3	1	0			
12	3	0	0			
12	46	1	0	*Doubled, comes* 93 *Miles run.*		

H.	K.	HK	F.	Courses.	Winds.	
2	3	1	0	S W b S	E b N	*March* the 30th. on *Sunday* was fair weather and smooth water, the Wind being *E b N*, the *Dutch* Ship left us. The true Course I judge to be *S W b S*, the distance 101 Miles, the difference of Latitude 84 Min. and the Meridian distance 56 Min.
4	3	1	0			
6	4	0	0			
8	4	0	0			
10	4	1	0			
12	4	1	0		E	
2	4	0	0			
4	4	0	0			Zenith distance . 34° 16'
6	5	0	0			Declination . . 08 2½
8	5	0	0			Lat. by Observation 42 49 N.
10	4	1	0			the 31th. of *March.*
12	4	0	0			
12	50	1	0	*Doubled, comes* 101 *Miles run.*		

H i.

H.	K.	HK	F.	Courses.	Winds.	
2	3	0	0	S W b S	N N W	*March* 31th. on *Monday* little Winds, fair weather, and smooth water, at five in the Morning we out with our Top-gallant-sails and Stay-sails. The true Course I make to be *S W* b *S*, the distance 78 Miles, the difference of Latitude 65 Min. and the Meridian distance 43 Min.
4	3	0	0			
6	3	0	0			
8	3	1	0			
10	3	1	0		N	
12	3	0	0			
2	3	0	0			
4	2	1	0			Zenith distance . 32° 39′
6	3	1	0			Declination . . 08 46
8	4	0	0			
10	4	0	0			Lat. by Observation 41 25 N.
12	3	0	0			the first of *April*.
12	39	0	0	*Doubled*, comes 78 *Miles run*.		

H.	K.	HK	F.	Courses.	Winds.	
2	5	1	0	S W b S	N b W	*April* the 1st. on *Tuesday* a fresh Gale, fair weather, and smooth water. The true Course I make to be *S W* 3° 15′ *Wly*, the distance 111 Miles, the difference of Latitude 74 Min. and the Meridian distance 83 Min.
4	5	0	0			
6	5	1	0			
8	5	1	0			
10	6	0	0			
12	6	0	0		N N W	
2	5	1	0			
4	5	1	0			Latitude . . . 41° 32′
6	5	0	0	W		Difference of Lat. 01 14
8	4	0	0			
10	4	1	0			Lat. by Dead Reck. 40 18 N.
12	4	1	0			the 2d. of *April*.
12	61	1	0	*Doubled*, comes 123 *Miles run*.		

H.

H.	K.	HK	F.	Courses.	Winds.	
2	4	1	0	W.	N N W	April the 2d. on *Wednesday* a fresh Gale and fair weather, at break of day we saw two Sails on our Larboard Bow, which fearing (or judging) to be two *Argerine* (or *Turks* men of War as without doubt they were) we clap close by a Wind, by which means we escaped; for they chased us 'till after Sun set, and then bore up, finding they did not get upon us. The Course I make to be S W 5° S*ly*, the distance 102 Miles, the difference of Latitude 78 Min. and the Meridian distance 66 Min.
4	4	1	0			
6	5	0	0			
8	5	0	0	S W ½ S*ly*	W N W	
10	5	1	0			
12	5	1	0			
2	5	1	0			
4	5	1	0	S ½ W*ly*	W b S	
6	6	0	0			
8	5	0	0			
10	4	0	0			
12	3	0	0			
12	59	0	0	*Doubled, comes 118 Miles run.*		Latitude . . 40° 18′ Diff. in Latitude 01 18 Lati. by Reckoning 39 00 N. the 30th. of *April.*

H.	K.	HK	F.	Courses.	Winds.	
2	4	0	0	S ¼ E*ly*	W S W	April the 3d. on *Thursday* a fresh Gale and fair weather, a great Sea from the South West. The Course is S b E 6° E*ly*, the distance 118 Miles, the difference of Latitude 111 Min. and the Meridian distance 39 Min.
4	4	0	0			
6	4	1	0			
8	4	1	0			
10	4	0	0			
12	4	0	0			
2	3	1	0			Zenith distance . . 26° 58′
4	3	1	0	S S E ½ E*ly*	S W	Declination, add. 9 51
6	3	0	0			
8	3	0	0			Lat. by Observation 36 49 N. the 4th. of *April.*
10	3	0	0			
12	3	0	0			
12	44	1	0	*Doubled, comes 89 Miles run.*		

H2.

H.	K.	HK	F.	Courses.	Winds.
2	3	1	0	S W	S S E
4	4	0	0		
6	4	0	0		
8	4	0	0		
10	4	0	0		
12	3	1	0		
2	3	1	0	S S W	S E
4	3	0	0		
6	3	1	0		
8	3	1	0		
10	3	1	0		
12	3	0	0		
12	43	0	0	*Doubled, comes 86 Miles run.*	

April the 4th. on *Friday* a fresh Gale and rainy weather. The Course is S W b S 2° 30' *Wly*, the distance 84 Miles, the difference of Latitude 68 Min. and the Meridian distance 50 Min.

Latitude . . 36° 49'
Diff. in Latitude 01 08
—————
Lat. by Reckoning 35 41 N.
the 5th. of *April*.

H.	K.	HK	F.	Courses.	Winds.
2	3	1	0	S W b S	S E b S
4	3	1	0		
6	4	0	0		
8	4	0	0		
10	4	0	0		
12	3	1	0		
2	3	1	0	W S W ½ Wly	S
4	3	1	0		
6	4	0	0		
8	4	1	0	S W b S	S E b E
10	4	1	0		
12	4	0	0		
12	46	1	0	*Doubled, comes 93 Miles run.*	

April the 5th. on *Saturday* a fresh Gale and fair weather. The Course is S W 2° S*ly*, the distance 88 Miles, the difference of Latitude 65 Min. and the Meridian distance 60 Min.

Zenith distance . 24° 00'
Declination, add   10 33
—————
Lat. by Observation 34 33 N.
the 6th. of *April*.

H

H.	K.	HK	F.	Courses.	Winds.	
2	3	0	0	S W	E S E	*April* the 6th. on *Sunday* little wind, fair weather, and smooth water. The true Course I allow to be *S W*, the distance 86 Miles, the difference of Latitude 61 Min. and the Meridian distance 61 Min.
4	3	0	0			
6	3	0	0			
8	3	1	0			
10	3	1	0			
12	4	0	0			Zenith distance . 22° 34'
2	4	0	0			Declination, add 10 54
4	4	0	0			Lat. by Observation 33 28 N.
6	4	1	0			the 7th. of *April.*
8	4	0	0			
10	3	1	0			
12	3	0	0			
12	43	0	0	*Doubled, comes* 86 *Miles run.*		

H.	K.	HK	F.	Courses.	Winds.	
2	4	0	0	S W	S E b E	*April* the 7th. on *Monday* a loom Gale, fair weather, and smooth water, which makes us believe the Wind will be more Northerly to morrow. The true Course I make to be *S W*, the distance 112 Miles, the difference of Latitude 93 Min. and the Meridian distance 93 Min.
4	4	0	0			
6	4	0	0			
8	4	0	0			
10	5	0	0			
12	5	0	0			
2	5	0	0			Zenith distance . 20° 37'
4	5	0	0			Declination, add : 11 15
6	5	0	0			Lat. by Observation 31 52 N.
8	5	0	0			the 8th. of *April.*
10	5	0	0			
12	5	0	0			
12	56	0	0	*Doubled, comes* 112 *Miles run.*		

H.	K.	HK	F.	Courses.	Winds.	
2	3	0	0	S W	E S E	*April* the 8th. on *Tuesday* little Wind fair weather, and smooth water. The true Course I judge to be *S W*, the distance 69 Miles. the difference in Latitude 49 Min. and the Meridian distance 49 Min.
4	3	0	0			
6	0	0	0			
8	3	1	0			
10	3	1	0			
12	3	0	0			Zenith distance . 19° 14'
2	3	0	0			Declination, add 11 54
4	3	0	0			Lat. by Observat. 31 03 N
6	3	1	0			the 9th. of *April*.
8	3	0	0			
10	3	0	0			
12	3	0	0			
12	34	1	0	*Doubled*, comes 69 *Miles run.*		

H.	K.	HK	F.	Courses.	Winds.	
2	5	1	0	S W	E S E	*April* the 9th. on *Wednesday* : very fresh Gale, rainy weather, and smooth water. The true Course I allow to be *S W*, the distance 15 Miles, the difference of Latitude 108 Min. and the Meridian distance 108 Min.
4	6	0	0			
6	6	1	0			
8	7	0	0		E b S	
10	7	0	0			
12	7	0	0		E S E	
2	6	1	0			Latitude . . 31° 13'
4	6	0	0			Diff. in Lat. Subtr. 01 48
6	6	1	0			Lat. by Dead Reck. 29 25 N
8	6	1	0			the 10th. of *April*.
10	6	0	0			
12	6	0	0			
12	76	1	0	*Doubled*, comes 153 *Miles run.*		

H

H.	K.	HK	F.	Courſes.	Winds.
2	6	0	0	S W	E b S
4	6	0	0		
6	6	1	0		
8	6	1	0		
10	6	1	0		
12	6	0	0		
2	6	0	0		
4	5	0	0		
6	5	1	0		
8	5	0	0		
10	5	0	0		
12	5	0	0		
12	69	0	0	*Doubled, comes* 138 *Miles run.*	

April the 10th. on *Thurſday* a freſh Gale, fair weather, and ſmooth water. The true Courſe I make to be *S W*, the diſtance 138 Miles, the difference of Latitude 98 Min. and the Meridian diſtance 98 Min.

Zenith diſtance . 15° 23'
Declination, add 12 15
Lat. by Obſervation 27 38 N.
the 11th. of *April.*

H.	K.	HK	F.	Courſes.	Winds.
2	5	1	0	S W b W	E b S
4	5	1	0		
6	6	0	0		
8	6	0	0		
10	6	0	0		E
12	6	1	0		
2	6	1	0		
4	6	1	0		
6	6	1	0		
8	6	0	0		
10	5	1	0		
12	5	0	0		
12	71	1	0	*Doubled, comes* 143 *Miles run.*	

April the 11th. on *Friday* a freſh Gale, cloſe weather, and ſmooth water. The true Courſe I allow to be *S W b W*, the diſtance 143 Miles, the difference of Latitude 79 Min. and the Meridian diſtance 119 Min.

Latitude . . 27° 47'
Diff. in Latitude 01 19
Lat. by Dead Reck. 26 28 N.
the 12th. of *April.*

H.	K.	HK	F.	Courfes.	Winds.	
2	5	1	0	S W b W	E b N	*April* the 12th. on *Saturday* a
4	6	0	0			freſh Gale, fair weather, and ſmooth
6	6	1	0			water. The true Courſe I allow to
8	7	0	0			be *S W* b *W* ½ *Wly*, the diſtance 149
10	7	0	0			Miles, the difference of Latitude 70
12	6	0	0			Min. and the Meridian diſtance 131
2	6	1	0			Min.
4	6	0	0			
6	6	0	0			Zenith diſtance . 12° 12'
8	6	0	0			Declinationin, add 12 54
10	5	1	0			Lat. by Obſervation 25 06 N.
12	5	1	0			
12	74	1	0	*Doubled, comes* 149 *Miles run.*		

H.	K.	HK	F.	Courfes.	Winds.	
2	5	1	0	S W b W	E N E	*April* the 13th. on *Sunday* a
4	5	1	0			freſh Gale, fair weather, and ſmooth
6	6	0	0			water. The true Courſe I allow to
8	6	0	0			be *S W* ½ *Wly*, the diſtance 131 Miles,
10	6	0	0			the difference of Latitude 83 Min.
12	6	0	0			and the Meridian diſtance 101 Min.
2	5	1	0			
4	5	1	0			Zenith diſtance . 10° 23'
6	5	1	0			Declination, add 13 14
8	5	0	0			Lat. by Obſervation 23 37 N.
10	5	0	0			the 4th. of *April*.
12	4	0	0			
12	55	1	0	*Doubled, comes* 131 *Miles run.*		

H.

H.	K.	HK	F.	Courses.	Winds.
2	3	1	0	SWbW	EbN
4	3	1	0		
6	4	0	0		
8	3	1	0		
10	3	0	0		
12	3	0	0		
2	3	1	0		
4	3	1	0		
6	4	0	0		
8	4	0	0		
10	4	0	0		
12	4	0	0		
12	43	1	0	Doubled, comes 87 Miles run.	

April the 14th. on *Monday* little Wind and smooth water; this day according to Custom we did Baptise those who never had passed the Tropick before. The true Course I make to be *SWbW ½ Wly*, the distance 87 Miles, the difference of Latitude 41 Min. and the Meridian distance 77 Min.

Latitude . . . 23° 37'
Diff. in Lat. Subtr. 00 41
Lat. by Reckoning 22 56 N

H.	K.	HK	F.	Courses.	Winds.
2	5	0	0	SWbW	E
4	5	0	0		
6	5	1	0		
8	5	1	0		
10	6	0	0		
12	6	0	0		
2	6	1	0		
4	6	1	0		
6	6	1	0		
8	6	0	0		
10	5	1	0		
12	5	0	0		
12	69	0	0	Doubled, comes 138 Miles run.	

April the 15th. on *Tuesday* a fresh Gale, fair weather, and smooth water. The true Course I make to be *SW ¼ Wly*, the distance 138 Miles, the difference in Latitude 83 Min. and the Meridian distance 107 Min.

Zenith distance . 07° 31'
Declination, add 13 53
Lat. by Observat. 21 24 N.

H3.

H.	K.	HK	F.	Courses.	Winds.
2	5	1	0	*S W b W*	*E b S*
4	5	1	0		
6	6	0	0		
8	6	0	0		
10	6	0	0		
12	5	1	0		
2	5	0	0		
4	5	0	0		
6	5	1	0		
8	5	0	0		
10	4	0	0		
12	3	1	0		
12	62	1	0	*Doubled, comes* 125 *Miles run.*	

April the 16th. on *Wednesday* a fresh Gale and fair weather. The true Course I make to be *S W b W.* ½ *Wly,* the diftance 125 Miles, the difference of Latitude 59 Min. and the Meridian diftance 110 Min.

Latitude . . . 21° 28'
Difference in Lat. 00   59
Lat. by Dead Reck. 20   29 N.

H.	K.	HK	F.	Courses.	Winds.
2	3	1	0	*S W b W*	*E b N*
4	3	1	0		
6	4	0	0		
8	4	0	0		
10	4	0	0		
12	4	0	0		*E N E*
2	3	1	0		
4	3	1	0		
6	4	0	0		
8	4	1	0		
10	4	1	0		
12	5	0	0		
12	48	0	0	*Doubled, comes* 96 *Miles run.*	

April the 17th. on *Thurfday* little Wind, fair weather, and fmooth water. The true Courfe I make to be *S W* ½ *Wly,* the diftance 96 Miles, the difference of Latitude 61 Min. and the Meridian diftance 74 Min.

Zenith diftance . 04° 51'
Declination, add   14   30
Lat. by Obfervation 19   21 N.

H.

H.	K.	HK	F.	Courſes.	Winds.
2	5	1	0	S W b W	E
4	5	1	0		
6	6	0	0		
8	6	0	0		
10	6	1	0		
12	7	0	0		
2	7	0	0		
4	7	0	0		
6	7	0	0		
8	6	1	0		
10	6	0	0		
12	6	0	0		
12	76	0	0	Doubled, comes 152 Miles run.	

April the 18th. on Friday a freſh Gale, fair weather, and ſmooth water. The true Courſe I make to be S W b W, the diſtance 152 Miles, the difference of Latitude 84 Min. and the Meridian diſtance 125 Min.

Latitude . . . 19° 28'
Diff. in Latit. Subtr. 01 24

Lat. by Dead Reck. 18 04

H.	K.	HK	F.	Courſes.	Winds.
2	6	0	0	S.W b W	E b N
4	6	0	0		
6	6	1	0		
8	6	1	0		
10	6	1	0		
12	6	0	0		
2	6	0	0		
4	6	0	0		
6	6	1	0		
8	6	0	0		
10	5	0	0		
12	4	0	0		
12	71	0	0	Doubled, comes 142 Miles run.	

April the 19th. on Saturday : freſh Gale, fair weather, and ſmooth water. The true Courſe I make to be S W b W ½ Wly, the diſtance 142 Miles, the difference of Latitude 67 Min. and the Meridian diſtance 125 Min.

Latitude . . . 18° 04'
Difference in Lat. . 01 07

Lat. by Reckoning . 16 57 N.

H.

H.	K.	HK	F.	Courses.	Winds.	
						April the 20th. on *Sunday* little
2	3	1	0	S W b W	E N E	Wind, fair weather, and smooth
4	3	1	0			water. The true Course I allow to
6	4	0	0			be *S W ¼ Wly*, the distance 90 Miles,
8	3	1	0			the difference of Latitude 57 Min.
10	3	1	0			and the Meridian distance 70 Min.
12	3	1	0			
2	3	1	0			Zenith distance . 00° 25′
4	3	0	0			Declination, add . 15 25
6	4	0	0			
8	4	0	0			Lat. by Observat. 15 50
10	4	1	0			
12	4	1	0			
12	45	0	0	*Doubled, comes* 90 *Miles run.*		

H.	K.	HK	F.	Courses.	Winds.	
						April the 21th. on *Monday* a fresh
2	5	1	0	S W b W	E b N	Gale, fair weather, and smooth
4	5	1	0			water. The true Course I make to
6	6	0	0			be *S W b W*, the distance 128 Miles,
8	5	1	0			the difference of Latitude 71 Min.
10	5	1	0			and the Meridian distance 106 Min.
12	5	0	0			
2	5	0	0			Latitude . . . 16° 00′
4	5	0	0			Difference in Lat. . 01 11
6	6	0	0			
8	6	0	0			Lat. by Dead Reck. . 14 49
10	5	0	0			
12	4	0	0			
12	64	0	0	*Doubled, comes* 128 *Miles run.*		

H.

H.	K.	HK	F.	Courses.	Winds.	
2	3	1	0	*SW bW*	*E*	*April* the 22d. on *Tuesday* little Wind, fair weather, & smooth water. The true Course I make to be *S W b W ½ Wly*, the distance 92 Miles, the difference of Latitude 43 Min. and the Meridian distance 81 Min.
4	3	1	0			
6	4	0	0			
8	4	0	0			
10	4	0	0			
12	4	0	0		*E b S*	Declination North 16° 00′
2	4	0	0			Zenith Dist. Subtr. 02 02
4	3	1	0.			
6	3	1	0			Lat. by Observat. 13 58 N.
8	4	0	0			
10	4	0	0			
12	4	0	0			
12	46	0	0	*Doubled, comes 92 Miles run.*		

H.	K.	HK	F.	Courses.	Winds.	
2	3	1	0	*SW bW*	*E b N*	*April* the 23d. on *Wednesday* little Wind, fair weather, and smooth water. The true Course I make to be *SW b W ½ Wly*, the distance 89 Miles, the difference in Latitude 4 Min. and the Meridian distance 78 Min.
4	3	1	0			
6	3	1	0			
8	3	1	0			
10	3	1	0			
12	4	0	0		*E N E*	Declination 16° 18′
2	4	0	0			Zenith distance 03 05
4	3	1	0			
6	3	1	0			Lat. by Observat. 13 13
8	3	1	0			
10	4	0	0			
12	4	1	0			
12	44	1	0	*Doubled, comes 89 Miles run.*		

R r   H.

H.	K.	HK	F.	Courses.	Winds.
2	6	o	o	W	E N E
4	6	o	o		
6	6	o	o		
8	6	o	o		
10	6	o	o		
12	5	o	o		
2	5	o	o		
4	5	o	o		
6	5	o	o		
8	6	o	o		
10	6	o	o		
12	6	o	o		
12	68	o	o	Doubled, comes 136 Miles run.	

April the 24th. on *Thursday* a fresh Gale, fair weather, and smooth water. The true Course I make to be *W ½ Sly*, the distance 136 Miles, the difference in Latitude 13 Min. and the Meridian distance 135 Min.

Latitude . . . 13° 13'
Difference in Lat. . 00 13
Lat. by Dead Reck. . 13 00 N.

H.	K.	HK	F.	Courses.	Winds.
2	3	I	o	W b N	E N E
4	3	I	o		
6	4	o	o		
8	4	o	o		
10	4	o	o		
12	3	I	o		
2	3	o	o		
4	3	o	o		
6	3	I	o		
8	3	i	o		
10	4	o	o		
12	4	I	o		
12	44	o	o	Doubled, comes 88 Miles run.	

April the 25th. on *Friday* a little Wind, fair weather, and smooth water. The true Course I make to be *W ½ Nly*, the distance 88 Miles, the difference of Latitude 9 Min. North, and the Meridian distance 88 Min.

Declination . . 16° 51'
Zenith distance . 03 44
Lat. by Observat. 13 07 N.

H.

H.	K.	HK	F.	Courses.	Winds.
2	5	0	0	W	E
4	5	0	0		
6	5	1	0		
8	5	1	0		
10	6	0	0		
12	6	0	0		
2	6	0	0		
4	5	1	0		
6	6	0	0		
8	6	0	0		
10	5	1	0		
12	5	0	0		
12	67	0	0	*Doubled, comes* 134 *Miles run.*	

April the 26th. on *Saturday* a loom Gale, fair weather, and smooth water. The true Course I make to be $W\frac{1}{2}Nly$, the distance 134 Miles, the difference of Latitude 13 Min. and the Meridian distance 133 Min.

Declination . . 17° 08'
Zenith dist. subtr. 03   50

Lat. by Observat. 13   18

H.	K.	HK	F.	Courses.	Winds.
2	4	0	0	WbS	EbN
4	3	1	0		
6	3	1	0		
8	3	1	0		
10	3	0	0		
12	3	0	0		
2	3	0	0	W	
4	3	1	0		
6	3	1	0		
8	3	1	0		
10	4	0	0		
12	4	0	0		
12	42	0	0	*Doubled, comes* 84 *Miles run.*	

April the 27th. on *Sunday* little Wind, fair weather, and smooth water. The true Course I make to be $W\frac{1}{2}Sly$, the distance 84 Miles, the difference of Latitude 8 Min. and the Meridian distance 84 Min.

Declination . . 17° 24'
Zenith distance . 04   12

Lat. by Observation 13   12 N.

R r 2   R.

H.	K.	HK	F.	Courses.	Winds.	
2	1	o	o	W	E b N	*April* the 28th. on *Monday* very little Wind, fair weather, and smooth water. The true Course I make to be *W*, the distance 41 Miles, the difference in Latitude oo Min. and the Meridian distance 41 Min.
4	1	o	o			
6	1	o	o			
8	1	1	o			
10	2	o	o			
12	2	o	o			Declination . . 17° 40'
2	2	o	o			Zenith distance . 04 28
4	2	o¹	o			Lat. by Dead Reck. 13 12 N.
6	2	o	o			
8	2	o	o			
10	2	o	o			
12	2	o	o			
12	20	1	o	*Doubled, comes* 41 *Miles run.*		

H.	K.	HK	F.	Courses.	Winds.	
2	2	o	o	W	E	*April* the 29th. on *Tuesday* little Wind, fair weather, and smooth water about Clock we saw the Island *Barbadoes* bear West about Leagues of us. The true Course is *W*, the distance 67 Miles, the difference of Latitude oo Min. and the Meridian distance 67 Min.
4	2	o	o			
6	3	o	o			
8	3	o	o			
10	3	1	o			
12	3	1	o			
2	3	1	o			
4	3	1	o			Latitude arrived . 13° 12' N.
6	3	1.	o			
8	3	o	o.			*April* the 30th. about 10 a Clock
10	3	o	o			we came to an Anchor in the Harbor.
10	33	1	o	*Doubled, comes* 67 *M. run. in* 20 *hours.*		

The JOURNAL to the former LOG-BOOK, for a Voyage intended by Gods assistance for the Island of Barbadoes, in the Ship The Prophet Daniel of Great Yarmouth. *Anno* 1684.

Month and Days.	Lat. by Observ. D. M	The Courses Corrected.	Correct. Dist. in M. or M	Nor-thing. M or M	Sou-thing. M or M	East-ing. M or M	West-ing. M or M	Lat. by Reckon. D. M	East Longit. M or M	West Longit. M or M
Mar. 26		S b W from the Lizard.	15		15		2	49.45		
27		S W b S	142		118		79	47.47		119
28	46.00	S W b S 3° 15′ Sl	112		97		57	46.10		84
		Correction by Observa.			10		6	46.00		08
29		S b E 4° Ely	58		56	15		45.04	22	
30		S W 2° Wly	93		63		68	44.01		95
31	42.40	S W b S	101		84		56	42.37		77
April 1	41.25	S W b S	78		65		43	41.32		59
2		S W 3° 15′ Wly	111		74		83	40.18		110
3		S W 5° Sly	102		78		66	39.00		85
4	36.49	S b E 6° Ely	118		113	35		37.07	45	
		Correction by Observa.			18		9	36.45		11
5		S W b S 2° 30′ Wly	84		68		50	35.41		62
6	34.33	S W 2° Sly	88		65		60	34.36		74
7	33.28	S W	86		61		61	33.35		74
8	31.52	S W	112		93		93	32.02		110
9	31.08	S W	69		49		45	31.13		58
10		S W	153		108		108	29.25		125
11	27.38	S W	138		98		98	27.47		112
12		S W b W	143		79		115	26.28		132
13	25.06	S W b W ¼ Wly	149		70		131	25.18		140
14	23.3	S W ¼ Wly	131		83		101	23.55		112
		Correction by Observa.			18		25	23.37		2c

Month and Days.	*Lat. by Observ.* D. M.	*The courses corrected.*	*Correct. Dist. in* M. or M	*Northing.* M or M	*Southing.* M. or M	*Easting.* M or M	*Westing.* M or M	*Lat. by Reckon.* D. M.	*East Longit.* M or M	*West Longit.* M or M
April	23.37	*Brought from the other side*			—	—	—	23.37	67	1677
15		*S W* b *W* ½ *Wly*	87		41		77	22.56		84
16	21.24	*S W* ½ *Wly*	138		88		107	21.28		116
17		*S W* b *W* ½ *Wly*	125		59		110	20.29		118
18	19.21	*S W* ½ *Wly*	96		61		74	19.28		79
19		*S W* b *W*	152		84		126	18.04		133
20		*S W* b *W* ½ *Wly*	142		67		125	16.57		131
21	15.50	*S W* ½ *Wly*	90		57		70	16.00		72
22		*S W* b *W*	128		71		106	14.49		111
23	13.58	*S W* b *W* ½ *Wly*	92		43		81	14.06		82
24	13.13	*S W* b *W* ½ *Wly*	89		42		78	13.24		82
		Correction by Oserva.			11		17	13.13		17
25		*W* ½ *Sly*	136		13		135	13.00		132
26	13.07	*W* ½ *Nly*	88	9			88	13.09		92
27	13.18	*W* ½ *Nly*	134	13			133	13.22		131
28	13.12	*W* ½ *Sly*	84		8		84	13.14		81
29	13.12	*W*	41				41	13.14		41
30	13.12	*W*	67				67	13.12		67

Total ———— 3246 Min.
Subtract East Longit. —— 67
Remaineth ——— 3179 Min.

60) 3179 (52
59

Which being divided by 60, comes 52 Degrees 159 Minutes, the whole difference of Longitude Westerly.

Note, That in the Use of the *Traverse Table* I neglect the parts of a Mile when they are under 50; but when they are above 50, I add a Minute for it.

THE

THE

EXPLANATION

OF THE

JOURNAL.

IN this JOURNAL there are Eleven Columns, the first contains the Month and Days of it ; the second the Latitude by Obfervation ; the third the Courfe Corrected by the allowance for Lee-way, or for the Variation of the Compafs if there be any ; the fourth the diftance failed ; the fifth, fixth, feventh and eighth, the Northing, Southing, Eafting and Wefting, being the difference of Latitude and departure of the feveral Courfes and Diftances ; the ninth the Latitude by Dead Reckoning ; the tenth the Eaft Longitude ; the eleventh the Weft Longitude.

Here I recommend to all that will keep a good account of their Reckoning, that they keep a particular account of that which they take off the *Log-Board* every day at Noon, as in the precedent Book, commonly call'd a *Log-Book.* Now the manner of proceeding in this *Journal*, by the help of the Table of Latitude and Departure, is very facile, as follows ; the 26th of *March* at Noon, I find the Lizard to bear. N b E, and to be diftant about 5 Leagues 15 Miles, or 15 Minutes ; therefore I am to the Southward of the Lizard 15 Minutes, which 15 Minutes I place in the South Column, and that makes my Latitude 49 Degrees 45 Minutes.

The 27th Days my Courfe is S W b S, and the diftance 142 Min. to find the difference of Latitude and Departure by the Traverfe Table according to Problem 1. The difference of Latitude is 118 Minutes, and the Departure 79, becaufe the Courfe is South Wefterly I place the difference of Latitude in the South Column, and my Departure in the Weft Column 118 Minutes, or 1 Degree 58 Minutes Subtracted from 49 Degrees 45 Minutes, giveth the Latitude 47 Degrees 47 Minutes.

Hom

How to find the difference of Longitude.

TO find the difference of Longitude, in the two laſt Columns you have both Latitudes 47 Degrees 47 Minutes, and 49 Degrees 45 Minutes, the preſent Latitude and the Latitude of the day before, and the Courſe is *S W b S*, by which you may find the difference of Longitude, ſaying:

As Radius,
   to the Meridional difference of Latitude;
So is the Tangent of the Courſe,
   to the difference of Longitude.

This Queſtion being wrought by the Logarithms, you will find the difference of Longitude 119 Minutes, which place in the Weſt Column, becauſe your Courſe is Weſterly.

The 28th. day is wrought after the manner of the 27th. having the Courſe and Diſtance given, to find the difference of Latitude, Departure, and difference of Longitude, as was ſhewed before; this queſtion may alſo be wrought with eaſe by the *Sinical Quadrant*, which may ſerve in this caſe.

How to Correct your Reckoning by Obſervation of the Latitude.

ON the 28th. of *March* by good Obſervation, I find my Latitude to be 46 Degrees, whereas by my Reckoning I ſhould be in the Latitude 46 Degrees 10 Minutes, ſo that the difference is 10 Minutes more Southerly: Therefore to Correct my Latitude, I place 10 Minutes in the South Column, which ſubtracted from 46 Degrees 10 Minutes, makes my Latitude by Reckoning to agree with the Obſervation. To Correct your Departure you muſt conſider, whether the fault may be imputed to your Courſe, or to your Diſtance; if your Courſe is well Steered, and you find no Current, nor any Variation of the Compaſs, then your Diſtance is faulty; but if you cannot truſt to the Courſe Steered, then your beſt way is to Correct your Latitude only, not medling with your Departure; if there be a Current, and you know which way the Current ſets, and how faſt, then find the Difference, Latitude, and Departure of the Current, and add or ſubtract that Latitude and Departure to or from the Ships Difference, Latitude, and Departure, according as the Current doth farther or hinder your Ship in her Courſe;
but

but if you only by fome propable reafon conjecture there is a Current, then give what allowance you think meet in Difference, Latitude, and Departure, and fee if that will reform your Reckoning in your Latitude, if fo, you have gueffed well; but if it will not, it's to be fuppofed that you are miftaken in your conjecture, or that there is fome other caufe of this error in your Reckoning.

If the Compafs varies (as moft commonly it doth) then finding what the Variation is, and which way it is, you muft allow it in the Ships Courfe; but if you cannot impute the error to any of thefe, then (as I faid before) the Diftance is faulty, and this is that which ufually makes the Difference between the Latitude obferved, and the Latitude by your Reckoning; and this I take to be the Caufe of the error this 28th. day of *March*, and generally in this Reckoning.

Now to Correct your Departure and Difference of Longitude, you muft add up the North, South, Eaft and Weft Columns, from the day that you Correct, to the beginning of your *Journal Tables*; if it be the firft Correction you have made, or from the day of Correction to the laft Correction, if it be the fecond, third or fourth Correction, &c. then fubtract the fums of the North and South Columns from each other, and likewife from the Eaft and Weft, and fay by the Rule of Proportion:

As the Difference of the North and South Columns,
　to the Difference of the Eaft and Weft Columns;
So is the Diff. between the Latitudes by Obfervation and Reckoning,
　to the Difference in Departure, and for the Diff. of Longitude.

As the Diff. between the Latitudes by Obfervation and Reckoning,
　to the Meridional Difference for thofe two Latitudes;
So is the Difference in the Departure,
　to the Difference in the Longitude.

Example.

The 28th. day you will find the fum of the North Column 60, the fum of the South Column (leaving out 10 Minutes the error) 230 Min. and therefore their Difference is 230 Minutes, the fum of the Eaft Column is 00 Minutes, of the Weft Column 138 Minutes, and their Difference 138 Minutes, then the Operation by the Logarithm will be,

Co. Ar.

As the Diff. of the North and South Columns 230 *Min.* *Logar.* 7.63827
to the Diff. of the East and West Columns 138 *Min.* *Logar.* 2.13987
So is the Diff. between the two Latitudes 10 *Min.* *Logar.* 1.00000

to the Diff. in the Departure . . . . . 6 ———— 10.77814

. Place this 6, Minutes in the West Column , becaufe the fum of the West Column exceeds the fum of the East Column.

The Operation for the Difference in the Longitude.

The two Latitudes are 46 Degrees 10 Minutes , and 46 Degrees 00 Minutes , by which in the Table of Meridional Parts, you will find the Meridional Difference of Latitude 8 Minutes.

Therefore ,

Co. Ar.

As the Diff. between the two Latitudes 10 *Min.* *Logar.* 9.00000
to the Meridional Diff. of thofe Lat. 14 *Min.* ——— 1.14612
So is the Diff. in the Departure . . . 6 *Min.* ——— 0.77815

to the Diff. in the Longitude . . . . . 8 ———— 10.92427

This 8 Minutes is placed in the West Column , becaufe the Departure is Wefterly.

After the fame manner are the Corrections made in this *Journal,* upon the 4th. 14th. and 24th. of *April*, the error being fuppofed to be in the Computation of the Diftance.

If your Ship fail feveral Courfes in 24 hours, you muft find your Difference, Latitude, and Departure , by working a Traverfe according to Problem   in the ufe of the Table of Latitude, and Departure ; your Difference of Latitude, will give you what Latitude the Ship is in, then have you two Latitudes, *viz.* the Latitude the Ship was in the day before at Noon , and the Latitude the Ship is now in , by which you may find the Meridional Difference of Latitude, by the Table of Meridional Parts.

Then

Then for your Difference of Longitude, say,

As the Diff. of Latitude found by the Traverse,
to the Diff. of Latitude in Meridional Parts;
So is the Departure found by the Traverse,
to the Diff. of Longitude for that Traverse.

To find the whole Difference of Longitude of the two Parts between which you make your Voyage: Add up the Columns of East and West Longitude, and subtract the one from the other, the Remainder reduced into Degrees and Minutes, is the Difference of Longitude sought. In this *Journal* the Difference of East and West Columns of Longitude is 3179 Minutes, which reduced into Degrees and Minutes, makes 52 Degrees 59 Minutes, the Difference of Longitude between the *Lizard* and *Barbadoes*.

The End of the Fourth Book.

THE

Book V

THE
Compleat ART
OF
NAVIGATION.

THE FIFTH BOOK.

Tables Useful in Navigation.

T. IN the first place I shall insert a Table readily shewing the *Dominical* or *Sunday* Letter for ever, which I have Transcribed out of that extraordinary useful Book entituled, *Vade-Mecum*, or the *Necessary Companion*, containing: 1. Sir *Samuel Morland*'s Perpetual Almanack. (Of which this Table is part.) 2. A Computation of Years. 3. Directions for Gardening. 4. Reduction of Weights, Measures and Coins. 5. Any Number of Farthings, Half-pence, Pence, and Shillings, ready cast up. 6. Interest and Rebate of Money, the forbearance, discount and purchase of Annuities. With several other very useful things.

This Table will make the following Tables of Months (so far as they respect the Day of the Week or Month for any Year past, present, or to come) perpetual. Also by this Table you may know the Leap-years for ever.

S. How shall I know by this Table what the *Sunday* Letter will be for the Year 1686?

T. That you may throughly understand how to find the *Sunday* Letter for that Year, or any other, take notice, That if you would know the *Dominical* Letter, for any Year, whose Number is Hundreds, you must look for that Number in the Top of the Table, and either over or

A a a   next

A TABLE shewing the Dominical Letter from the first Year of Our Lord to the Year 3400, and may be continued for ever.

			DC	ED	FE	GF	AG	BA	CB
			000	100	200	300	400	500	600
			700	800	900	1000	1100	1200	1300
			1400	1500	1600	1700	1800	1900	2000
			2100	2200	2300	2400	2500	2600	2700
			2800	2900	3000	3100	3200	3300	3400
			&c.	&c.	&c.	&c.	&c.	&c.	&c.
0	28	56 84	DC	ED	FE	GF	AG	BA	CB
1	29	57 85	B	C	D	E	F	G	A
2	30	58 86	A	B	C	D	E	F	G
3	31	59 87	G	A	B	C	D	E	F
4	32	60 88	FE	GF	AG	BA	CB	DC	ED
5	33	61 89	D	E	F	G	A	B	C
6	34	62 90	C	D	E	F	G	A	B
7	35	63 91	B	C	D	E	F	G	A
8	36	64 92	AG	BA	CB	DC	ED	FE	GF
9	37	65 93	F	G	A	B	C	D	E
10	38	66 94	E	F	G	A	B	C	D
11	39	67 95	D	E	F	G	A	B	C
12	40	68 96	CB	DC	ED	FE	GF	AG	BA
13	41	69 97	A	B	C	D	E	F	G
14	42	70 98	G	A	B	C	D	E	F
15	43	71 99	F	G	A	B	C	D	E
16	44	72	ED	FE	GF	AG	BA	CB	DC
17	45	73	C	D	E	F	G	A	B
18	46	74	B	C	D	E	F	G	A
19	47	75	A	B	C	D	E	F	G
20	48	76	GF	AG	BA	CB	DC	ED	FE
21	49	77	E	F	G	A	B	C	D
22	50	78	D	E	F	G	A	B	C
23	51	79	C	D	E	F	G	A	B
24	52	80	BA	CB	DC	ED	FE	GF	AG
25	53	81	G	A	B	C	D	E	F
26	54	82	F	G	A	B	C	D	E
27	55	83	E	F	G	A	B	C	D

next under the Hundreds in each Column, you have two Letters, for the _Dominical_ Letters, those Years bring all Leap-years, the first Letter serving only for _January_ and _February_, but the latter, the rest of the Year, as the _Dominical_ Letters for 700, 1400, &c. are D C; for 100, 800, &c. E D. If you would know the _Dominical_ Letter for any Year under 100, then you must seek the Number in the Columns on the left hand of the Table, and in the first Column of _Dominical_ Letters, and right against the Number sought, you have the proper _Dominical_ Letter or Letters, as for the Years 28, 56, 84, (Leap-years) you have D C; for 1, 29, 57, 85, (first Year after Leap-years) B, &c. But if you would know the _Dominical_ Letter of a Years whose Number consists of Hundreds and Parts, as your Question for 1686 does, then you must seek the Hundreds in the Top and the Parts in the side, and underneath the Hundreds and against the Parts is the _Dominical_ Letter, as under 1600 and against 86, you have C, which is the _Dominical_ Letter for 1686, and is the second after Leap-year.

The Moveable Feasts and Terms Calculated for Sixteen Years.

Year	Shrove Sunday.	Easter Day.	Ascen. Day.	Whit Sunday.	Easter Term		Trinity Term	
					Begins,	Ends,	Begins,	Ends,
1685	March 1	April 19	May 28	June 7	May 6	June 1	June 19	July 8
86	Febr. 14	April 4	May 13	May 23	Apr. 21	May 17	June 4	June 23
87	Febr. 6	Mar. 27	May 5	May 15	Apr. 13	May 9	May 27	June 15
88	Febr. 26	April 15	May 24	June 3	May 2	May 28	June 15	July 4
89	Febr. 10	Mar. 31	May 9	May 19	Apr. 17	May 13	May 31	June 19
90	March 2	April 20	May 29	June 8	May 7	June 2	June 20	July 9
91	Febr. 23	April 12	May 21	May 31	Apr. 29	May 25	June 12	July 1
92	Febr. 7	Mar. 27	May 5	May 15	Apr. 13	May 9	May 27	June 15
93	Febr. 26	April 16	May 25	June 4	May 3	May 29	June 16	July 5
94	Febr. 18	April 8	May 17	May 27	Apr. 25	May 21	June. 8	June 27
95	Febr. 3	Mar. 24	May 2	May 12	Apr. 10	May 6	May 24	June 12
96	Febr. 23	April 12	May 21	May 31	Apr. 29	May 25	June 12	July 1
97	Febr. 14	April 4	May 13	May 23	Apr. 21	May 17	June 4	June 23
98	March 6	April 2	June 2	June 12	May 11	June 6	June 24	July 13
99	Febr. 19	April 9	May 18	May 28	Apr. 26	May 22	June. 9	June 28
1700	Febr. 11	Mar. 31	May 9	May 19	Apr. 17	May 13	May 31	June 19

Next follows Twelve Tables for the Twelve Months in the Year, shewing the Days in each Month, the fixed Feasts and remarkable Days and Things, the Southing of the Stars at Midnight, the Suns true Place in the Ecliptick, and his Declination newly Calculated for the Years 1684, 1685, 1686, 1687.

Aaa 2

The Month of *JANUARY* hath XXXI Days.

Month	Week	Festivals, remarkable Days, and Things, Southing of the Stars at Midnight, and Suns Rising and Setting.	Leap-year.		First-year.		Second-year.		Third-year.	
			☽ Pla.	☉ Dec.	☽ Pla.	☉ Dec.	☽ Pla.	☉ Dec.	☽ Pla.	☉ Dec.
			D. M.	D. M.	D. M.	D. M.	D. M.	D. M.	D. M.	D. M.
			♈.	South.	♈.	South.	♈.	South.	♈.	South.
1	A	New-years-day, or Circumcision.	21 17	21 49	22 02	21 42	21 47	21 44	21 32	21 46
2	B		22 19	21 39	23 03	21 32	22 48	21 34	23 35	21 37
3	C	4. 164⅞. General *Monk* dyed	23 20	21 29	24 04	21 21	23 45	21 24	23 35	21 26
4	D	5. 164⅔. *The Jews petition'd*	24 21	21 18	25 05	21 10	24 51	21 13	24 36	21 16
5	E	*or admittance into England, having been banish'd thence by*	25 22	21 07	26 07	20 55	25 52	21 02	25 37	21 05
6	F	Twelf-day or Epiphany.	26 23	20 56	27 08	20 47	26 53	20 50	26 38	20 54
7	G	K. *Edw. I.* 1290.	27 24	20 44	28 05	20 35	27 54	20 38	27 35	20 41
8	A	☉ Rises at 8 Sets at 4.	28 25	20 32	29 10	20 23	28 55	20 26	28 40	20 29
9	B		29 26	20 19	♈.	20 11	29 56	20 13	29 41	20 16
10	C	10. 164⅔. *A.B. Laud beheaded.*	♈.	20 07	01 12	19 57	♈.	19 59	♈.	20 03
11	D	11. 166⅔. 152 English *Slaves*	01 28	19 54	02 13	19 43	01 58	19 46	01 43	19 49
12	E	*redeem'd from Argiers, by the*	02 30	19 40	03 14	19 29	02 55	19 32	02 44	19 36
13	F	*Charity of the English Clergy.*	03 31	19 26	04 15	19 14	04 00	19 18	03 45	19 22
14	G		04 32	19 11	05 16	19 00	05 01	19 03	04 46	19 07
15	A		05 31	18 56	06 17	18 45	06 02	18 48	05 47	18 52
16	B	16.164⅔. Scots *enter England.*	06 34	18 41	07 18	18 30	07 03	18 33	06 48	18 37
17	C	17. 166⅔. *The French declar'd*	07 35	18 26	08 15	18 14	08 04	18 17	07 49	18 21
18	D	*a war with England.*	08 35	18 09	09 20	17 58	09 05	18 02	08 50	18 06
19	E	19. 164⅔. Bodmin *Fight.*	09 36	17 54	10 21	17 41	10 06	17 45	09 51	17 49
20	F		10 37	17 37	11 22	17 24	11 07	17 29	10 52	17 33
21	G	☉ Rises 39 Min. after 7,	11 38	17 21	12 23	17 08	12 08	17 12	11 53	17 16
22	A	and Sets 21 Min. after 4	12 39	17 03	13 23	16 45	13 08	16 54	12 54	16 58
23	B	Term begins.	13 40	16 45	14 24	18 31	14 05	16 36	13 55	16 40
24	C	Hydra's Heart, S. at mid.	14 41	16 27	14 24	16 13	15 06	18 18	14 55	16 23
25	D	Conv. of St. Paul.	15 41	16 10	16 25	15 55	16 11	16 00	15 56	16 05
26	E	☉ Ri. at ½ hour after 7,	16 42	15 51	17 26	15 37	17 12	15 42	16 57	15 46
27	F	and Sets ½ hour after 4	17 43	15 32	18 27	15 19	18 12	15 23	17 58	15 28
28	G	29. 166⅔. *Sir Christopher*	18 44	15 14	19 28	15 00	19 13	15 06	18 58	15 05
29	A	*Mins. set sail with his Squadron for the Downs.*	19 44	14 55	20 28	14 40	20 14	14 45	19 59	14 50
30	B	Martyrdom of KING	20 45	14 36	21 28	14 21	21 14	14 26	21 00	14 31
31	C	*CHARLES I.* 164⅔	21 46	14 16	22 30	14 02	22 15	14 06	22 00	14 11

January.	1685.			1686.			1687.			1688.			1689.			1690.		
Changes of the ☽	D.	H.	M.	D.	H.	M.	D.	H.	M.	D.	H.	M.	D.	H.	M.	D.	H.	M.
New Moon	24	03	38	13	11	59	03	07	55	22	10	12	10	22	10	29	20	17
First Quarter	31	11	18	20	20	30	10	00	24	29	00	57	17	20	20	07	16	53
Full Moon	10	03	34	28	22	12	18	00	27	07	01	12	24	22	12	14	15	38
Last Quarter	17	07	05	07	01	12	26	01	28	15	04	50	03	01	02	21	18	08

The Month of *FEBRUARY* hath XXVIII or XXIX Days.

Month Days.	Week Days.	Festivals, remarkable Days and Things, Southing of the Stars at Midnight, and Suns Rising and Setting.	Leap-year.		Firſt-year.		Second-year.		Third-year.	
			⊙ Pla.	⊙ Dec.	⊙ Pla.	⊙ Dec.	⊙ Pla.	⊙ Dec.	☉ Pla.	☉ Dec
			D. M.	D. M.	D. M.	D. M.	D. M.	D. M.	D. M.	D. M.
			♒	South.	♒	South.	♒	South.	♒	ːouth.
1	D		22 46	13 56	23 30	13 42	23 16	13 46	23 01	13 51
2	B	Candlemas, or Puri.B.V.	23 47	13 36	24 31	13 21	24 16	13 26	24 01	13 31
3	F	Lions Heart South at	24 47	13 16	25 31	13 01	25 17	13 06	25 02	13 11
4	G	Midnight.	25 48	12 56	26 32	12 40	26 17	12 46	26 02	12 50
5	A	2. 1668. Sir Chr.Mins with his Squadron arriv'd in the Downs	26 48	12 35	27 32	12 19	27 18	12 26	27 03	12 25
6	B	Lions Neck.	27 49	12 15	28 33	11 59	28 18	12 04	28 03	12 05
7	C	and chaſed the Dutch Fleet into	28 49	11 53	29 33	11 37	29 19	11 43	29 04	11 48
8	D	the Wielings.	29 50	11 32	♓ 34	11 16	♓ 19	11 22	♓ 04	11 27
9	E	8. 1659. G. Monk entred Lon.	♓ 50	11 11	01 34	10 55	01 19	11 00	01 05	11 05
10	F	☉Ri. 1' aſt.7.S.59'aſt. 4. 5. 1665. The S. attrick deſer	01 50	10 45	02 34	10 33	02 20	10 36	02 05	10 43
11	G	ted by her Fireſhip was taken, for	02 51	10 28	03 35	10 11	03 20	10 16	03 05	10 22
12	A	Lent ends.	03 51	10 06	04 35	09 49	04 20	09 54	04 06	10 00
13	B	which the Capt. of the Fireſhip	04 51	09 44	05 35	09 27	05 20	09 32	05 06	09 38
14	C	Valentine's Day.	05 51	09 22	06 35	09 05	06 21	09 10	06 06	09 16
15	D	Seely, was ſhot to death the 7th. of May following.	06 52	08 55	07 35	08 42	07 21	08 48	07 06	08 53
16	E	Loweſt of the two firſt	07 52	08 37	08 36	08 20	08 21	08 25	08 06	08 31
17	F	in ♌ Great Bear.	08 52	08 14	09 36	07 57	09 21	08 03	09 07	08 08
18	G	6. 1684. K. Charles II. dyed.	09 52	07 52	10 36	07 34	10 21	07 35	10 07	07 45
19	A	10. 1665. War decl.r'd with the Dutch.	10 52	07 29	11 36	07 11	11 21	07 17	11 07	07 22
20	B	19 and 20. 1653. Monk, Dean	11 52	07 06	12 36	06 48	12 21	06 54	12 07	07 00
21	C	and Blake fought Van Trump, and beat him; he loſing 11 Men	12 52	06 43	13 36	06 26	13 21	06 31	13 07	06 37
22	D	of War; 30 Merchant Ships, and	13 52	06 20	14 36	06 02	14 21	06 08	14 07	06 13
23	E	had 1500 Men Killed:	14 52	05 57	15 36	05 33	15 21	05 45	15 07	05 50
24	F	St. Matthias.	15 52	05 34	16 36	05 16	16 21	05 21	16 07	05 27
25	G	☉Ri.30'aſ.6. S. 30'aſ.5.	16 52	05 10	17 35	04 52	17 21	04 58	17 06	05 03
26	A	26.1665. The Rupert, a Third Rate Launched at Harwich.	17 52	04 46	18 35	04 29	18 21	04 34	18 06	04 40
27	B		18 52	04 23	19 35	04 05	19 21	04 11	19 06	04 17
28	C		19 51	04 00	20 35	03 41	20 20	03 48	20 06	03 55
29	D	Note, That in Leap-years only Feb. has 29 days.	20 51	03 36						

February.	1685.			1686.			1687.			1688.			1689.			1690.		
Changes of the ☽	D.	H.	M.	D.	H.	M.	D.	H.	M.	D.	H.	M.	D.	H.	M.	D.	H.	M.
New Moon	22	18	19	12	01	26	01	18	13	20	20	24	09	11	21	28	11	04
Firſt Quarter	No firſt Qu.			20	15	37	08	12	38	27	12	39	16	03	41	06	01	40
Full Moon	08	17	52	16	20	16	19	26	19	19	04	13	14	12	11	20	39
Laſt Quarter	15	12	50	08	07	57	22	22	47	13	22	20	01	11	58	20	11	1

The Month of *MARCH* hath XXXI Days.

Mon. days.	Week days.	*Festivals, remarkable Days and Things, Southing of the Stars at Midnight, and Suns Rising and Setting.*	*Leap-year.* ☉ Pla.	*Leap-year.* ☉ Dec.	*First-year.* ☉ Pla	*First-year.* ☉ Dec.	*Second-year.* ☉ Pla.	*Second-year.* ☉ Dec.	*Third-year.* ☉ Pla.	*Third-year.* ☉ Dec.
			D. M.	*D. M.*	*D. M.*	*D. M.*	*D. M.*	*D. M.*	*D. M.*	*D. M.*
			♓.	South.	♓.	South.	♓.	South.	♓.	South.
1	D	S. *David*, Welch Patron.	21 51	03 12	21 34	03 18	21 20	03 24	21 06	03 30
2	E	Lions Tail.	22 50	02 48	22 34	02 55	22 20	03 00	22 05	03 06
3	F	Lowest of the two last in	23 50	02 25	23 34	02 31	23 19	02 39	23 05	02 42
4	G	☐ of the Great Bear.	24 50	02 02	24 33	02 07	24 19	02 13	24 05	02 18
5	A	5. 166⅔. Charles *the Second* *Launched at* Deptford.	25 49	01 38	25 33	01 44	25 19	01 49	25 04	01 55
6	B	3. 167⅘. *Sir* Jnº Narborough	26 49	01 14	26 33	01 20	26 18	01 26	26 04	01 32
7	C	*burnt* 4 Tripoline *Men of War*	27 48	00 50	27 32	00 56	27 18	01 02	27 03	01 08
8	D	9. Upper of two lowest	28 48	00 27	28 32	00 32	28 17	00 38	28 03	00 44
9	E	in ☐ of Gr. Bear.	29 47	00 03	29 31	00 8	29 17	00 14	29 02	00 20
10	F	☽ rises at 6. and sets at 6. *in Harbour.*	♈. 47	*No.* 21	♈. 30	*No.* 16	♈. 16	*No.* 10	♈. 02	*No.* 04
11	G	7. 166⅘. London Frigot *ac-*	01 46	00 45	01 30	00 40	01 15	00 35	01 01	00 28
12	A	*cidentally Blown up.*	02 45	01 9	02 29	01 04	02 15	00 57	02 00	00 51
13	B		03 45	01 32	03 28	01 27	03 14	01 21	03 00	01 15
14	C	14. 167½. *Sir* R. Holms *took*	04 44	01 56	04 28	01 50	04 13	01 44	03 59	01 38
15	D	*Five of the* Dutch *Smirna Fleet.*	05 43	02 19	05 27	02 14	05 12	02 08	04 58	02 02
16	E		06 42	02 43	06 26	02 37	06 12	02 31	05 57	02 25
17	F	S. *Patrick*, Irish Patron.	07 41	03 06	07 25	03 00	07 11	02 55	06 56	02 49
18	G	17. 167⅔. *Royal* Charles *Laun-ched at* Portsmouth.	08 40	03 30	08 24	03 24	08 10	03 18	07 56	03 12
19	A		09 39	03 43	09 23	03 47	09 09	03 42	08 55	03 36
20	B	☉ Ri. 40' aft. 5. S. 20' aft. 6.	10 38	04 16	10 22	04 10	10 08	04 05	09 54	03 59
21	C	Last but two in Great	11 37	04 39	11 21	04 34	11 07	04 28	10 53	04 22
22	D	Bears Tail.	12 36	05 02	12 20	04 57	12 06	04 51	11 52	04 46
23	E	22. Sweepstakes *Launched at* Yarmouth.	13 35	05 25	13 19	05 20	13 05	05 14	12 51	05 08
24	F		14 34	05 49	14 18	05 43	14 04	05 37	13 50	05 32
25	G	**Lady-day** or Annuntiation of B. V.	15 33	06 11	15 17	06 05	15 03	06 00	14 49	05 55
26	A	27. 1666. Defiance *Launched*	16 32	06 34	16 16	06 28	16 02	06 23	15 47	06 17
27	B	*at* Deptford, *and Sir* R. Holms	17 31	06 56	17 15	06 51	17 00	06 46	16 46	06 40
28	C	*Knighted.*	18 30	07 18	18 14	07 13	17 59	07 08	17 45	07 02
29	D	Virgins Spike,	19 28	07 41	19 12	07 36	18 58	07 30	18 44	07 25
30	E	Last but one in Great	20 27	08 04	20 11	07 58	19 57	07 53	19 43	07 47
31	F	ears Tail.	21 26	08 25	21 10	08 19	20 55	08 15	20 41	08 10

March.	1685.			1686.			1687.			1688.			1689.			1690.		
Changes of the ☽	*D.*	*H.*	*M.*	*D.*	*H.*	*M.*	*D.*	*H.*	*M.*	*D.*	*H.*	*M.*	*D.*	*H.*	*M.*	*D.*	*H.*	*M.*
New Moon	24	09	45	13	14	02	03	04	03	21	05	15	10	21	57	29	22	08
First Quarter	02	20	04	21	11	01	10	08	49	28	02	29	17	12	17	07	08	06
Full Moon	10	05	49	29	06	56	18	12	17	06	13	58	25	16	04	14	10	14
Last Quarter	16	20	58	06	18	39	25	22	47	14	13	14	03	17	08	22	09	46

The Month of *APRIL* hath XXX Days.

M.	W.	Festivals, remarkable Days and Things, Southing of the Stars at Midnight, and Suns Rising and Setting.	Leap-year.		First-year.		Second-year.		Third-year.	
			⊙ Pla.	⊙ Dec.	⊙ Pla.	⊙ Dec.	⊙ Pla.	⊙ Dec.	⊙ Pla.	⊙ Dec.
			D. M.	D. M.	D. M.	D. M.	D. M.	D. M.	D. M.	D. M.
			♈.	North	♈.	North	♈.	North	♈.	North
1	G		22 24	08 47	22 08	08 42	21 54	08 37	21 40	08 31
2	A		23 23	09 09	23 07	09 04	22 53	08 59	22 38	08 53
3	B		24 21	09 31	24 05	09 26	23 51	09 21	23 37	09 15
4	C	Last in Gr. Bears Tail.	25 20	09 52	25 04	09 47	24 50	09 42	24 36	09 37
5	D		26 18	10 14	26 02	10 08	25 48	10 03	25 34	09 58
6	E		27 17	10 34	27 01	10 30	26 47	10 25	26 33	10 19
7	F		28 15	10 56	27 59	10 51	27 45	10 45	27 31	10 41
8	G		29 14	11 17	28 58	11 12	28 44	11 07	28 25	11 01
9	A		♉. 12	11 38	29 56	11 33	29 42	11 27	29 28	11 23
10	B	⊙ Rises at 5. Sets at 7.	01 10	11 58	♉. 55	11 53	♉. 40	11 47	♉. 26	11 43
11	C	Dragons Tail.	02 09	12 18	01 53	12 13	01 39	12 08	01 25	12 03
12	D	Arcturus.	03 07	12 38	02 51	12 33	02 37	12 28	02 23	12 23
13	E		04 05	12 58	03 49	12 53	03 35	12 48	03 21	12 43
14	F	15. 1665. *Young Everston and*	05 03	13 17	04 48	13 13	04 33	13 08	04 15	13 03
15	G	*3 Dutch Frigots after a sharp Dispute, taken by the Diamond and Mermaid.*	06 02	13 37	05 46	13 33	05 32	13 28	05 18	13 23
16	A		07 00	13 56	06 44	13 52	06 30	13 47	06 13	13 42
17	B		07 58	14 15	07 42	14 10	07 28	14 06	07 14	14 01
18	C	South Ballance.	08 56	14 34	08 40	14 29	08 26	14 24	08 12	14 20
19	D	20. 1665. *De Ruyter beaten off from Barbadoes.*	09 54	14 52	09 38	14 48	09 24	14 43	09 10	14 39
20	E		10 52	15 10	10 36	15 07	10 22	15 02	10 08	14 57
21	F	21. 1665. *His present Majesty (then Lord H. Admiral) set Sail with the whole Fleet, and the 28*	11 50	15 28	11 34	15 24	11 20	15 20	11 06	15 15
22	G	*came before the Texel.*	12 48	15 46	12 32	15 42	12 18	15 38	12 04	15 33
23	A	St. George.	13 46	16 04	13 30	15 59	13 16	15 55	13 02	15 51
24	B		14 44	16 21	14 28	16 17	14 14	16 13	14 00	16 08
25	C	St. Mark.	15 42	16 37	15 26	16 33	15 12	16 29	14 58	16 24
26	D	Upper of the two first in □ of L. Bear.	16 40	16 54	16 24	16 50	16 10	16 46	15 56	16 42
27	E	23. 1666. *Prince Rupert and Gen. Monk set forth to Command the Fleet.*	17 38	17 10	17 22	17 06	17 08	17 02	16 54	16 58
28	F		18 36	17 26	18 20	17 22	18 06	17 19	17 52	17 15
29	G		19 33	17 42	19 18	17 38	19 04	17 34	18 50	17 31
30	A	North Ballance.	20 31	17 58	20 15	17 54	20 01	17 50	19 47	17 46

April.	1685.			1686.			1687.			1688.			1689.			1690.		
Changes of the ☽	D.	H.	M.	D.	H.	M.	D.	H.	M.	D.	H.	M.	D.	H.	M.	D.	H.	M.
New Moon	23	01	35	12	04	32	01	14	08	19	15	27	09	06	32	28	06	59
First Quarter	01	09	44	20	05	13	09	02	25	26	18	14	15	22	42	05	14	02
Full Moon	08	14	58	27	15	01	17	02	18	05	06	04	23	23	00	13	21	34
Last Quarter	15	05	25	5	00	07	24	04	2	13	00	04	02	09	31	21	03	07

The Month of *MAY* hath XXXI Days.

Mon. days.	Week days.	Festivals, remarkable Days and Things, Southing of the Stars at Midnight, and Suns Rising and Setting.	Leap-year. ⊙ Pla.	☽ Dec.	First-year. ⊙ Pla.	☽ Dec.	Second-year. ⊙ Pla.	☽ Dec.	Third-year. ⊙ Pla.	☽ Dec.
			D. M.	D. M.	D. M.	D. M.	D. M.	D. M.	D. M.	D. M.
			♉	North	♉	North	♉	North	♉	North
1	B	May-day, or S. Ph. & Jas.	21 29	18 13	21 13	18 09	20 59	18 06	20 45	18 02
2	C	Brightest in the Crown.	22 27	18 28	22 11	18 24	21 57	18 21	21 43	18 17
3	D	Bright. in the Serp. neck.	23 24	18 42	23 09	18 39	22 55	18 35	22 41	18 32
4	E		24 22	18 57	24 00	18 53	23 52	18 50	23 38	18 46
5	F	☉ Ri. 12' aft. 4. S. 48'aft. 7.	25 20	19 11	25 04	19 07	24 50	19 04	24 36	19 00
6	G	8. 1660. K. Ch. 2. proclaim'd in London.	26 17	19 25	26 02	19 21	25 48	19 18	25 34	19 14
7	♈		27 15	19 38	26 59	19 34	26 45	19 31	26 31	19 28
8	B	6. 1671. Sir E. Spragg burns	28 13	19 51	27 57	19 47	27 43	19 44	27 29	19 41
9	C	Scorpions Forehead.	29 10	20 03	28 54	20 00	28 40	19 57	28 26	19 54
10	D	10 Algerine *Men of War* at Bugia.	♊ 08	20 16	29 52	20 13	29 38	20 10	29 24	20 07
11	E	12. 1641. Earl of Strafford beheaded.	01 05	20 28	♊ 49	20 25	♊ 36	20 22	♊ 22	20 19
12	F		02 03	20 39	01 47	20 37	01 33	20 34	01 19	20 31
13	G		03 00	20 51	02 44	20 48	02 31	20 45	02 17	20 43
14	A	☉ Rises at 4. Sets at 8.	03 58	21 02	03 42	20 59	03 28	20 56	03 14	20 54
15	B	Scorpion's Heart.	04 55	21 12	04 39	21 10	04 26	21 07	04 12	21 04
16	C	16. 1667. Capt. Utbert arriv'd at Plimouth from the Streights with Seven Prizes.	05 53	21 23	05 71	21 18	05 23	21 18	05 09	21 15
17	D		06 50	21 32	06 34	21 30	06 20	21 28	06 07	21 25
18	E	19. 1652. Sea Fight between the English and Dutch, Blake and Van Trump.	07 47	21 42	07 32	21 39	07 18	21 37	07 04	21 35
19	F		08 45	21 51	08 29	21 49	08 15	21 47	08 01	21 44
20	G	28. 1672. His present Majesty (then Lord High Admiral) fought the Dutch, and after 8 hours sharp dispute made them run, and pursued them to their own Coasts. In this Fight we lost the Earl of Sandwich.	09 42	22 00	09 27	21 58	09 13	21 55	08 59	21 53
21	A		10 40	22 08	10 24	22 06	10 10	22 04	09 56	22 02
22	B		11 37	22 16	11 21	22 14	11 07	22 12	10 53	22 10
23	C		12 34	22 24	12 19	22 22	12 05	22 20	11 51	22 17
24	D		13 32	22 31	13 16	22 30	13 02	22 27	12 48	22 26
25	E		14 29	22 38	14 13	22 36	13 59	22 34	13 45	22 33
26	F	☉ Ri. 47' aft. 3. S. 13' af. 8.	15 26	22 44	15 11	22 42	14 57	22 41	14 43	22 40
27	G	28. 1673. Prince Rupert beat the Dutch.	16 24	22 50	16 08	22 49	15 54	22 47	15 40	22 46
28	A	29. K. Ch. 2. Born 1630. and restor'd to His Kingdoms 166.	17 21	22 55	17 05	22 53	16 51	22 52	16 37	22 51
29	B	30. 1667. The S. David Laun-	18 18	23 00	18 02	22 59	17 48	22 57	17 35	22 56
30	C	ched in the Forest of Dean.	19 15	23 05	19 00	23 04	18 46	23 02	18 32	23 00
	D		20 13	23 09	19 56	23 08	19 43	23 07	19 29	23 06

May.	1685.			1685.			1686.			1687.			1689.			1689.		
Changes of the ☽	D.	H.	M.	D.	H.	M.	D.	H.	M.	D.	H.	M.	D.	H.	M.	D.	H.	M.
New Moon	22	18	01	11	16	09	01	12	40	18	16	10	08	13	34	27	14	31
Frst Quarter	30	03	39	19	21	18	08	19	55	26	10	41	15	11	21	04	21	00
Full Moon	07	22	48	26	23	00	16	14	04	04	21	36	23	14	19	12	15	19
Last Quarter	14	15	59	04	05	41	23	08	12	12	06	56	01	07	40	20	17	53

The Month of *JUNE* hath XXX Days.

Mon. days	Week days	Festivals, remarkable Days and Things, Southing of the Stars at Midnight, and Suns Rising and Setting.	Leap-year.				First-year.				Second-year.				Third-year.			
			⊙ Pla.		⊙ Dec.		⊙ Pla.		⊙ Dec.		☽ Pla.		☽ Dec.		☽ Pla.		☽ Dec.	
			D. M.		D. M.		D. M.		D. M.		D. M.		D. M.		D. M.		D. M.	
			♊		North		♊		North		♊		North		♊		North	
1	E	2 and 3, 1653. Monk and Dean, beat Van Trump.	21	10	23	13	20	54	23	12	20	40	23	11	20	26	23	10
2	F	3. 1665. His present Majesty then D. of Y. and Lord H. Admiral, engaged the whole Dutch Fleet, and took and destroyed above Thirty Capital Ships and 8000 Men.	22	07	23	17	21	51	23	16	21	37	23	15	21	24	23	14
3	G		23	04	23	20	22	49	23	15	22	35	23	19	22	21	23	18
4	A		24	01	23	23	23	46	23	22	23	32	23	22	23	18	23	21
5	B		24	59	23	25	24	43	23	25	24	25	23	24	24	15	23	24
6	C		25	56	23	27	25	40	23	27	25	26	23	26	25	12	23	26
7	D	1,2,3. 1666. Monk and Prince Rupert fought the Dutch, the two first days Monk with only 50 Ships fought above 80.	26	53	23	29	26	37	23	28	26	23	23	28	26	10	23	28
8	E		27	50	23	30	27	34	23	30	27	21	23	29	27	07	23	29
9	F		28	47	23	31	28	32	23	31	28	18	23	30	28	04	23	30
10	G	⊙ Ri. 41 aft. 3. S. 19. aft 8.	29	45	23	31	29	25	23	31	29	15	23	31	29	01	23	31
11	A	8. 1658. Sir H. Slingsby and Dr. Hewit beheaded.	♋	42	23	31	♋	26	23	31	♋	12	23	31	29	58	23	31
12	B	11. 1666. The Loyal London of 100 Guns Launched at Deptford, and the Warspight at Black-wall.	01	39	23	31	01	23	23	31	01	09	23	31	♋	56	23	31
13	C		02	36	23	30	02	20	23	30	02	06	23	30	01	53	23	30
14	D		03	33	23	28	03	18	23	29	03	04	23	29	02	50	23	29
15	E	12. 1660. Mr. Oughtred, the Mathematitian, dyed.	04	30	23	26	04	15	23	27	04	01	23	27	03	47	23	28
16	F		05	27	23	24	05	12	23	25	04	58	23	25	04	44	23	26
17	G	Brightest in the Harp.	06	25	23	22	06	09	23	23	05	55	23	23	05	41	23	24
18	A	14. 1645. Naseby Battel.	07	22	23	19	07	06	23	20	06	52	23	21	06	38	23	21
19	B	15. 1658. Dunkirk surrendred to the French, and by them delivered to the English.	08	19	23	16	08	03	23	17	07	49	23	17	07	36	23	18
20	C		09	16	23	12	09	00	23	12	08	47	23	14	08	33	23	15
21	D	21. 1665. Capt. Minns and Capt. Smith were Knighted as on the 29th. were Rere-Admiral.	10	13	23	08	09	58	23	09	09	44	23	09	09	30	23	11
22	E		11	10	23	03	10	55	23	04	10	41	23	05	10	27	23	06
23	F	⊙ Ri. 44 aft. 3. S. 16 aft. 8.	12	08	22	58	11	52	22	59	11	38	23	00	11	24	23	02
24	G	St. *John* Baptist.	13	05	22	53	12	49	22	54	12	35	22	55	12	21	22	57
25	A	24. Admiral Tiddiman, Capt. Spragg, Capt. Jordan, and Capt. Cuttings.	14	02	22	47	13	46	22	48	13	32	22	50	13	19	22	51
26	B		14	59	22	39	14	43	22	41	14	30	22	43	14	16	22	44
27	C	25. 1667. Sir John Harman with 16 Sail engaged about 30	15	56	22	33	15	41	22	35	15	27	22	36	15	13	22	38
28	D		16	53	22	26	16	38	22	28	16	24	22	30	16	10	22	32
29	E	St. *Peter.*	17	51	22	19	17	35	22	21	17	21	22	22	17	07	22	24
30	F	French Men of War, and destroy'd most of them.	18	48	22	11	18	32	22	13	18	18	22	15	18	04	22	17

June.	1685.			1686.			1687.			1688.			1689.			1690.		
Changes of the ☽	D.	H.	M.	D.	H.	M.	D.	H.	M.	D.	H.	M.	D.	H.	M.	D.	H.	M.
New Moon	21	07	49	10	09	16	29	01	29	17	08	05	06	22	08	25	21	21
First Quarter	28	23	10	18	11	41	07	12	56	25	03	34	14	01	51	03	06	30
Full Moon	06	05	00	25	05	55	14	22	30	03	09	56	22	04	18	11	04	13
Last Quarter	13	04	40	02	12	50	21	12	52	10	05	46	29	13	56	19	05	43

The Month of *JULY* hath XXXI Days.

Mon. days	Fest. day	Festivals, remarkable Days, and Things, Something of the Stars at Midnight, and Suns Rising and Setting.	Leap-year.				First-year.				Second-year.				Third-year.			
			☉ Pla.		☉ Dec.		☉ Pla.		☉ Dec.		☉ Pla.		☉ Dec.		☉ Pla.		☉ Dec	
			D.	M.	D.	M.	D.	M.	D.	M.	D.	M.	D.	M.	D.	M.	D.	M.
			♋		North		♋		North		♋		North		♋		North	
1	G		19	45	22	03	19	25	22	05	19	15	22	07	19	02	22	09
2	A		20	42	21	54	20	26	21	57	20	15	21	59	19	59	22	01
3	B		21	39	21	46	21	24	21	48	21	10	21	50	20	56	21	52
4	C	Brighteft between the	22	37	21	37	22	21	21	39	22	07	21	41	21	53	21	43
5	D	Eagles fhoulders.	23	34	21	27	23	18	21	30	23	04	21	32	22	50	21	34
6	E	5. 1643. *Mr. Tomkins and Mr. Chaloner Executed*	24	31	21	17	24	15	21	20	24	02	21	22	23	48	21	24
7	F	7. 1648. *Francis Lord Villers flain.*	25	28	21	07	25	13	21	09	24	55	21	12	24	45	21	14
8	G	8. 1640. *D. of Gloucefter Born*	26	26	20	56	26	10	20	59	25	56	21	01	25	42	21	04
9	A		27	23	20	45	27	07	20	48	26	53	20	51	26	39	20	53
10	B	☉ Rifes at 4. Sets at 8.	28	20	20	34	28	04	20	37	27	51	20	38	27	37	20	42
11	C	10. 1669. *The Sea ebb'd & flow'd about Weymouth 7 times in 3 ho.*	29	17	20	22	29	02	20	25	28	48	20	28	28	34	20	31
12	D	12. 1642. *Earl of Effex Voted General by the Parliament againft the King.*	♌.	15	20	10	29	59	20	13	29	45	20	16	29	31	20	19
13	E		01	12	19	58	♌.	56	20	01	♌.	42	20	04	♌.	29	20	07
14	F	25, 26. 1666. *The Prince and Monk beat the Dutch Fleet into their Harbours, and on Aug. 7.*	02	09	19	45	01	54	19	48	01	40	19	51	01	26	19	54
15	G		03	07	19	32	02	51	19	35	02	37	19	38	02	23	19	41
16	A	*Sir Robert Holms burns above*	04	04	19	19	03	48	19	22	03	34	19	25	03	21	19	28
17	B	150 *Sail of Ships in the Fly, and*	05	01	19	05	04	46	19	08	04	32	19	11	04	18	19	14
18	C	*Swans Tail.*	05	55	18	51	05	43	18	54	05	25	18	58	05	15	19	01
19	D	*Dog-days begin.*	06	56	18	37	06	40	18	40	06	27	18	45	06	13	18	47
20	E	☉R. 1. 14. aft. 4. S. 46. aft. 7 *the Town of Bandaris on the ifland Scheling.*	07	54	18	22	07	38	18	25	07	24	18	29	07	10	18	32
21	F		08	51	18	07	08	35	18	11	08	21	18	14	08	08	18	18
22	G		09	48	17	50	09	33	17	55	09	19	17	59	09	05	18	03
23	A		10	46	17	36	10	30	17	40	10	16	17	44	10	02	17	47
24	B		11	43	17	20	11	28	17	24	11	14	17	28	11	00	17	32
25	C	*St. James.*	12	41	17	04	12	25	17	08	12	11	17	12	11	57	17	16
26	D		13	38	16	48	13	24	16	52	13	09	16	56	12	55	17	00
27	E	29. 1668. *The Edgar of 1100 Tun Launched at Briftol.*	14	36	16	31	14	20	16	35	14	06	16	40	13	52	16	44
28	F	29 and 30 1653. *A Sea Fight betwixt Monk and Van Trump,*	15	24	16	14	15	18	16	18	15	04	16	22	14	50	16	27
29	G	*the Dutch beaten, Van Trump*	16	31	15	57	16	15	16	02	16	01	16	06	15	48	16	10
30	A	*flain, and 33 Ships funk.*	17	29	15	39	17	13	15	43	16	59	15	47	16	45	15	51
31	B		18	26	15	21	18	11	15	25	17	57	15	30	17	43	15	34

July.	1685.			1686.			1687.			1688.			1689.			1690.		
Changes of the ☽	D.	H.	M.	D.	H.	M.	D.	H.	M.	D.	H.	M.	D.	H.	M.	D.	H.	M.
New Moon	20	21	22	09	23	59	28	16	04	16	19	47	06	05	28	25	04	42
Firft Quarter	28	04	50	17	21	57	07	04	51	24	20	34	17	13	54	02	18	27
Full Moon	05	13	04	24	12	40	14	06	24	02	20	38	21	16	58	10	21	28
Laft Quarter	12	19	41		11		20	20	12	09	15	10	28	18	07	18	14	45

The Month of AUGUST hath XXXI Days.

Mon. days		Festivals, remarkable Days and Things, Something of the Stars at Midnight, and Suns Rising and Setting.	Leap-year.				First-year.				Second-year.				Third-year.			
			☉ Pla.		☉ Dec.		☉ Pla.		☉ Dec.		☉ Pla.		☉ Dec.		☉ Pla.		☉ Dec.	
			D. M.		D. M.		D. M.		D. M.		D. M.		D. M.		D. M.		D. M.	
			♌		North		♌		North		♌		North		♌		North	
1	C	Lammas.	19	24	15	03	19	08	15	07	18	54	15	12	18	40	15	10
2	D	Pegasus Mouth.	20	22	14	45	20	06	14	50	19	52	14	54	19	38	14	58
3	E	11. 1673. Prince Rupert beat the Dutch on their own Coasts,	21	19	14	26	21	04	14	31	20	50	14	35	20	36	14	40
4	F	but loft Sir E. Spragg in the	22	17	14	08	22	01	14	13	21	47	14	17	21	33	14	21
5	G	Engagement.	23	15	13	49	22	59	13	54	22	45	13	58	22	31	14	03
6	A	15. 1667. De Ruyter set upon the Virginia Fleet in Foy, but	24	13	13	30	23	57	13	35	23 · 43		13	39	23	29	13	44
7	B	without success.	25	10	13	11	24	55	13	15	24	41	13	20	24	27	13	25
8	C	16. 1652. Sea Fight between the English and Dutch, Sir George	26	08	12	52	25	52	12	56	25	38	13	00	25	24	13	05
9	D	Ayscue and De Ruyter.	27	06	12	33	26	50	12	36	26	36	12	41	26	22	12	46
10	E	16. 1665. The Dutch attaqu'd	28	04	12	12	27	48	12	16	27	34	12	21	27	20	12	26
11	F	by Rere-Admiral Tidduman in	29	02	11	52	28	46	11	57	28	32	12	01	28	18	12	06
12	G	Bergen, without success.	30	00	11	31	29	44	11	36	29	30	11	41	29	16	11	46
13	A	17. 1670. Capt. Beach run bers destroy'd Six stout Alge-	♍.	58	11	11	♍.	42	11	16	♍.	28	11	21	♍.	14	11	26
14	B	rine Ships.	01	56	10	50	01	40	10	55	01	26	11	00	01	12	11	05
15	C	☉ Rises at 5. Sets 7.	02	54	10	29	02	38	10	34	02	24	10	35	02	10	10	44
16	D	18. 1648. The Great Seal bro- ken by the Rebels.	03	52	10	08	03	36	10	13	03	22	10	18	03	08	10	23
17	E	20. 1658. Gravelin deliver'd	04	50	09	47	04	34	09	52	04	20	09	57	04	06	10	03
18	F	to the French.	05	48	09	26	05	32	09	31	05	18	09	36	05	04	09	41
19	G	22. 1642. King Charles I. set up his Standard.	06	46	09	04	06	30	09	09	06	16	09	15	06	02	09	20
20	A	23. 1628. Duke of Bucking-	07	44	08	43	07	28	08	48	07	14	08	53	07	00	08	58
21	B	Fomahant.	08	42	08	21	08	27	08	26	08	12	08	32	07	58	08	36
22	C	ham stabb'd by Felton.	09	41	07	59	09	25	08	04	09	11	08	9	08	57	08	15
23	D	24. 1667. Six of our Fleet fought a Squadron of Dutch and took	10	39	07	37	10	23	07	42	10	09	07	47	09	55	07	52
24	E	St. Bartholomew.	11	37	07	14	11	21	07	20	11	07	07	26	10	53	07	30
25	F	First in Pegasus Wing.	12	35	06	52	12	19	06	58	12	05	07	03	11	51	07	09
26	G	3 Men of War, and a Merchant man; and that day Peace pro-	13	34	06	30	13	18	06	35	13	04	06	41	12	50	06	46
27	A	claim'd.	14	32	06	8	14	16	06	13	14	02	06	18	13	48	06	24
28	B	Dog-days end.	15	30	05	45	15	15	05	51	15	00	05	56	14	46	06	01
29	C	28. 1648. Sir Charles Lucas	16	29	05	22	16	13	05	28	15	59	05	33	15	45	05	38
30	D	and Sir George Lisle shot to	17	27	04	59	17	11	05	05	16	57	05	11	16	43	05	15
31	E	death at Colchester.	18	26	04	37	18	10	04	42	17	56	04	48	17	42	04	53

August.	1685.			1686.			1687.			1688.			1689.			1690.		
Changes of the ☽	D.	H.	M.	D.	H.	M.	D.	H.	M.	D.	H.	M.	D.	H.	M.	D.	H.	M.
New Moon	19	10	10	08	15	07	27	07	59	5	09	10	04	15	21	23	13	21
First Quarter	26	09	19	16	04	15	05	18	38	23	12	34	12	11	13	01	01	22
Full Moon	03	21	41	22	20	50	12	15	11		20	04	17	09	11	45
Last Quarter	11	12	37	30	04	02	19	02	56	07	20	45	26	22	38	16	21	32

The Month of *SEPTEMBER* hath XXX Days.

Moon days	Week days	Festivals, remarkable Days and Things, Southing of the Stars at Midnight, and Suns Rising and Setting.	Leap-year.				First-year.				Second-year.				Third-year.			
			☉ Pla.		☉ Dec.		☉ Pla.		☉ Dec.		☉ Pla.		☉ Dec.		☽ Pla.		☽ Dec.	
			D. M.	D. M.	D. M.	D. M.	D. M.	D. M.	D. M.	D. M.	D. M.	D. M.	D. M.	D. M.	D. M.	D. M.	D. M.	D. M.
			♍.		North	♍.		North	♍.		North	♍.		North
1	F	2. 1666. London Burnt.	19 24	04 13	19 08	04 19	18 54	04 25	18 40	04 30
2	G	3. 1658. Oliv. Cromwell dyed.	20 23	03 50	20 07	03 57	19 53	04 01	19 39	04 07
3	A	3, 4. 1665. The English Fleet	21 22	03 27	21 06	03 33	20 51	03 39	20 37	03 44
4	B	under the Command of the Earl of Sandwich, took 4 Dutch	22 20	03 04	22 04	03 10	21 50	03 15	21 36	03 21
5	C	Men of War and two East India ships, with the loss only of a small vessel called the Hecto.	23 19	02 41	23 03	02 46	22 48	02 52	22 34	02 58
6	D		24 17	02 17	24 01	02 23	23 47	02 29	23 33	02 34
7	E	5. 1652. French Fleet beaten	25 16	01 55	25 00	02 00	24 46	02 06	24 32	02 11
8	F	by Blake.	26 15	01 31	25 59	01 37	25 45	01 42	25 30	01 48
9	G	6 1669. Sir T. Allen with a Fleet before Argiers declares war.	27 14	01 08	26 58	01 14	26 43	01 19	26 29	01 25
10	A	Andromeda's Head.	28 13	00 44	27 57	00 50	27 42	00 55	27 28	01 01
11	B	Tip of Pegasus's Wing.	29 11	00 21	28 55	00 26	28 41	00 32	28 27	00 38
12	C	9. 1664. Part of our Fleet en-	♎. 10	Sou. 03	29 54	00 03	29 40	00 08	29 26	00 14
13	D	☉ Rises at 6. Sets at 6.	01 09	00 27	♎. 53	Sou. 21	♎. 39	Sou. 25	♎. 25	Sou. 10
14	E	gaged 18 Dutch Ships, took	02 08	00 50	01 52	00 45	01 38	00 35	01 24	00 33
15	F	most of them (whereof 4 Men of War) and about 1000 Prisoners.	03 07	01 14	02 51	01 08	02 37	01 02	02 23	00 57
16	G	13. 1660. The Duke of Glocester dyed.	04 06	01 37	03 50	01 32	03 36	01 26	03 22	01 20
17	A	17. 1643. Auburn Fight.	05 05	02 01	04 49	01 55	04 35	01 50	04 21	01 44
18	B	18. 1666. Capt. De Roch and	06 04	02 24	05 48	02 18	05 34	02 13	05 20	02 07
19	C	a French Ship of 54 Brass	07 04	02 48	06 47	02 42	06 33	02 36	06 19	02 30
20	D	Guns taken.	08 03	03 11	07 47	03 05	07 32	03 00	07 18	02 54
21	E	23. 1642. Worcester Fight. St. Matthew.	09 02	03 35	08 46	03 28	08 31	03 23	08 17	03 18
22	F	Pole-star.	10 01	03 58	09 45	03 52	09 31	03 46	09 16	03 41
23	G	25. 1658. Queen Consort Born.	11 00	04 21	10 44	04 15	10 30	04 10	10 16	04 04
24	A	26. 1670. Capt. Pierce and his Lieutenant executed for cow-	12 00	04 44	11 44	04 39	11 29	04 33	11 15	04 28
25	B	ardly losing the Saphire Frigot.	12 59	05 08	12 43	05 02	12 29	04 56	12 14	04 51
26	C	Southermost in Andro-	13 59	05 31	13 42	05 25	13 28	05 20	13 14	05 14
27	D	meda's Girdle.	14 58	05 54	14 42	05 48	14 27	05 43	14 13	05 37
28	E	30. 1669. The St. Michael of	15 57	06 17	15 41	06 11	15 27	06 06	15 12	06 00
29	F	Michaelmas or S. Michael	16 57	06 40	16 41	06 34	16 26	06 29	16 12	06 23
30	G	100 Guns Launched at Portsmouth.	17 56	07 03	17 40	06 57	17 26	06 52	17 11	06 46

September.	1685.			1686.			1687.			1688.			1689.			1690.		
Changes of the ☽	D.	H.	M.	D.	H.	M.	D.	H.	M.	D.	H.	M.	D.	H.	M.	D.	H.	M.
New Moon	17	21	53	07	05	58	26	00	25	14	00	49	03	03	39	22	00	32
First Quarter	24	14	57	14	13	15	04	06	02	22	03	42	11	05	17	29	21	12
Full Moon	02	08	39	21	06	46	10	21	30	28	23	00	18	17	46	08	01	42
Last Quarter	10	00	55	28	02	32	17	21	16	04	04	28	25	08	13	15	03	08

The Month of *OCTOBER* hath XXXI Days.

Mon. day	Week days	Festivals, remarkable Days, and Things, Southing of the Stars at Midnight, and Suns Rising and Setting.	Leap-year ☉ Pla.	☉ Dec.	First-year ☉ Pla.	☉ Dec.	Second-year ☉ Pla.	☉ Dec.	Third-year ☉ Pla.	☉ Dec.
			♎ D. M.	South. D. M.	♎ D. M.	South. D. M.	♎ D. M.	South. D. M.	♎ D. M.	South. D. M.
1	A	2. 1662. *Capt.* Mins *took Saint* Jago, *with the Castle and Block-houses, and six Ships.*	18 56	07 26	18 40	07 20	18 25	07 15	18 11	07 09
2	B		19 56	07 48	19 39	07 43	19 25	07 37	19 10	07 32
3	C	2. 1662. *Cuba attack'd by the*	20 55	08 11	20 39	08 05	20 25	08 00	20 10	07 54
4	D	*English from Jamaica, the*	21 55	08 33	21 39	08 28	21 24	08 22	21 10	08 17
5	E	*Spaniards routed, and some Towns destroyed.*	22 55	08 56	22 38	08 50	22 24	08 45	22 09	08 39
6	F	3. 1656. *The Thames ebb'd and flow'd twice in 3 Hours.*	23 54	09 18	23 38	09 12	23 24	09 07	23 09	09 02
7	G		24 54	09 40	24 38	09 35	24 23	09 29	24 09	09 24
8	A	5. 1662. *Sir* John Lawson *concluded a Peace with Argiers.*	25 54	10 02	25 38	09 56	25 23	09 51	25 09	09 46
9	B	9. 1666. *War declar'd against*	26 54	10 23	26 57	10 18	26 23	10 13	26 08	10 08
10	C	Denmark.	27 54	10 45	27 37	10 40	27 23	10 34	27 08	10 29
11	D	South foot of Andromeda	28 54	11 07	28 37	11 01	28 23	10 56	28 08	10 51
12	E	☉ Rifes at 7. Sets at 5.	29 53	11 28	29 37	11 23	29 23	11 17	29 08	11 12
13	F	13. 1633. *His present Majesty*	m. 53	11 49	m. 37	11 44	m. 23	11 39	m. 08	11 34
14	G	K. J A M E S *the Second, Born.*	01 53	12 10	01 37	12 05	01 23	12 00	01 08	11 55
15	A	15. 1651. *Earl of Derby be-headed.*	02 53	12 31	02 37	12 25	02 23	12 21	02 08	12 16
16	B		03 54	12 51	03 37	12 46	03 23	12 41	03 08	12 36
17	C		04 54	13 11	04 37	13 06	04 23	13 02	04 08	12 57
18	D	St. **Luke.**	05 54	13 32	05 37	13 27	05 23	13 22	05 08	13 17
19	E	23. 1641. *Irish Rebellion. And*	06 54	13 51	06 38	13 47	06 23	13 42	06 08	13 37
20	F	Edgehill *Battel*, 1642. *And*	07 54	14 11	07 38	14 07	07 23	14 02	07 05	13 57
21	G	Mardike *surrendred to the* Fren h, *and by them after de-*	08 54	14 31	08 38	14 26	08 23	14 21	08 05	14 17
22	A	*liver'd to the* English, 1657.	09 55	14 50	09 38	14 46	09 24	14 41	09 05	14 36
23	B	**Term begins.**	10 55	15 09	10 39	15 05	10 24	15 00	10 05	14 55
24	C	26. 1664. *The Royal* Catherine	11 55	15 28	11 39	15 24	11 24	15 19	11 05	15 14
25	D	*Launched.*	12 56	15 46	12 39	15 42	12 25	15 37	12 10	15 33
26	E	Whales Jaw.	13 56	16 05	13 40	16 00	13 25	15 55	13 10	15 51
27	F	28. 1652. De Wit, *beaten by*	14 56	16 22	14 40	16 18	14 25	16 14	14 11	16 05
28	G	Blake.	15 57	16 40	15 40	16 36	15 26	16 31	15 11	16 27
29	A	Perseus bright side.	16 57	16 57	16 41	16 53	16 26	16 49	16 12	16 44
30	B	30. 1664. *Sir* Thoms Allen	17 58	17 15	17 41	17 11	17 27	17 06	17 12	17 01
31	C	*concluded a Peace with* Argiers.	18 58	17 31	18 42	17 27	18 27	17 23	18 13	17 10

October.	1685. D. H. M.	1686. D. H. M.	1687. D. H. M.	1688. D. H. M.	1689. D. H. M.	1690. D. H. M.
Changes of the ☽	D. H. M.	D. H. M.	D. H. M.	D. H. M.	D. H. M.	D. H. M.
New Moon	17 08 45	06 20 01	25 10 50	13 18 40	02 18 55	21 13 10
First Quarter	23 22 50	13 19 18	03 14 10	21 16 59	10 23 00	29 15 19
Full Moon	01 23 16	20 19 22	10 06 49	28 08 12	18 00 40	07 14 26
Last Quarter	10 01 31	28 18 17	17 14 53	05 16 56	24 14 21	14 10 35

The Month of *NOVEMBER* hath XXX Days.

Mon. day.	Week day.	Festivals, remarkable Days and Things, Southing of the Stars at Midnight, and Suns Rising and Setting.	Leap-year. ☉ Pla. ☉ Dec. ♍. South.		First-year. ☉ Pla. ☉ Dec. ♍. South.		Second-year. ☉ Pla. ☉ Dec. ♍. South.		Third-year. ☉ Pla. ☉ Dec. ♍. South.	
1	D	All-Saints.	19 59 17 47	19 42	17 43 19 28	17 39	19 13	17 35		
2	E		21 00 18 03	20 43	17 59 20 28	17 56	20 14	17 52		
3	F	3. 1640. Long-Parliament began.	22 00 18 19	21 44	18 15 21 29	18 11	21 14	18 08		
4	G		23 01 18 35	22 44	18 31 22 30	18 27	22 15	18 24		
5	A	5. 1605. Gunpowder-Plot.	24 01 18 50	23 45	18 46 23 30	18 42	23 15	18 39		
6	B	☉ Ri. 16' aft. 7 S. 14' aft. 4	25 02 19 05	24 46	19 01 24 31	18 58	24 16	18 54		
7	C		26 03 19 19	25 46	19 16 25 32	19 12	25 17	19 09		
8	D	8. 1666. The Vice-Admiral of	27 04 19 34	26 47	19 30 26 32	19 27	26 18	19 23		
9	E	Denmark of 52 Guns reign.	28 04 19 48	27 48	19 44 27 33	19 41	27 18	19 38		
10	F	9. 1664. His present Majesty then D. of York, went to Ports-	29 05 20 01	28 49	19 58 28 34	19 54	28 19	19 51		
11	G	mouth to command the Fleet and arriv'd at St. Helens the	♐. 06 20 14	29 49	20 11 29 35	20 08	29 20	20 04		
12	A	11th. and return'd to Whitehall	01 07 20 27	♐. 50	20 24 ♐. 35	20 21	♐. 20	20 17		
13	B	Dec. 4.	02 08 20 39	01 51	20 36 01 36	20 32	01 22	20 30		
14	C	12. 1642. Brainford Fight.	03 08 20 51	02 52	20 48 02 37	20 45	02 22	20 42		
15	D	Q. Catherine Born.	04 09 21 03	03 53	21 00 03 38	20 57	03 23	20 54		
16	E		05 10 21 14	04 54	21 11 04 39	21 09	04 24	21 06		
17	F	Bulls Eye. Q. El. B. 1534.	06 11 21 25	05 55	21 22 05 40	21 20	05 25	21 17		
18	G		07 12 21 35	06 56	21 33 06 41	21 30	06 26	21 28		
19	A	K. Charles I. Born.	08 13 21 45	07 57	21 43 07 42	21 40	07 27	21 38		
20	B		09 14 21 54	08 58	21 53 08 43	21 50	08 28	21 48		
21	C	19. 1671. Sir R. Spragg made Peace with Algiers.	10 15 21 59	09 59	22 02 09 44	21 59	09 29	21 57		
22	D		11 16 22 08	11 00	22 11 10 45	22 08	10 30	22 06		
23	E	☉ Ri. 10' aft. 8. S. 50' aft. 3	12 17 22 17	12 01	22 19 11 46	22 17	11 31	22 15		
24	F		13 18 22 25	13 02	22 26 12 47	22 25	12 32	22 23		
25	G	29. 1652. Blake beaten by Van Trump.	14 19 22 33	14 03	22 34 13 48	22 33	13 33	22 31		
26	A	The Goat.	15 20 22 40	15 04	22 41 14 49	22 40	14 34	22 38		
27	B	Orion's left Foot.	16 22 22 47	16 05	22 48 15 50	22 47	15 35	22 45		
28	C	End of the Bulls Horn.	17 23 22 52	17 06	22 54 16 51	22 52	16 36	22 51		
29	D	28. Termends.	18 24 22 58	18 07	23 00 17 52	22 58	17 37	22 57		
30	E	St. Andrew.	19 25 23 03	19 08	23 04 18 53	23 03	18 39	23 02		

November.	1685. D. H. M.	1686. D. H. M.	1687. D. H. M.	1688. D. H. M.	1689. D. H. M.	1690. D. H. M.
Changes of the D						
New Moon	15 20 16	05 09 16	24 08 38	12 12 54	01 12 35	20 07 34
First Quarter	22 10 52	12 02 13	02 00 22	04 08 09	09 15 35	28 09 51
Full Moon	30 10 24	19 11 24	08 18 26	26 19 05	16 10 30	06 00 25
Last Quarter	08 19 42	27 14 30	16 10 55	04 02 51	23 05 22	12 16 52

The Month of *DECEMBER* hath XXXI Days.

Mon. days	Week days	Festivals, remarkable Days and Things, Southing of the Stars at Midnight, and Suns Rising and Setting.	Leap-year. ☉ Pla.	☉ Dec.	First-year. ☉ Pla.	☉ Dec.	Second-year. ☉ Pla.	☉ Dec.	Third-year. ☉ Pla.	☉ Dec.
			D. M.	D. M.	D. M.	D. M.	D. M.	D. M.	D. M.	D. M.
			⊋. South.	⊋. South.	⊋. South.	⊋. South.
1	F	First in Orion's Belt.	20 26	23 08	20 09	23 05	19 55	23 08	19 40	23 07
2	G		21 27	23 12	21 11	23 13	20 56	23 12	20 41	23 11
3	A	Last in Orion's Belt.	22 28	23 16	22 12	23 17	21 57	23 16	21 42	23 15
4	B	6. 1667. *The Resolution, a 3d.*	23 30	23 20	23 13	23 21	22 58	23 20	22 43	23 19
5	C	*Rate, Launched at Harwich.*	24 31	23 23	24 14	23 24	23 59	23 23	23 44	23 22
6	D	Orion and Auriga's right	25 32	23 25	25 15	23 26	25 00	23 25	24 46	23 25
7	E	shoulders.	26 35	23 27	26 17	23 28	26 02	23 27	25 47	23 27
8	F	11. 1640. *Petition against*	27 34	23 29	27 18	23 29	27 03	23 29	26 48	23 29
9	G	*B.ſhops.*	28 36	23 30	28 19	23 30	28 04	23 30	27 49	23 30
10	A	12. 1666. *Capt. Robinſon took and ſunk Three Dutch Men of War near the Texel.*	29 37	23 31	29 20	23 31	29 05	23 31	28 50	23 31
11	B	22. 1668. *The Nonſuch Launched at Portſmouth.*	♈. 38	23 31	♈. 21	23 31	♈. 07	23 31	29 52	23 31
12	C		01 39	23 31	01 23	23 31	01 08	23 31	♈. 53	23 31
13	D	Foot of the great Dog.	02 40	23 30	02 24	23 30	02 05	23 30	01 54	23 30
14	E	Gemini's bright Foot.	03 42	23 29	03 25	23 28	03 10	23 29	02 55	23 29
15	F	22. 1672. *Tabago taken from the Dutch.*	04 43	23 27	04 26	23 27	04 11	23 27	03 57	23 28
16	G	☉Ri.18' af.8.S. 42' af.3.	05 44	23 25	05 28	23 24	05 13	23 25	04 58	23 26
17	A	Mouth of the great Dog.	06 45	23 22	06 29	23 21	06 14	23 22	05 59	23 23
18	B	26. 1664. *Admiral Allen took four Prizes from the Dutch in the Streights, and much about*	07 47	23 19	07 30	23 18	07 15	23 19	07 00	23 20
19	C		08 48	23 16	08 31	23 15	08 16	23 16	08 02	23 16
20	D		09 49	23 12	09 33	23 11	09 18	23 12	09 03	23 12
21	E	St. Thomas.	10 50	23 07	10 34	23 06	10 19	23 07	10 04	23 08
22	F	*this time His Majeſty's Fleet off 'ortſmouth took near One hundred more, ſmall and great.*	11 52	23 04	11 35	22 59	11 20	23 01	11 05	23 02
23	G	23.1644.*S. A. Carew beheaded.*	12 53	22 58	12 36	22 54	12 21	22 55	12 07	22 57
24	A		13 54	22 51	13 37	22 47	13 23	22 49	13 08	22 50
25	B	Chriſtmas, or Na. of Chr.	14 55	22 45	14 39	22 40	14 24	22 42	14 09	22 43
26	C	St. Steven.	15 57	22 38	15 40	22 33	15 25	22 35	15 10	22 36
27	D	St. John the Evangeliſt.	16 58	22 31	16 41	22 26	16 26	22 28	16 11	22 29
28	E	Innocents.	17 59	22 24	17 42	22 18	17 27	22 20	17 13	22 22
29	F	27. Head of Caſtor.	19 00	22 16	18 44	22 05	18 29	22 11	18 14	22 13
30	G	Head of Pollux.	20 01	22 07	19 45	22 00	19 30	22 02	19 15	22 05
31	A		21 03	21 58	20 46	21 51	20 31	21 54	20 16	21 56

December.	1685.			1686.			1687.			1688.			1689.			1690.		
Changes of the ☽	D.	H.	M.	D.	H.	M.	D.	H.	M.	D.	H.	M.	D.	H.	M.	D.	H.	M.
New Moon	15	05	27	04	21	04	23	22	30	11	06	08	01	01	11	20	02	42
Firſt Quarter	22	02	20	11	13	05	09	11	17	19	13	02	09	06	00	28	05	05
Full Moon	30	04	56	19	05	15	08	08	31	26	07	42	15	22	02	05	13	11
Laſt Quarter	08	08	05	27	09	21	16	08	05	04	06	15	22	22	42	12	04	25

A TABLE *shewing what time* Aldebaran *or the* Bulls Eye *comes to the Meridian throughout the Year.*

Days.	Janua.	Febru.	March	April.	May.	June.	July.	Aug.	Sept.	Octo.	Nov.	Dec.
	Even.	Even.	Even.	Even.	Even.	Morn.	Morn.	Morn.	Morn.	Morn.	Morn.	Even.
1	8 42	5 35	4 49	2 56	1 03	10 58	8 54	6 52	4 58	3 09	1 10	11 02
2	8 38	5 31	4 45	2 52	0 55	10 54	8 50	6 48	4 54	3 05	1 06	10 57
3	8 34	5 27	4 41	2 48	0 55	10 50	8 46	6 44	4 51	3 02	1 02	10 52
4	8 30	5 23	4 38	2 44	0 51	10 46	8 42	6 40	4 47	2 58	0 58	10 48
5	8 26	5 19	3 34	2 41	0 47	10 41	8 38	6 37	4 44	2 55	0 54	10 43
6	8 21	5 15	4 31	2 37	0 43	10 37	8 34	6 33	4 40	2 51	0 50	10 39
7	8 17	5 11	4 27	2 33	0 39	10 33	8 30	6 29	4 36	2 47	0 46	10 35
8	8 13	5 07	4 24	2 30	0 35	10 29	8 26	6 26	4 33	2 43	0 41	10 30
9	8 08	5 03	4 20	2 26	0 31	10 25	8 22	6 22	4 29	2 39	0 37	10 26
10	8 04	5 00	4 16	2 23	0 27	10 21	8 18	6 15	4 26	2 36	0 32	10 21
11	8 00	4 56	4 13	2 19	0 23	10 17	8 14	6 15	4 22	2 32	0 28	10 17
12	7 55	4 52	4 09	2 15	0 19	10 13	8 10	6 11	4 18	2 28	0 24	10 12
13	7 51	4 48	4 05	2 11	0 15	10 09	8 06	6 07	4 15	2 24	0 19	10 08
14	7 47	4 44	4 02	2 07	0 11	10 05	8 02	6 03	4 11	2 20	0 15	10 03
15	7 43	4 41	3 58	2 04	0 07	10 00	7 58	6 00	4 08	2 17	0 10	9 58
16	7 39	4 37	3 54	2 00	0 03	9 56	7 54	5 56	4 04	2 13	0 06	9 53
17	7 35	4 33	5 51	1 56	11m59	9 52	7 50	5 52	4 00	2 09	0 02	9 49
18	7 31	4 29	4 47	1 52	11 55	9 48	7 46	5 49	3 57	2 05	11 58	9 44
19	7 27	4 25	4 44	1 48	11 51	9 44	7 42	5 45	3 53	2 01	11 54	9 40
20	7 23	4 22	3 40	1 45	11 47	9 39	7 38	5 42	3 50	1 57	11 49	9 36
21	7 19	4 18	3 36	1 41	11 43	9 35	7 34	5 38	3 46	1 53	11 45	9 32
22	7 14	4 14	3 33	1 37	11 39	9 31	7 30	5 34	3 42	1 49	11 41	9 28
23	7 10	4 11	3 29	1 33	11 35	9 27	7 26	5 31	3 39	1 45	11 37	9 23
24	7 06	4 07	3 25	1 29	11 31	9 24	7 22	5 27	3 35	1 41	11 33	9 19
25	7 02	4 04	3 22	1 26	11 27	9 15	7 19	5 24	3 32	1 38	11 28	9 14
26	6 58	4 00	3 18	1 22	11 23	9 15	7 15	5 20	3 28	1 34	11 24	9 10
27	6 54	3 56	3 14	1 18	11 19	9 11	7 11	5 16	3 24	1 30	11 20	9 06
28	6 50	3 52	3 11	1 14	11 15	9 07	7 07	5 12	3 20	1 26	11 15	9 01
29	6 46		3 07	1 10	11 11	9 03	7 03	5 09	3 16	1 22	11 11	8 57
30	6 42		3 03	1 07	11 06	8 58	6 59	5 05	3 13	1 18	11 06	8 52
31	6 39		3 00		11 02		6 56	5 02		1 14		8 47

A TABLE

A TABLE *shewing what time* 31 *chief Stars come upon the Meridian, before or after the Bulls Eye.*

Names of the Stars.	H.	M.	
Draco's Eye	10	28	
Bright Star in the Harp	9	51	
Vulters Heart	8	42	
Swans Tail	7	47	
Pegasus's Mouth	6	49	
Fomahant	5	38	
Andromada's Head	4	25	
Southermost in the Whales Tail	3	49	South before the Bulls Eye.
Andromada's Girdle	3	26	
—— In her Foot	2	33	
Bright Star in *Aries*	2	28	
Whales Jaw	1	32	
Brightest in the *Seven* Stars	0	49	
Capella	0	35	
Orion's left Foot	0	42	
—— The middle in his Belt	1	02	
—— In his right shoulder	1	10	
Foot of *Castor*	2	02	
Great Dog	2	13	
Head of *Castor*	2	55	
Head of *Pollux*	3	08	
Hydra's Heart	4	54	
Lion's Heart	5	34	South after the Bulls Eye.
—— In his Tail	7	15	
Virgin's Spike	8	51	
Last in the Great Bears Tail	9	18	
Arcturus	9	44	
South Ballance	10	16	
North Ballance	10	43	
Brightest in the Crown	11	05	
Scorpion's Heart	11	53	

A Table *of the Right Ascentions and Declinations of the chiefest and most known Stars in the Firmament, with their Magnitude, Latitude, Longitude, and Distance from the Pole; exactly Calculated for the Year* 1684.

Northern Constellations.	*Mag.*	*Latitude.*		*Longitud.*		*Declinati.*		*Distance*		*Right Ascention*			
		D.	M.	D.	M.	D.	M.	D.	M.	D.	M.	H.	M.
The North Star	2	56	2 N	24	14 ♊	87	38 N	2	22	8	44	0	35
The brightest of the Guards .	2	72	51½ N	8	27 ♌	75	32¼ N	14	27¾	222	47½	14	51
The upper of the two foremost of the □ in the Great Bear.	2	49	40 N	10	45 ♌	63	27⅓ N	26	32⅓	161	1	10	44
The lower of the two foremost of the □ in the Great Bear.	2	45	3½ N	14	55 ♌	58	3¼ N	31	56¼	160	32¼	10	42
The lower of the two latter of the □ in the Great Bear . .	2	47	6½ N	25	56 ♌	55	28½ N	34	31¼	174	11½	11	37
The upper of the two latter of the □ in the Great Bear .	2	51	37 N	26	37 ♌	58	46¼ N	31	13¼	180		12	
The first in the great Bears Tail	2	54	18 N	4	21 ♍	57	43 N	32	17	189	58	12	40
The middlemost in her Tail .	2	56	22 N	11	7 ♍	56	37¼ N	33	22¼	197	41	13	11
The end of her Tail . . .	2	54	25 N	22	23 ♍	50	56½ N	39	3½	203	45	13	35
The Dragons Tail . . .	2	66	36 N	3	22 ♍	65	52 N	24	8	209	44½	13	59
Arcturus (in *Bootes*) . .	1	31	2 N	19	50 ♎	20	54 N	69	6	210	22	14	1½
In the Breast of *Cassiopea* . .	3	46	35½ N	3	29 ♉	54	49⅔ N	35	10⅔	5	44	0	23
In *Cassiopea*'s Knee . . .	3	46	22 N	13	32 ♉	58	34¾ N	31	25¼	16	22½	1	5½
Brightest in *Cassiopea*'s Chair .	3	51	14½ N	0	47 ♉	57	26¼ N	32	33¼	358	9	23	52½
The brightest in the Crown .	2	44	23 N	7	50 ♏	27	48 N	62	12	230	21	15	21½
The brightest in the Harp . .	1	61	47 N	10	54 ♑	38	31 N	51	29	276	34	18	26
The Swans Tail	2	59	57 N	1	5 ♓	44	12 N	45	48	307	41	20	31
The brightest in *Perseus*'s right side	2	30	5 N	27	28 ♉	48	39 N	41	21	45	13	3	1
Wagoners left shoulder . .	1	22	50½ N	17	27 ♊	45	38 N	44	22	73	21	4	53½
Wagoners right shoulder .	2	21	28 N	25	39 ♊	44	53¼ N	45	6¼	84	17	5	37
The brightest in the Serpents Neck	2	25	36 N	17	41 ♏	7	28⅘ N	82	31⅓	232	14	15	29
The Eagles Heart . . .	2	29	21½ N	27	20 ♑	8	5 N	81	55	293	50½	19	35⅓
The brightest in the Dolphins Tail	3	29	8 N	9	43 ♒	10	16 N	79	44	304	33½	20	18
The first in *Pegasus*'s Wing .	2	19	26 N	19	8 ♓	13	32¼ N	76	27¼	342	17	22	49
The end of *Pegasus*'s Wing .	2	12	35 N	4	49 ♈	13	26½ N	76	33½	359	20	23	57⅓
In the beginning of *Pegasus*'s right Leg	2	31	7½ N	25	1 ♓	26	22¾ N	63	37¼	342	10	22	49
Andromeda's Head . . .	2	25	42 N	9	58 ♈	27	23 N	62	37	358	3	23	52
The Southermost in *Andromeda*'s Girdle	2	25	59 N	26	0 ♈	33	59 N	56	1	13	0	0	52
The brightest in *Andromeda*'s Southermost Foot . . .	2	27	46½ N	9	50 ♉	40	48 N	49	12	26	10	1	45

Signs of the Zodiack.	Mag.	Latitude. D.	M.	Longitude D.	M.	Declinati. D.	M.	Distance D.	M.	Right Ascensions. D.	M.	H.	M.
Brighteſt in the Rams Head .	3	9	57 N	3	17 ♉	21	58⅙ N	68	1⅙	27	23¼	1	49½
The Bulls Eye, or *Aldebaran* .	1	5	31 S	5	24 ♊	15	50½ N	74	9½	64	29	4	18
The end of the Bulls Northermoſt Horn	2	5	20 N	18	10 ♊	28	18 N	61	42	76	35½	5	6⅓
The brighteſt in the *Pleiades* .	3	4	00 N	25	35 ♉	23	6 N	66	54	52	12	3	29
The upper or the Northermoſt head of the Twins *Caſtor* .	2	10	2 N	15	52 ♋	32	32 N	57	28	108	37⅓	7	14½
The lower head *Pollux* . .	2	6	38 N	18	54 ♋	28	44⅔ N	61	15½	111	31¼	7	26
Bright Star, left Foot of *Pollux*	2	6	48½ S	4	42 ♋	16	38 N	73	22	94	52	6	19½
The brighteſt in the Lions neck	2	8	47 N	25	10 ♌	21	25½ N	68	34½	150	37½	10	2½
The Lions heart, *Regulus* . .	1	0	26½ N	25	28 ♌	13	29⅓ N	76	30⅓	147	54	9	52
Brighteſt Star in the Lions back	2	14	20 N	6	52 ♍	22	16 N	67	44	164	19	10	57
The Lions Tail	1	12	18 N	17	14 ♍	16	21 N	73	39	173	14	11	33
Virgins Spike	1	1	59 S	19	27 ♎	9	28 S	80	32	197	11	13	9
Southermoſt Scale of Libra .	2	0	26 N	10	42 ♏	14	40 S	75	20	218	24	14	34
Northermoſt Scale of Libra .	2	8	35 N	14	59 ♏	8	10 S	81	50	225	2¼	15	0
Scorpion's heart	1	4	27 S	5	24 ♐	25	39 S	64	21	242	35¼	16	10⅓
Fomahant	1	21	3 S	29	23 ♒	31	16 S	58	44	339	59	22	40
Southern Conſtellations.													
Bright Star in the Whales Jaw	2	12	37 S	9	58 ♉	2	51½ N	87	8½	41	30⅔	2	46
Brighteſt in the Whales Tail	2	20	47 S	28	7 ♓	19	44 S	70	16	6	55	0	28
Orion's right ſhoulder . .	2	16	6 S	24	23 ♊	7	18⅔ N	82	41⅓	84	33⅔	5	38
Orion's left ſhoulder . . .	2	16	53 S	16	33 ♊	6	1⅔ N	83	58⅓	77	4	5	8¼
The firſt or Northermoſt in *Orion*'s belt	2	23	38 S	18	2 ♊	0	34 S	89	26	79	3	5	16
The ſecond or middlemoſt in *Orion*'s belt	2	24	33½ S	19	5 ♊	1	25 S	88	35	80	4⅔	5	20⅓
The third or Southermoſt in *Orion*'s belt	2	25	21½ S	20	18 ♊	2	8 S	87	52	81	14¼	5	25
Orion's left Foot, or *Regel* .	1	31	11½ S	12	28 ♊	8	35½ S	81	24½	74	53½	4	59½
The brighteſt Star in the mouth of the great Dog, *Sirius*. . .	1	39	30 S	9	47 ♋	16	15 S	73	45	97	51	6	31⅓
The great Dog's fore Foot .	2	41	18½ S	2	54 ♋	17	49 S	72	11	92	17¼	6	9
The little Dog's Thigh, *Procyon*	2	15	57 S	21	30 ♋	6	1½ N	83	58½	110	45	7	23
Hydra's heart	1	22	24 S	22	57 ♌	7	18⅓ S	82	41⅔	138	4	9	12¼
Bright Star in *Argos* Helm . .	1	75	48 S	10	38 ♋	52	26½ S	37	33½	94	15½	6	17
End of the *Centaur*'s right Foot	1	42	23 S	25	31 ♏	59	26 S	30	34 S	214	41	14	19
In the left Knee of the *Centaur*.	2	44	2 S	18	24 ♏	58	25 S	31	35	204	19	13	37
The foremoſt of the *Croſiers* .	3	50	18 S	1	21 ♏	56	56½ S	33	3½ S				
The Foot of the *Croſiers* . .	2	52	45 S	7	32 ♏	61	17 S	28	43 S	182	34	12	10
The head of the *Croſiers* . . .	2	47	41 S	2	22 ♏	55	16 S	34	44	183	36	12	14

A TABLE of the Sun's RIGHT ASCENSION.

Days	Jan. Afcen. H. M.	Feb. Afcen. H. M.	Mar. Afcen. H. M.	April. Afcen. H. M.	May. Afcen. H. M.	June. Afcen. H. M.	July. Afcen. H. M.	Aug. Afcen. H. M.	Sept. Afcen. H. M.	Octob. Afcen. H. M.	Nov. Afcen. H. M.	Dec. Afcen. H. M.
1	19 34	21 42	23 28	01 21	03 14	05 19	07 23	09 25	11 19	13 07	15 07	17 15
2	19 38	21 46	23 32	01 24	03 18	05 23	07 27	09 25	11 22	13 11	15 11	17 20
3	19 42	21 50	23 35	01 28	03 22	05 27	07 31	09 32	11 26	13 15	15 15	17 24
4	19 47	21 54	23 39	01 32	03 26	05 31	07 35	09 36	11 29	13 18	15 20	17 29
5	19 51	21 58	23 43	01 36	03 29	05 36	07 39	09 40	11 33	13 22	15 24	17 33
6	19 55	22 02	23 46	01 39	03 33	05 40	07 43	09 44	11 37	13 26	15 28	17 38
7	20 00	22 06	23 50	01 43	03 37	05 44	07 47	09 47	11 40	13 30	15 32	17 42
8	20 04	22 09	23 53	01 47	03 41	05 48	07 51	09 51	11 44	13 33	15 36	17 46
9	20 08	22 13	23 57	01 50	03 45	05 52	07 55	09 55	11 47	13 37	15 40	17 51
10	20 12	22 17	00 01	01 54	03 49	05 56	07 59	09 58	11 51	13 41	15 45	17 55
11	20 16	22 21	00 04	01 58	03 53	06 00	08 03	10 02	11 55	13 45	15 49	18 00
12	20 21	22 25	00 08	02 02	03 57	06 04	08 07	10 06	11 58	13 49	15 53	18 04
13	20 25	22 29	00 12	02 05	04 01	06 09	08 11	10 10	12 02	13 52	15 57	18 09
14	20 29	22 32	00 15	02 09	04 05	06 13	08 15	10 13	12 05	13 56	16 01	18 13
15	20 33	22 36	00 19	02 13	04 09	06 17	08 19	10 17	12 09	14 00	16 06	18 18
16	20 37	22 40	00 23	02 16	04 13	06 21	08 23	10 21	12 13	14 04	16 10	18 22
17	20 42	22 44	00 26	02 20	04 17	06 25	08 27	10 24	12 16	14 08	16 14	18 27
18	20 46	22 47	00 30	02 24	04 21	06 29	08 31	10 28	12 20	14 12	16 19	18 31
19	20 50	22 51	00 33	02 28	04 26	06 33	08 35	10 32	12 23	14 16	16 23	18 35
20	20 54	22 55	00 37	02 32	04 30	06 38	08 39	10 35	12 27	14 19	16 27	18 40
21	20 58	22 58	00 41	02 35	04 34	06 42	08 43	10 39	12 31	14 23	16 32	18 44
22	21 02	23 02	00 44	02 39	04 38	06 46	08 46	10 42	12 34	14 27	16 36	18 48
23	21 06	23 06	00 48	02 43	04 42	06 50	08 50	10 46	12 38	14 31	16 40	18 53
24	21 10	23 10	00 52	02 47	04 46	06 54	08 54	10 50	12 42	14 35	16 45	18 58
25	21 14	23 13	00 55	02 51	04 50	06 58	08 58	10 53	12 45	14 39	16 49	19 02
26	21 18	23 17	00 59	02 55	04 54	07 02	09 02	10 57	12 49	14 43	16 53	19 06
27	21 22	23 21	01 03	02 59	04 58	07 06	09 06	11 01	12 53	14 47	16 58	19 11
28	21 26	23 24	01 06	03 03	05 02	07 10	09 10	11 04	12 56	14 51	17 02	19 15
29	21 30		01 10	03 06	05 06	07 15	09 13	11 08	13 00	14 55	17 06	19 19
30	21 34		01 13	03 10	05 11	07 19	09 17	11 11	13 04	14 59	17 11	19 24
31	21 38		01 17		05 15		09 21	11 15		15 03		19 28

A

A *Traverse-Table* for every *Point*, *Half-Point*, and *Quarter-Point*, of the *Compass*, to the 100 part of a *League* or *Mile*.

Dist. in Legg. or Miles fail'd	02 deg. 49 min.		05 deg. 37 min.		08 deg. 26 min.		11 deg. 15 min.		Dist. in Legg. or Miles fail'd
	0 *Point* ¼		0 *Point* ½		0 *Point* ¾		1 *Points.*		
	N S	E W	N S	E W	N S	E W	N S	E W	
1	01.00	00.05	01.00	00.10	00.98	00.14	00.98	00.20	1
2	02.00	00.10	01.99	00.20	01.97	00.29	01.96	00.39	2
3	03.00	00.15	02.98	00.29	02.96	00.44	02.94	00.58	3
4	04.00	00.20	03.98	00.39	03.95	00.58	03.92	00.78	4
5	04.99	00.25	04.97	00.49	04.94	00.73	04.90	00.98	5
6	05.99	00.29	05.97	00.59	05.93	00.88	05.88	01.17	6
7	06.99	00.34	06.97	00.69	06.92	01.02	06.86	01.37	7
8	07.99	00.39	07.96	00.78	07.91	01.17	07.85	01.56	8
9	08.99	00.44	08.96	00.88	08.90	01.32	08.83	01.76	9
10	09.99	00.49	09.95	00.98	09.89	01.46	09.81	01.95	10
11	10.98	00.54	10.95	01.08	10.88	01.61	10.79	02.15	11
12	11.98	00.59	11.94	01.18	11.87	01.76	11.77	02.34	12
13	12.98	00.63	12.94	01.27	12.86	01.91	12.75	02.54	13
14	13.98	00.68	13.93	01.37	13.85	02.05	13.73	02.73	14
15	14.98	00.73	14.93	01.47	14.84	02.20	14.71	02.93	15
16	15.98	00.78	15.92	01.57	15.83	02.34	15.65	03.12	16
17	16.98	00.83	16.92	01.67	16.82	02.49	16.67	03.32	17
18	17.97	00.88	17.91	01.76	17.80	02.64	17.65	03.51	18
19	18.97	00.93	18.91	01.86	18.79	02.79	18.64	03.71	19
20	19.97	00.98	19.90	01.96	19.78	02.93	19.62	03.90	20
21	20.97	01.03	20.90	02.06	20.77	03.08	20.60	04.10	21
22	21.97	01.08	21.89	02.16	21.76	03.22	21.58	04.29	22
23	22.97	01.13	22.89	02.25	22.75	03.37	22.56	04.49	23
24	23.97	01.17	23.88	02.35	23.74	03.52	23.54	04.68	24
25	24.97	01.22	24.88	02.45	24.73	03.66	24.52	04.88	25
26	25.96	01.27	25.87	02.55	25.71	03.81	25.50	05.07	26
27	26.96	01.32	26.87	02.65	26.70	03.96	26.48	05.27	27
28	27.96	01.37	27.86	02.75	27.69	04.10	27.46	05.46	28
29	28.96	01.42	28.86	02.84	28.68	04.25	28.44	05.66	29
30	29.96	01.47	29.86	02.94	29.67	04.40	29.42	05.85	30
31	30.96	01.52	30.85	03.04	30.66	04.55	30.40	05.05	31
32	31.96	01.57	31.85	03.14	31.65	04.69	31.38	06.24	32
	E W	N S	E W	N S	E W	N S	E W	N S	
	7 *Points* ¾		7 *Points* ½		7 *Point* ¼		7 *Points.*		
	87 deg. 11 min.		84 deg. 22 min.		81 deg. 34 min.		78 deg. 45 min.		

A TRAVERSE TABLE.

Dist. in Leag. or Miles lat/dep.	02 deg. 49 min. 0 Points ¼		05 deg. 37 min. 0 Point ½		08 deg. 26 min. 0 Points ¾		11 deg. 15 min. 1 Point.		Dist. in Leag. or Miles lat/dep.
	N S	E W	N S	E W	N S	E W	N S	E W	
33	32.96	01.61	32.84	03.23	32.64	04.84	32.37	06.44	33
34	33.95	01.66	33.84	03.33	33.63	04.98	33.35	06.63	34
35	34.95	01.71	34.83	03.43	34.62	05.13	34.33	06.83	35
36	35.95	01.75	35.83	03.53	35.61	05.28	35.31	07.02	36
37	36.95	01.81	36.82	03.63	36.60	05.42	36.29	07.22	37
38	37.95	01.86	37.82	03.73	37.59	05.57	37.27	07.41	38
39	38.95	01.91	38.81	03.82	38.58	05.72	38.25	07.61	39
40	39.95	01.96	39.81	03.92	39.57	05.87	39.23	07.80	40
41	40.95	02.01	40.80	04.02	40.55	06.02	40.21	08.00	41
42	41.95	02.06	41.80	04.12	41.54	06.16	41.19	08.19	42
43	42.95	02.11	42.79	04.21	42.53	06.31	42.17	08.35	43
44	43.94	02.15	43.79	04.31	43.52	06.45	43.15	08.58	44
45	44.94	02.20	44.78	04.41	44.51	06.60	44.14	08.78	45
46	45.94	02.25	45.78	04.51	45.50	06.48	45.12	08.77	46
47	46.94	02.30	46.77	04.61	46.49	06.63	46.10	09.17	47
48	47.94	02.35	47.77	04.70	47.48	07.77	47.08	09.36	48
49	48.94	02.40	48.76	04.80	48.47	07.92	48.06	09.56	49
50	49.94	02.45	49.76	04.90	49.46	07.07	49.04	09.75	50
51	50.93	02.50	50.75	05.00	50.44	07.48	50.02	09.95	51
52	51.93	02.55	51.75	05.10	51.43	07.63	51.00	10.14	52
53	52.93	02.60	52.74	05.20	52.42	07.77	51.98	10.34	53
54	53.93	02.65	53.74	05.29	53.41	07.92	52.96	10.53	54
55	54.93	02.70	54.73	05.39	54.40	08.07	53.94	10.73	55
56	55.93	02.75	55.73	05.49	55.30	08.21	54.92	10.92	56
57	56.92	02.75	56.72	05.59	56.38	08.36	55.90	11.02	57
58	57.93	02.84	57.72	05.68	57.37	08.51	56.89	11.31	58
59	58.92	02.89	58.71	05.78	58.36	08.65	57.87	11.51	59
60	59.92	02.94	59.71	05.88	59.35	08.80	58.85	11.70	60
70	69.91	03.43	69.66	06.86	69.24	10.27	68.65	13.65	70
80	79.90	03.92	79.61	07.84	79.13	11.73	78.46	15.60	80
90	89.89	04.41	89.56	08.82	89.02	13.20	88.27	17.55	90
100	99.87	04.90	99.51	09.80	98.91	14.67	98.08	19.50	100
200	199.76	09.80	199.02	19.90	197.82	29.34	196.16	39.00	200
	E W	N S	E W	N S	E W	N S	E W	N S	
	7 Points ¼		7 Points ½		7 Points ¾		7 Points.		
	87 deg. 11 min.		84 deg. 11 min.		81 deg. 34 min.		78 deg. 45 min.		

A TRAVERSE-TABLE.

Dist. in Long. or Miles (sail'd)	14 deg. 04 min. 1 Point ¼		16 deg. 52 min. 1 Point ½		19 deg. 41 min. 1 Point ¾		22 deg. 30 min. 2 Points		Dist. in Long. or Miles (sail'd)
	N S	E W	N S	E W	N S	E W	N S	E W	
1	00.97	00.24	00.96	00.29	00.94	00.33	00.92	00.38	1
2	01.94	00.48	01.91	00.58	01.88	00.67	01.85	00.76	2
3	02.91	00.72	02.87	00.87	02.82	01.01	02.77	01.75	3
4	03.88	00.97	03.83	01.16	03.77	01.34	03.70	01.53	4
5	04.85	01.21	04.78	01.45	04.71	01.68	04.62	01.91	5
6	05.82	01.45	05.74	01.74	05.65	02.02	05.54	02.30	6
7	06.79	01.70	06.70	02.03	06.59	02.35	06.47	02.68	7
8	07.76	01.94	07.66	02.32	07.53	02.70	07.35	03.16	8
9	08.73	02.18	08.61	02.61	08.47	03.03	08.31	03.44	9
10	09.70	02.43	09.57	02.90	09.41	03.37	09.24	03.83	10
11	10.67	02.67	10.53	03.19	10.36	03.71	10.16	04.21	11
12	11.64	02.91	11.48	03.48	11.30	04.04	11.05	04.59	12
13	12.61	03.15	12.44	03.77	12.24	04.36	12.01	04.97	13
14	13.58	03.40	13.40	04.06	13.18	04.72	12.93	05.36	14
15	14.55	03.64	14.35	04.35	14.12	05.05	13.86	05.74	15
16	15.52	03.88	15.31	04.64	15.06	05.39	14.78	06.12	16
17	16.49	04.13	16.27	04.93	16.00	05.73	15.71	06.51	17
18	17.46	04.37	17.22	05.22	16.95	06.06	16.63	06.89	18
19	18.43	04.61	18.18	05.51	17.89	06.40	17.55	07.27	19
20	19.40	04.86	19.14	05.81	18.83	06.74	18.48	07.65	20
21	20.37	05.10	20.10	06.10	19.77	07.08	19.40	08.04	21
22	21.34	05.34	21.05	06.39	20.71	07.41	20.32	08.42	22
23	22.34	05.48	22.01	06.68	21.66	07.75	21.25	08.80	23
24	23.28	05.83	22.97	06.97	22.60	08.08	22.17	09.18	24
25	24.25	06.07	23.91	07.26	23.54	08.42	23.10	09.57	25
26	25.22	06.31	24.88	07.55	24.48	08.76	24.02	09.95	26
27	26.19	06.56	25.84	07.84	25.42	09.10	24.94	10.33	27
28	27.16	06.80	26.75	08.13	26.36	09.43	25.87	10.71	28
29	28.13	07.04	27.75	08.42	27.30	09.77	26.75	11.10	29
30	29.10	07.28	28.71	08.71	28.25	10.11	27.72	11.48	30
31	30.07	07.53	29.66	09.00	29.19	10.44	28.64	11.86	31
32	31.04	07.77	30.62	09.29	30.13	10.74	29.56	12.25	32
	E W	N S	E W	N S	E W	N S	E W	N S	
	6 Points ¼		6 Points ½		6 Point ¾		6 Points		
	75 deg. 56 min.		73 deg. 07 min.		70 deg. 19 min.		67 deg. 30 min.		

A TRAVERSE-TABLE.

Dist. in Leag. or Miles fast.	14 deg. 04 min. 1 Point ¼		16 deg. 52 min. 1 Point ½		19 deg. 41 min. 1 Point ¾		22 deg. 30 min. 2 Points.		Dist. in Leag. or Miles fast.
	N S	E W	N S	E W	N S	E W	N S	E W	
33	32.01	08.01	31.58	09.58	31.07	11.12	30.49	12.63	33
34	32.98	08.26	32.54	09.87	32.01	11.45	31.44	13.01	34
35	33.95	08.50	33.45	10.16	32.95	11.79	32.34	13.39	35
36	34.92	08.74	34.45	10.45	33.89	12.13	33.26	13.78	36
37	35.80	08.99	35.41	10.74	34.84	12.47	34.18	14.16	37
38	36.26	09.23	36.36	11.03	35.78	12.80	35.11	14.54	38
39	37.83	09.47	37.32	11.32	36.72	13.14	36.03	14.92	39
40	38.80	09.71	38.28	11.61	37.66	13.48	36.96	15.31	40
41	39.77	09.96	39.23	11.90	38.60	13.81	37.88	15.69	41
42	40.74	10.20	40.19	12.19	39.34	14.15	38.80	16.07	42
43	41.71	10.44	41.15	12.48	40.49	14.49	39.73	16.45	43
44	42.68	10.69	42.12	12.77	41.43	14.72	40.65	16.84	44
45	43.65	10.93	43.06	13.06	42.37	15.16	41.57	17.22	45
46	44.62	11.17	44.62	13.35	43.41	15.50	42.50	17.60	46
47	45.59	11.42	44.98	13.64	44.25	15.83	43.42	17.99	47
48	46.56	11.66	45.95	13.93	45.19	16.17	44.35	18.37	48
49	47.33	11.90	46.89	14.22	46.13	16.37	45.27	18.75	49
50	48.50	12.14	47.85	14.51	47.08	16.85	46.19	19.13	50
51	49.47	12.39	48.80	14.80	48.01	17.18	47.12	19.92	51
52	50.44	12.63	49.76	15.09	48.96	17.31	48.04	19.89	52
53	51.41	12.87	50.72	15.38	49.90	17.85	48.97	20.28	53
54	52.38	13.12	51.67	15.67	50.84	18.15	49.89	20.66	54
55	53.35	13.36	52.63	15.96	51.78	18.53	50.81	21.04	55
56	54.32	13.60	53.55	16.26	52.73	18.87	51.73	21.15	56
57	55.29	13.85	54.55	16.55	53.67	19.20	52.66	21.43	57
58	56.26	14.09	55.50	16.84	54.61	19.54	53.58	22.81	58
59	57.23	14.33	56.46	17.23	55.55	19.88	54.21	22.20	59
60	58.20	14.57	57.42	17.42	56.45	20.21	55.43	22.58	60
70	67.90	17.00	66.90	20.31	65.90	23.58	64.67	26.78	70
80	77.60	19.43	76.55	23.22	75.32	26.95	73.91	30.61	80
90	87.30	21.86	86.10	26.12	84.73	30.39	83.14	34.44	90
100	97.00	24.29	95.69	29.02	94.15	33.68	92.38	38.26	100
200	194.00	48.58	191.38	58.04	188.30	67.36	184.76	76.52	200
	E W	N S	E W	N S	E W	N S	E W	N S	
	6 Points ¾		6 Points ½		6 Points ¼		6 Points.		
	75 deg. 56 min.		73 deg. 07 min.		70 deg. 19 min.		67 deg. 30 min.		

A TRAVERSE-TABLE.

Dist. in Leag. or Miles failed	25 deg. 19 min.		28 deg. 07 min.		30 deg. 56 min.		33 deg. 45 min.		Dist. in Leag. or Miles failed
	2 Points ¼		2 Points ½		2 Points ¾		3 Points.		
	N S	E W	N S	E W	N S	E W	N S	E W	
1	00.90	00.43	00.88	00.47	00.86	00.51	00.83	00.56	1
2	01.81	00.85	01.76	00.94	01.71	01.03	01.66	01.55	2
3	02.71	01.28	02.65	01.41	02.57	01.54	02.49	01.67	3
4	03.61	01.71	03.53	01.89	03.43	02.06	03.32	02.22	4
5	04.52	02.14	04.41	02.36	04.29	02.57	04.16	02.78	5
6	05.42	02.56	05.29	02.83	05.15	03.08	04.99	03.33	6
7	06.33	02.99	06.17	03.30	06.00	03.60	05.82	03.89	7
8	07.23	03.42	07.05	03.77	06.86	04.11	06.65	04.44	8
9	08.14	03.85	07.94	04.24	07.72	04.63	07.48	05.00	9
10	09.04	04.28	08.82	04.71	08.58	05.14	08.31	05.55	10
11	09.94	04.70	09.70	05.18	09.44	05.66	09.15	06.11	11
12	10.85	05.13	10.58	05.66	10.29	06.17	09.98	06.67	12
13	11.75	05.56	11.46	06.13	11.15	06.68	10.81	07.22	13
14	12.66	05.99	12.35	06.60	12.00	07.20	11.64	07.78	14
15	13.56	06.41	13.23	07.07	12.87	07.71	12.47	08.33	15
16	14.46	06.84	14.11	07.54	13.72	08.23	13.30	08.89	16
17	15.37	07.27	14.99	08.01	14.58	08.74	14.13	09.44	17
18	16.27	07.70	15.87	08.48	15.44	09.25	14.97	10.00	18
19	17.18	08.12	16.76	08.96	16.30	09.77	15.80	10.56	19
20	18.08	08.55	17.64	09.43	17.16	10.28	16.63	11.11	20
21	18.98	08.98	18.52	09.90	18.01	10.80	17.46	11.67	21
22	19.89	09.41	19.40	10.37	18.87	11.31	18.29	12.22	22
23	20.79	09.83	20.28	10.84	19.73	11.82	19.12	12.78	23
24	21.70	10.26	21.17	11.31	20.59	12.34	19.95	13.33	24
25	22.60	10.69	22.05	11.78	21.44	12.85	20.79	13.89	25
26	23.50	11.12	22.93	12.26	22.30	13.57	21.62	14.44	26
27	24.41	11.54	23.81	12.73	23.16	13.88	22.45	15.15	27
28	25.31	11.97	24.69	13.20	24.02	14.39	23.28	15.56	28
29	26.22	12.40	25.58	13.67	24.87	14.91	24.11	16.11	29
30	27.12	12.83	26.46	14.14	25.73	15.42	24.94	16.67	30
31	28.02	13.25	27.34	14.61	26.59	15.94	25.78	17.22	31
32	28.93	13.68	28.22	15.08	27.45	16.45	26.61	17.78	32
	E W	N S	E W	N S	E W	N S	E W	N S	
	5 Points ¾		5 Points ½		5 Points ¼		5 Points.		
	64 deg. 42 min.		61 deg. 52 min.		59 deg. 04 min.		56 deg. 15 min.		

A TRAVERSE-TABLE.

Dist. in Long. or Miles sail'd	25 deg. 19 min. 2 Points ¼		28 deg. 07 min. 2 Points ½		30 deg. 56 min. 2 Points ¾		33 deg. 45 min. 3 Points.		Dist. in Long. or Miles sail'd
	N S	E W	N S	E W	N S	E W	N S	E W	
33	29.83	14.11	29.10	15.56	28.31	16.97	27.44	18.33	33
34	30.74	14.54	29.98	16.03	29.16	17.48	28.27	18.89	34
35	31.64	14.96	30.87	16.50	30.02	17.99	29.10	19.44	35
36	32.54	15.39	31.75	16.97	30.88	18.51	29.93	20.00	36
37	33.45	15.82	32.63	17.44	31.74	19.02	30.76	20.56	37
38	34.35	16.25	33.57	17.91	32.59	19.54	31.60	21.11	38
39	35.26	16.68	34.40	18.38	33.45	20.05	32.43	21.67	39
40	36.16	17.10	35.28	18.86	34.31	20.56	33.26	22.22	40
41	37.06	17.53	36.16	19.34	35.17	21.08	34.09	22.78	41
42	37.67	17.96	37.04	19.81	36.02	21.59	34.92	23.34	42
43	38.87	18.38	37.92	20.27	36.88	22.11	35.75	23.89	43
44	39.78	18.81	38.80	20.74	37.74	22.62	36.58	24.44	44
45	40.68	19.24	39.69	21.21	38.60	23.14	37.42	25.00	45
46	41.51	19.67	40.57	21.68	39.46	23.65	38.25	25.56	46
47	42.49	20.09	41.41	22.16	40.31	24.16	39.08	26.11	47
48	43.39	20.52	42.34	22.63	41.17	24.68	39.91	26.67	48
49	44.30	20.95	43.21	23.10	42.03	25.19	40.74	27.22	49
50	45.20	21.38	44.10	23.57	42.89	25.71	41.57	27.78	50
51	46.10	21.61	44.98	24.04	43.74	26.22	42.40	28.33	51
52	47.01	22.23	45.86	24.51	44.60	26.73	43.24	28.89	52
53	47.91	22.66	46.74	24.98	45.46	27.25	44.07	29.44	53
54	48.82	23.08	47.62	25.46	46.32	27.76	44.90	30.00	54
55	49.72	23.52	48.51	25.93	47.17	28.28	45.73	30.56	55
56	50.62	23.94	49.39	26.40	48.03	28.79	46.56	31.11	56
57	51.53	24.37	50.27	26.87	48.89	29.30	47.39	31.67	57
58	52.43	24.79	51.15	27.34	49.75	29.82	48.22	32.22	58
59	53.33	25.23	52.03	27.81	50.61	30.33	48.06	32.78	59
60	54.24	25.65	52.91	28.28	51.46	30.84	49.89	33.33	60
70	63.27	29.92	61.73	32.99	60.04	35.98	58.22	38.88	70
80	72.31	34.20	70.55	37.71	68.61	41.12	66.51	44.44	80
90	81.35	38.47	79.37	42.43	77.19	46.26	74.85	50.00	90
100	90.39	42.75	88.19	47.13	85.77	51.41	83.14	55.55	100
200	180.78	85.50	176.38	94.26	171.54	102.82	166.28	111.11	200
	E W	N S	E W	N S	E W	N S	E W	N S	
	5 Points ¼		5 Points ½		5 Points ¾		5 Points.		
	64 deg. 42 min.		61 deg. 52 min.		59 deg. 04 min.		56 deg. 15 min.		

A TRAVERSE-TABLE.

Dist. in Leag. or Miles fail'd	36 deg. 34 min.		39 deg. 22 min.		42 deg. 11 min.		45 deg. 00 min.		Dist. in Leag. &c.
	3 Point ¼		3 Point ½		3 Point ¾		4 Points.		
	N S	E W	N S	E W	N S	E W	N S	E W	
1	00.80	00.60	00.77	00.63	00.74	00.67	00.71	00.71	1
2	01.61	01.15	01.55	01.27	01.48	01.34	01.41	01.41	2
3	02.41	01.41	02.32	01.90	02.22	02.01	02.12	02.12	3
4	03.21	02.38	03.09	02.54	02.96	02.69	02.83	02.83	4
5	04.02	02.98	03.86	03.17	03.70	03.36	03.54	03.54	5
6	04.82	03.57	04.64	03.81	04.44	04.03	04.24	04.24	6
7	05.62	04.17	05.41	04.44	05.18	04.78	04.95	04.95	7
8	06.43	04.70	06.18	05.07	05.93	05.37	05.66	05.66	8
9	07.23	05.36	06.96	05.71	06.67	06.04	06.36	06.36	9
10	08.03	05.96	07.73	06.34	07.41	06.72	07.07	07.07	10
11	08.83	06.55	08.50	06.98	08.15	07.39	07.78	07.78	11
12	09.64	07.15	09.28	07.61	08.89	08.06	08.49	08.49	12
13	10.44	07.74	10.05	08.25	09.63	08.73	09.19	09.19	13
14	11.24	08.34	10.82	08.88	10.37	09.40	09.90	09.90	14
15	12.05	08.94	11.60	09.52	11.11	10.07	10.61	10.61	15
16	12.85	09.53	12.37	10.15	11.85	10.74	11.31	11.31	16
17	13.66	10.13	13.14	10.78	12.70	11.42	12.02	12.02	17
18	14.46	10.72	13.91	11.42	13.34	12.09	12.73	12.73	18
19	15.26	11.32	14.69	12.04	14.08	12.76	13.44	13.44	19
20	16.06	11.91	15.46	12.69	14.82	13.43	14.14	14.14	20
21	16.87	12.51	16.23	13.32	15.56	14.10	14.85	14.85	21
22	17.67	13.11	17.01	13.96	16.30	14.77	15.56	15.56	22
23	18.47	13.70	17.78	14.59	17.04	15.45	16.26	16.26	23
24	19.28	14.30	18.55	15.22	17.78	16.12	16.97	16.97	24
25	20.08	14.89	19.32	15.86	18.52	16.79	17.67	17.68	25
26	20.88	15.49	20.10	16.49	19.26	17.46	18.38	18.38	26
27	21.69	16.08	20.87	17.13	20.00	18.13	19.48	0905	27
28	22.49	16.68	21.64	17.76	20.75	18.79	19.46	80.80	28
29	23.29	17.27	22.42	18.40	21.45	19.44	20.44	51.51	29
30	24.10	17.87	23.19	19.03	22.23	20.12	21.42	21.21	30
31	24.90	18.47	23.96	19.67	22.97	20.82	21.40	92.92	31
32	25.70	19.06	24.74	20.30	23.71	21.49	22.38	63.63	32
	E W	N S	E W	N S	E W	N S	E W	N S	
	4 Points ¼		4 Points ½		4 Points ¾		4 Points.		
	53 deg. 26 min.		50 deg. 37 min.		47 deg. 49 min.		45 deg. 00 min.		

A TRAVERSE-TABLE.

Dist. in Leag. or Miles fail'd	35 deg. 34 min. 3 Points ¼		39 deg. 22 min. 3 Points ¼		42 deg. 11 min. 3 Points ¼		45 deg. 00 min. 4 Points		Dist. in Leag. or Miles fail'd
	N S	E W	N S	E W	N S	E W	N S	E W	
33	26.51	19.66	25.51	20.93	24.45	22.16	23.33	23.33	33
34	27.31	20.25	26.28	21.57	25.15	22.83	24.04	24.04	34
35	28.11	20.85	27.06	22.20	25.93	23.50	24.75	24.75	35
36	28.91	21.46	27.83	22.84	26.67	24.17	25.46	25.46	36
37	29.72	22.04	28.60	23.47	27.41	24.85	26.16	26.16	37
38	30.52	22.64	29.37	24.11	28.16	25.52	26.87	26.87	38
39	31.33	23.23	30.15	24.74	28.90	26.19	27.56	27.56	39
40	32.13	23.83	30.92	25.38	29.64	26.86	28.28	28.28	40
41	32.93	24.42	31.69	26.01	30.38	27.53	28.99	28.99	41
42	33.73	25.02	32.47	26.64	31.12	28.21	29.10	29.10	42
43	34.54	25.61	33.24	27.28	31.86	28.88	30.41	30.41	43
44	35.34	26.21	34.01	27.91	32.60	29.55	31.11	31.11	44
45	36.14	26.81	34.78	28.55	33.34	30.22	31.82	31.82	45
46	36.94	27.40	35.56	29.18	34.08	30.85	32.53	32.53	46
47	37.75	28.00	36.33	29.82	34.82	31.56	33.23	33.23	47
48	38.55	28.59	37.10	30.45	35.57	32.23	33.94	33.94	48
49	39.36	29.15	37.88	31.08	36.31	32.91	34.65	34.65	49
50	40.17	29.78	38.65	31.72	37.05	33.58	35.35	35.35	50
51	40.96	30.38	39.42	32.35	37.79	34.25	36.06	36.06	51
52	41.77	30.98	40.20	32.99	38.53	34.92	36.77	36.77	52
53	42.57	31.57	40.97	33.62	39.27	35.55	37.48	37.48	53
54	43.37	32.17	41.74	34.26	40.01	36.26	38.14	38.14	54
55	44.18	32.76	42.52	34.85	40.75	36.94	38.89	38.89	55
56	44.98	33.36	43.25	35.53	41.45	37.61	39.60	39.60	56
57	45.78	33.96	44.06	36.16	42.23	38.28	40.30	40.30	57
58	46.59	34.55	44.83	36.79	43.07	38.95	41.01	41.01	58
59	47.39	35.15	45.61	37.43	43.72	39.62	41.72	41.72	59
60	48.19	35.74	46.38	38.06	44.45	40.25	42.43	42.43	60
70	56.22	41.69	54.11	44.41	51.85	47.00	49.49	49.49	70
80	64.25	47.65	61.84	50.75	59.26	53.72	56.56	56.56	80
90	72.28	53.61	69.57	57.09	66.67	60.44	63.63	63.63	90
100	80.32	59.56	77.30	63.43	74.08	67.15	70.71	70.71	100
200	160.64	119.12	154.60	126.86	148.16	134.30	141.41	141.41	200
	E W	N S	E W	N S	E W	N S	E W	N S	
	4 Points ¼		4 Points ¼		4 Points ¼		4 Points.		
	54 deg. 26 min.		50 deg. 37 min.		47 deg. 49 m.		45 deg. 00 min.		

A
TABLE
O F
Meridional Parts.

A TABLE of

L.	0	1	2	3	4	5	6
M.	Min.	Min.	Min.	Min.	Min.	Min.	Min.
0	0	600	1200	1801	2402	3004	3607
1	10	610	1210	1811	2412	3014	3617
2	20	620	1220	1821	2422	3024	3627
3	30	630	1230	1831	2432	3034	3637
4	40	640	1240	1841	2442	3044	3647
5	50	650	1250	1851	2452	3054	3657
6	60	660	1260	1861	2462	3064	3667
7	70	670	1270	1871	2472	3074	3677
8	80	680	1280	1881	2482	3084	3687
9	90	690	1290	1891	2492	3094	3697
10	100	700	1300	1901	2502	3104	3707
11	110	710	1310	1911	2512	3114	3717
12	120	720	1320	1921	2522	3124	3727
13	130	730	1330	1931	2532	3134	3737
14	140	740	1340	1941	2542	3144	3747
15	150	750	1350	1951	2552	3154	3758
16	160	760	1360	1961	2562	3165	3768
17	170	770	1370	1971	2572	3175	3778
18	180	780	1380	1981	2582	3185	3788
19	190	790	1390	1991	2593	3195	3798
20	200	800	1400	2001	2603	3205	3808
21	210	810	1410	2011	2613	3215	3818
22	220	820	1420	2021	2623	3225	3828
23	230	830	1430	2031	2633	3235	3838
24	240	840	1440	2041	2643	3245	3848
25	250	850	1450	2051	2653	3255	3858
26	260	860	1460	2061	2663	3265	3868
27	270	870	1470	2071	2673	3275	3878
28	280	880	1481	2081	2683	3285	3888
29	290	890	1491	2091	2693	3295	3898

MERIDIONAL PARTS.

L.	0	1	2	3	4	5	6
M.	Min.	Min.	Min.	Min.	Min.	Min.	Min.
30	300	900	1501	2101	2702	3305	3908
31	310	910	1511	2111	2713	3315	3919
32	320	920	1521	2121	2723	3325	3929
33	330	930	1531	2131	2733	3335	3939
34	340	940	1541	2141	2743	3345	3949
35	350	950	1551	2151	2753	3355	3959
36	360	960	1561	2161	2763	3365	3969
37	370	970	1571	2171	2773	3375	3979
38	380	980	1581	2181	2783	3386	3989
39	390	990	1591	2191	2793	3396	3999
40	400	1000	1601	2202	2803	3406	4009
41	410	1010	1611	2212	2813	3416	4019
42	420	1020	1621	2222	2823	3426	4029
43	430	1030	1631	2232	2833	3436	4039
44	440	1040	1641	2242	2843	3446	4049
45	450	1050	1651	2252	2853	3456	4059
46	460	1060	1661	2262	2863	3466	4070
47	470	1070	1671	2272	2873	3476	4080
48	480	1080	1681	2282	2883	3486	4090
49	490	1090	1691	2292	2893	3496	4100
50	500	1100	1701	2302	2903	3506	4110
51	510	1110	1711	2312	2914	3516	4120
52	520	1120	1721	2322	2924	3526	4130
53	530	1130	1731	2332	2934	3536	4140
54	540	1140	1741	2342	2944	3546	4150
55	550	1150	1751	2352	2954	3556	4160
56	560	1160	1761	2362	2964	3566	4170
57	570	1170	1771	2372	2974	3576	4180
58	580	1180	1781	2382	2984	3587	4190
59	590	1190	1791	2392	2994	3597	4200

A Table of

L.	7	8	9	10	11	12	13
M.	Min.	Min.	Min.	Min.	Min.	Min.	Min.
0	4211	4816	5422	6031	6641	7253	7868
1	4221	4826	5433	6041	6651	7264	7879
2	4231	4836	5443	6051	6661	7274	7889
3	4241	4846	5453	6061	6671	7284	7899
4	4251	4856	5463	6071	6681	7294	7909
5	4261	4866	5473	6082	6692	7305	7920
6	4271	4876	5483	6092	6702	7315	7930
7	4281	4886	5493	6102	6712	7325	7940
8	4291	4896	5503	6112	6722	7335	7950
9	4301	4907	5514	6122	6732	7346	7961
10	4311	4917	5524	6132	6743	7356	7971
11	4321	4927	5534	6142	6753	7366	7981
12	4331	4937	5544	6153	6763	7376	7991
13	4342	4947	5554	6163	6773	7387	8002
14	4352	4957	5564	6173	6783	7397	8012
15	4362	4967	5574	6183	6794	7407	8022
16	4372	4977	5584	6193	6804	7417	8032
17	4382	4987	5594	6203	6814	7428	8043
18	4392	4998	5605	6213	6824	7438	8053
19	4402	5008	5615	6224	6834	7448	8063
20	4412	5018	5625	6234	6845	7458	8073
21	4422	5028	5635	6244	6855	7469	8084
22	4432	5038	5645	6254	6865	7479	8094
23	4442	5048	5655	6264	6875	7489	8104
24	4452	5058	5666	6274	6885	7499	8114
25	4463	5068	5676	6285	6896	7510	8125
26	4473	5078	5686	6295	6906	7520	8135
27	4483	5089	5696	6305	6916	7530	8145
28	4493	5099	5706	6315	6926	7540	8155
29	4503	5100	5716	6325	6936	7551	8166

MERIDIONAL PARTS.

L.	7	8	9	10	11	12	13
M.	*Min.*	*Min.*	*Min.*	*Min.*	*Min.*	*Min.*	*Min.*
30	4513	5119	5726	6335	6947	7561	8176
31	4523	5129	5737	6346	6957	7571	8186
32	4533	5139	5747	6356	6967	7581	8196
33	4543	5149	5757	6366	6977	7592	8207
34	4553	5159	5767	6376	6987	7602	8217
35	4563	5169	5777	6386	6998	7612	8227
36	4573	5180	5787	6396	7008	7622	8237
37	4584	5190	5797	6406	7018	7633	8248
38	4594	5200	5808	6417	7028	7643	8258
39	4604	5210	5818	6427	7038	7653	8268
40	4614	5220	5828	6437	7049	7663	8279
41	4624	5230	5838	6447	7059	7674	8289
42	4634	5240	5848	6457	7069	7684	8299
43	4644	5250	5858	6467	7079	7694	8310
44	4654	5260	5868	6477	7089	7704	8320
45	4664	5271	5879	6488	7100	7715	8330
46	4674	5281	5889	6498	7110	7725	8341
47	4684	5291	5899	6508	7120	7735	8351
48	4695	5301	5909	6518	7130	7745	8361
49	4705	5311	5919	6528	7141	7756	8372
50	4715	5321	5929	6539	7151	7766	8382
51	4725	5331	5939	6549	7161	7776	8392
52	4735	5341	5950	6559	7171	7786	8403
53	4745	5351	5960	6569	7182	7797	8413
54	4755	5362	5970	6579	7192	7807	8423
55	4765	5372	5980	6590	7202	7817	8434
56	4775	5382	5990	6600	7212	7827	8444
57	4785	5392	6000	6610	7223	7838	8454
58	4796	5402	6010	6620	7233	7848	8465
59	4806	5412	6021	6630	7243	7858	8475

A T A B L E of

L.	14	15	16	17	18	19	20
M	Min.	Min.	Min.	Min.	Min.	Min.	Min.
0	8485	9105	9728	10353	10982	11615	12251
1	8495	9115	9738	10363	10993	11625	12262
2	8506	9126	9748	10374	11003	11636	12273
3	8516	9136	9759	10384	11014	11647	12283
4	8526	9146	9769	10395	11024	11657	12294
5	8537	9157	9780	10405	11035	11668	12304
6	8547	9167	9790	10416	11045	11678	12315
7	8557	9177	9800	10426	11056	11689	12326
8	8568	9188	9811	10437	11066	11700	12336
9	8578	9198	9821	10447	11077	11710	12347
10	8589	9208	9832	10458	11087	11721	12358
11	8599	9219	9842	10468	11098	11731	12368
12	8610	9229	9852	10479	11108	11742	12379
13	8620	9239	9863	10489	11119	11752	12390
14	8630	9250	9873	10499	11129	11763	12400
15	8641	9260	9884	10510	11140	11774	12411
16	8651	9270	9894	10520	11150	11784	12422
17	8661	9281	9904	10531	11161	11795	12432
18	8672	9291	9915	10541	11171	11805	12443
19	8682	9301	9925	10552	11182	11816	12454
20	8692	9312	9936	10562	11192	11827	12464
21	8703	9322	9946	10573	11203	11837	12475
22	8713	9332	9956	10583	11213	11848	12486
23	8723	9343	9967	10593	11224	11858	12496
24	8734	9353	9977	10604	11234	11869	12507
25	8744	9363	9988	10614	11245	11880	12518
26	8754	9374	9998	10625	11255	11890	12528
27	8765	9384	10008	10635	11266	11901	12539
28	8775	9394	10019	10646	11276	11911	12550
29	8785	9405	10029	10656	11287	11922	12560

MERIDIONAL PARTS.						
L.	14	15	16	17	18	19	20
M.	*Min.*	*Min.*	*Min.*	*Min.*	*Min.*	*Min.*	*Min.*
30	8796	9415	10040	10667	11297	11932	12571
31	8806	9425	10050	10677	11308	11943	12582
32	8816	9436	10061	10688	11318	11954	12592
33	8827	9446	10071	10698	11329	11964	12603
34	8837	9456	10081	10709	11340	11975	12614
35	8847	9467	10092	10720	11351	11985	12624
36	8858	9477	10102	10730	11361	11996	12635
37	8868	9487	10113	10741	11372	12007	12646
38	8878	9498	10123	10751	11382	12017	12656
39	8889	9508	10134	10762	11393	12028	12667
40	8899	9519	10144	10772	11403	12039	12678
41	8909	9529	10154	10783	11414	12049	12688
42	8920	9539	10165	10793	11424	12060	12699
43	8930	9550	10175	10804	11435	12071	12710
44	8940	9560	10186	10814	11446	12081	12721
45	8951	9571	10196	10825	11456	12092	12731
46	8961	9581	10206	10835	11467	12102	12742
47	8971	9592	10217	10846	11477	12113	12753
48	8982	9602	10227	10856	11488	12124	12763
49	8992	9613	10238	10867	11498	12134	12774
50	9002	9623	10248	10877	11509	12145	12785
51	9012	9634	10259	10888	11520	12155	12795
52	9023	9644	10269	10898	11530	12166	12806
53	9033	9655	10280	10909	11541	12177	12817
54	9043	9665	10290	10919	11551	12187	12827
55	6054	9676	10301	10930	11562	12198	12838
56	9064	9686	10311	10940	11572	12209	12849
57	9074	9696	10322	10951	11583	12219	12860
58	9084	9707	10332	10961	11594	12230	12870
59	9095	9717	10343	10972	11604	12241	12881

A TABLE of

L.	21	22	23	24	25	26	27
M.	*Min.*	*Min.*	*Min.*	*Min.*	*Min.*	*Min.*	*Min.*
0	12852	13537	14187	14841	15500	16165	16836
1	12902	13548	14197	14852	15511	16176	16847
2	12913	13558	14208	14863	15522	16187	16858
3	12924	13569	14219	14873	15533	16198	16869
4	12935	13580	14230	14884	15544	16209	16880
5	12945	13590	14241	14895	15555	16220	16891
6	12956	13601	14251	14906	15566	16232	16903
7	12967	13612	14262	14917	15577	16243	16914
8	12978	13623	14273	14928	15588	16254	16925
9	12988	13633	14284	14939	15599	16265	16936
10	12999	13644	14295	14950	15610	16276	16948
11	13010	13655	14306	14961	15621	16287	16959
12	13020	13666	14317	14972	15632	16298	16970
13	13031	13676	14328	14983	15643	16310	16981
14	13042	13687	14339	14994	15654	16320	16993
15	13053	13698	14349	15005	15665	16332	17004
16	13063	13709	14360	15016	15676	16343	17015
17	13074	13720	14371	15027	15687	16354	17026
18	13085	13731	14382	15038	15698	16365	17038
19	13096	13742	14393	15049	15710	16377	17049
20	13106	13753	14404	15060	15721	16388	17060
21	13117	13764	14415	15071	15732	16399	17071
22	13128	13774	14426	15082	15743	16410	17083
23	13138	13785	14437	15093	15754	16421	17094
24	13149	13796	14448	15104	15765	16432	17105
25	13160	13807	14458	15115	15776	16443	17116
26	13171	13818	14469	15126	15787	16455	17128
27	13181	13828	14480	15137	15798	16466	17139
28	13192	13839	14491	15148	15889	16477	17150
29	13203	13850	14502	15159	15880	16488	17161

MERIDIONAL PARTS.

L.	21	22	23	24	25	26	26
M.	Min.	Min.	Min.	Min.	Min.	Min.	Min.
30	13214	13861	14513	15170	15832	16499	17173
31	13224	13872	14524	15181	15843	16510	17184
32	13235	13883	14535	15192	15854	16522	17195
33	13246	13894	14546	15203	15865	16533	17207
34	13257	13904	14557	15214	15876	16544	17218
35	13267	13915	14568	15225	15887	16555	17229
36	13278	13926	14579	15235	15898	16566	17240
37	13289	13937	14589	15247	15909	16578	17252
38	13300	13948	14600	15258	15920	16589	17263
39	13310	13958	14611	15269	15931	16600	17274
40	13321	13969	14622	15280	15943	16611	17286
41	13332	13980	14633	15291	15954	16622	17297
42	13342	13991	14644	15302	15965	16634	17308
43	13353	14002	14655	15313	15976	16645	17319
44	13364	14013	14666	15324	15987	16656	17331
45	13375	14024	14677	15335	15998	16667	17342
46	13386	14034	14688	15346	16009	16678	17353
47	13397	14045	14698	15357	16020	16690	17365
48	13407	14056	14709	15368	16031	16701	17376
49	13418	14067	14720	15379	16043	16712	17387
50	13429	14078	14731	15390	16054	16723	17399
51	13440	14088	14742	15401	16065	16734	17410
52	13450	14099	14753	15412	16076	16746	17421
53	13461	14110	14764	15423	16087	16757	17432
54	13472	14121	14775	15434	16098	16769	17444
55	13483	14132	14786	15445	16109	16780	17455
56	13494	14143	14797	15456	16120	16791	17466
57	13504	14154	14808	15467	16131	16802	17478
58	13515	14165	14819	15478	16143	16813	17489
59	13526	14176	14830	15489	16154	16824	17500

A TABLE of

L.	28	29	30	31	32	33	34
M.	*Min.*	*Min.*	*Min.*	*Min.*	*Min.*	*Min.*	*Min.*
0	17512	18195	18884	19581	20284	20996	21715
1	17523	18206	18895	19592	20296	21007	21727
2	17534	18217	18907	19604	20307	21019	21739
3	17546	18229	18919	19615	20319	21031	21751
4	17557	18240	18930	19627	20331	21043	21763
5	17568	18252	18942	19639	20343	21055	21775
6	17580	18263	18953	19650	20355	21067	21787
7	17591	18275	18965	19662	20367	21079	21800
8	17602	18286	18976	19674	20378	21091	21812
9	17614	18297	18988	19685	20390	21103	21824
10	17625	18309	18999	19697	20402	21115	21836
11	17636	18320	19011	19709	20414	21127	21848
12	17648	18332	19023	19720	20426	21139	21860
13	17659	18343	19034	19732	20438	21151	21872
14	17670	18355	19046	19744	20449	21163	21884
15	17682	18366	19057	19756	20461	21175	21896
16	17693	18378	19069	19768	20473	21187	21908
17	17705	18389	19081	19779	20485	21198	21920
18	17716	18401	19092	19791	20497	21210	21933
19	17727	18412	19104	19803	20508	21222	21945
20	17739	18424	19115	19814	20520	21234	21957
21	17750	18435	19127	19826	20532	21246	21969
22	17761	18446	19138	19837	20544	21258	21981
23	17772	18458	19150	19849	20550	21270	21993
24	17784	18469	19162	19861	20568	21282	22005
25	17795	18481	19173	19873	20580	21294	22017
26	17806	18492	19185	19884	20591	21306	22030
27	17818	18504	19196	19896	20603	21318	22042
28	17830	18515	19208	19908	20615	21330	22054
29	17841	18527	19219	19920	20627	21342	22066

MERIDIONAL PARTS.

L.	28	29	30	31	32	33	34
M.	Min.	Min.	Min.	Min.	Min.	Min.	Min.
30	17852	18538	19231	19931	20639	21354	22078
31	17864	18550	19243	19943	20651	21366	22090
32	17875	18561	19254	19955	20662	21378	22102
33	17886	18573	19266	19966	20674	21390	22114
34	17898	18584	19278	19978	20686	21402	22127
35	17909	18596	19289	19990	20698	21414	22139
36	17921	18607	19301	20002	20710	21426	22151
37	17932	18619	19313	20013	20722	21438	22163
38	17943	18630	19324	20025	20734	21450	22175
39	17955	18642	19336	20037	20746	21462	22187
40	17966	18653	19347	20049	20757	21474	22199
41	17978	18665	19359	20060	20769	21486	22212
42	17989	18676	19371	20072	20781	21498	22224
43	18000	18688	19382	20084	20793	21510	22236
44	18012	18699	19394	20096	20805	21522	22248
45	18023	18711	19405	20107	20817	21534	22260
46	18035	18722	19417	20119	20829	21546	22272
47	18046	18734	19429	20131	20841	21458	22285
48	18057	18745	19440	20143	20853	21570	22297
49	18069	18757	19452	20154	20865	21582	22309
50	18080	18768	19464	20166	20877	21594	22321
51	18092	18780	19475	20178	20889	21607	22333
52	18103	18792	19487	20190	20901	21619	22346
53	18114	18803	19499	20202	20913	21631	22358
54	18126	18815	19510	20213	20925	21643	22370
55	18137	18826	19522	20225	20937	21655	22382
56	18149	18838	19534	20237	20949	21667	22394
57	18160	18849	19545	20249	20961	21679	22407
58	18172	18861	19557	20260	20973	21691	22419
59	18183	18872	19569	20272	20985	21703	22431

A TABLE of

L.	35	36	37	38	39	40	41
Al.	Min.	Min.	Min.	Min.	Min.	Min.	Min.
0	22443	23180	23927	24683	25450	26227	27016
1	22455	23193	23939	24696	25462	26240	27029
2	22468	23205	23952	24708	25475	26253	27043
3	22480	23217	23964	24721	25488	26266	27056
4	22492	23230	23977	24734	25501	26279	27069
5	22504	23242	23989	24746	25514	26292	27083
6	22516	23254	24002	24759	25527	26305	27096
7	22529	23267	24014	24771	25540	26319	27109
8	22541	23279	24027	24785	25553	26332	27122
9	22553	23292	24039	24797	25566	26345	27136
10	22565	23304	24052	24810	25578	26358	27149
11	22578	23316	24064	24823	25591	26371	27162
12	22590	23329	24077	24835	25604	26384	27175
13	22602	23341	24090	24848	25617	26397	27189
14	22614	23353	24102	24861	25630	26410	27202
15	22627	23366	24115	24874	25643	26423	27215
16	22639	23378	24127	24886	25656	26436	27229
17	22651	23390	24140	24899	25669	26449	27242
18	22663	23403	24152	24912	25682	26463	27255
19	22676	23415	24165	24925	25695	26476	27269
20	22688	23428	24178	24937	25707	26489	27282
21	22700	23440	24190	24950	25720	26502	27295
22	22712	23453	24203	24963	25733	26515	27308
23	22725	23465	24215	24976	25746	26528	27322
24	22737	23478	24228	24988	25759	26541	27335
25	22749	23490	24240	25001	25772	26554	27348
26	22761	23502	24253	25014	25785	26568	27362
27	22774	23515	24265	25027	25798	26581	27375
28	22786	23527	24278	25034	25811	26594	27388
29	22798	23540	24291	25052	25824	26607	27402

MERIDIONAL PARTS.

L.	35	36	37	38	39	40	41.
M.	Min.	Min.	Min.	Min.	Min.	Min.	Min.
30	22810	23552	24303	25065	25837	26620	27415
31	22823	23565	24316	25078	25850	26633	27429
32	22835	23577	24329	25090	25863	26646	27442
33	22847	23589	24341	25103	25876	26660	27455
34	22860	23602	24354	25116	25889	26673	27469
35	22872	23614	24367	25129	25902	26686	27482
36	22884	23627	24379	25142	25915	26699	27495
37	22897	23639	24392	25154	25928	26712	27509
38	22909	23652	24404	25167	25941	26725	27522
39	22921	23664	34417	25180	25954	26739	27535
40	22933	23677	24430	25193	25967	26751	27549
41	22946	23689	24442	25206	25980	26765	27562
42	22958	23702	24455	25218	25993	26778	27576
43	22970	23714	24468	25231	26006	26791	27589
44	22983	23727	24480	25244	26019	26805	27602
45	22995	23739	24493	25257	26032	26818	27615
46	23007	23752	24506	25270	26045	26831	27629
47	23020	23764	24518	25283	26058	26844	27643
48	23032	23777	24531	25295	26071	26857	27656
49	23044	23789	24543	25308	26084	26871	27669
50	23057	23801	24556	25321	26097	26884	27683
51	23069	23813	24569	25334	26110	26898	27696
52	23081	23826	24581	25347	26123	26910	27710
53	23094	23839	24594	25360	26136	26923	27723
54	23106	23851	24607	25372	26149	26937	27737
55	23118	23864	24616	25385	26162	26950	27750
56	23131	23876	24632	25398	26175	26963	27764
57	23143	23889	24645	25411	26188	26976	27777
58	23155	23902	24658	25424	26201	26990	27790
59	23167	23914	24670	25437	26214	27003	27804

Fff

A TABLE of

L.	42	43	44	45	46	47	48
M.	Min.	Min.	Min.	Min.	Min.	Min.	Min.
0	27817	28631	29459	30300	31156	32028	32916
1	27831	28645	29472	30314	31170	32042	32931
2	27844	28658	29486	30328	31185	32057	32946
3	27858	28672	29500	30342	31199	32072	32961
4	27871	28685	29514	30356	31214	32086	32975
5	27885	28700	29528	30370	31228	32101	32990
6	27898	28713	29542	30384	31242	32116	33005
7	27912	28727	29556	30398	31257	32130	33020
8	27925	28741	29570	30413	31271	32145	33035
9	27939	28754	29584	30427	31286	32160	33050
10	27951	28768	29598	30441	31300	32174	33065
11	27966	28782	29611	30455	31315	32189	33080
12	27979	28795	29625	30470	31329	32204	33095
13	27993	28809	29639	30484	31343	32219	33110
14	28006	28823	29653	30498	31358	32233	33125
15	28020	28837	29667	30512	31372	32248	33140
16	28033	28850	29681	30526	31387	32263	33155
17	28047	28864	29695	30541	31401	32277	33170
18	28060	28878	29709	30555	31416	32292	33185
19	28074	28892	29723	30569	31430	32307	33200
20	28087	28905	29737	30583	31445	32322	33215
21	28101	28919	29751	30597	31459	32336	33231
22	28114	28933	29765	30612	31474	32351	33246
23	28128	28947	29779	30626	31488	32366	33261
24	28141	28960	29793	30640	31503	32381	33276
25	28155	28974	29807	30654	31517	32395	33291
26	28168	28988	29821	30669	31532	32410	33306
27	28182	29002	29835	30683	31546	32425	33321
28	28195	29015	29849	30697	31561	32440	33336
29	28209	29029	29863	30711	31575	32455	33351

MERIDIONAL PARTS.

L.	42	43	44	45	46	47	48
M.	*Min.*	*Min.*	*Min.*	*Min.*	*Min.*	*Min.*	*Min.*
30	28223	29043	29877	30726	31590	32469	33366
31	28236	29057	29891	30740	31604	32484	33381
32	28250	29071	29905	30754	31619	32499	33396
33	28263	29084	29919	30769	31633	32514	33411
34	28277	29097	29933	30783	31648	32529	33427
35	28290	29112	29947	30797	31662	32544	33442
36	28304	29126	29961	30811	31677	32558	33457
37	28318	29140	29975	30826	31691	32573	33472
38	28331	29153	29989	30840	31706	32588	33487
39	28345	29167	30003	30854	31721	32603	33501
40	28358	29181	30018	30869	31735	32618	33517
41	28372	29195	30032	30883	31750	32633	33532
42	28386	29209	30046	30897	31764	32647	33548
43	28399	29223	30060	30912	31779	32662	33563
44	28413	29236	30074	30926	31793	32677	33578
45	28426	29250	30088	30940	31808	32692	33593
46	28440	29264	30102	30955	31823	32707	33608
47	28454	29278	30116	30969	31837	32722	33623
48	28467	29292	30130	30983	31852	32737	33639
49	28481	29306	30144	30998	31866	32752	33654
50	28495	29320	30158	31012	31881	32766	33669
51	28508	29333	30172	31026	31896	32781	33684
52	28522	29347	30187	31041	31910	32796	33699
53	28536	29361	30201	31056	31925	32811	33715
54	28549	29375	30215	31070	31940	32826	33730
55	28563	29389	30229	31084	31954	32841	33745
56	28577	29403	30243	31098	31969	32856	33760
57	28591	29417	30257	31112	31984	32871	33776
58	28605	29431	30271	31127	31998	32886	33791
59	28618	29444	30285	31141	32013	32901	33806

A TABLE of

L.	49	50	51	52	53	54	55
M	Min.	Min.	Min.	Min.	Min.	Min.	Min.
0	33821	34745	35688	36652	37638	38647	39680
1	33836	34761	35704	36669	37655	38664	39697
2	33852	34776	35720	36685	37671	38681	39715
3	33867	34792	35736	36701	37688	38698	39732
4	33882	34807	35752	36717	37704	38715	39750
5	33897	34823	35768	36734	37721	38732	39767
6	33913	34839	35784	36750	37738	38749	39785
7	33928	34854	35800	36766	37754	38766	39802
8	33943	34870	35816	36782	37771	38783	39820
9	33959	34885	35832	36799	37788	38800	39837
10	33974	34901	35848	36815	37804	38817	39855
11	33989	34917	35864	36831	37821	38834	39872
12	34004	34932	35880	36848	37838	38851	39890
13	34020	34948	35895	36864	37855	38868	39907
14	34035	34963	35911	36880	37871	38886	39925
15	34050	34979	35927	36897	37888	38903	39942
16	34066	34995	35943	36913	37905	38920	39960
17	34081	35010	35959	36929	37921	38937	39977
18	34096	35026	35975	36946	37938	38954	39995
19	34112	35042	35991	36962	37955	38971	40013
20	34127	35057	36007	36978	37972	38988	40030
21	34142	35073	36023	36995	37988	39005	40048
22	34158	35089	36039	37011	38005	39023	40065
23	34173	35104	36055	37027	38022	39040	40083
24	34188	35120	36071	37044	38039	39057	40100
25	34204	35136	36087	37060	38055	39074	40118
26	34219	35151	36103	37077	38072	39091	40136
27	34235	35167	36119	37093	38089	39109	40153
28	34250	35183	36136	37109	38106	39126	40171
29	34265	35198	36152	37126	38123	39143	40189

MERIDIONAL PARTS.

L.	49	50	51	52	53	54	55
M.	Min.	Min.	Min.	Min.	Min.	Min.	Min.
30	34281	35214	36168	37142	38139	39160	40206
31	34296	35230	36184	37159	38156	39177	40224
32	34312	35246	36200	37175	38173	39195	40242
33	34327	35261	36216	37192	38190	39212	40259
34	34342	35277	36232	37208	38207	39229	40277
35	34358	35293	36248	37224	38223	39246	40295
36	34373	35309	36264	37241	38240	39264	40312
37	34389	35324	36280	37257	38257	39281	40330
38	34404	35340	36296	37274	38274	39298	40348
39	34420	35356	36313	37290	38291	39315	40366
40	34435	35372	36329	37307	38308	39333	40383
41	34450	35388	36345	37323	38325	39350	40401
42	34466	35403	36361	37340	38342	39367	40419
43	34481	35419	36377	37356	38358	39385	40436
44	34497	35425	36393	37373	38375	39402	40454
45	34512	35451	36409	37389	38392	39419	40472
46	34528	35467	36425	37406	38409	39437	40490
47	34543	35482	36442	37422	38426	39454	40508
48	34559	35498	36458	37439	38443	39471	40525
49	34574	35514	36474	37456	38460	39489	40543
50	34590	35530	36490	37472	38477	39506	40561
51	34605	35546	36506	37489	38494	39523	40579
52	34621	35561	36523	37505	38511	39541	40597
53	34636	35577	36539	37522	38528	39558	40614
54	34652	35593	36555	37538	38545	39576	40632
55	34667	35609	36571	37555	38562	39593	40650
56	34683	35625	36587	37572	38579	39610	40668
57	34698	35641	36604	37588	38596	39628	40686
58	34714	35657	36620	37605	38613	39645	40704
59	34730	35673	36636	37621	38630	39663	40722

A TABLE of

L.	56	57	58	59	60	61	62
M.	Min.	Min.	Min.	Min.	Min.	Min.	Min.
0	40739	41827	42943	44092	45274	46493	47750
1	40757	41845	42962	44111	45294	46513	47771
2	40775	41863	42981	44131	45314	46534	47793
3	40793	41882	43000	44150	45334	46555	47814
4	40811	41900	43019	44170	45354	46575	47835
5	40829	41918	43038	44189	45374	46596	47857
6	40847	41937	43057	44208	45394	46617	47878
7	40865	41955	43076	44228	45414	46637	47900
8	40883	41974	43095	44247	45434	46658	47921
9	40901	41992	43114	44267	45454	46679	47942
10	40919	42011	43132	44286	45475	46699	47964
11	40937	42029	43151	44306	45495	46720	47985
12	40955	42047	43170	44325	45515	46741	48007
13	40973	42066	43189	44345	45535	46762	48028
14	40991	42084	43208	44364	45555	46782	48049
15	41009	42103	43227	44384	45575	46803	48071
16	41027	42121	43246	44404	45595	46824	48092
17	41045	42140	43265	44423	45615	46845	48114
18	41063	42158	43284	44443	45636	46866	48135
19	41081	42177	43303	44462	45656	46886	48157
20	41099	42195	43322	44482	45676	46907	48178
21	41117	42214	43342	44502	45696	46928	48200
22	41135	42232	43361	44521	45716	46949	48222
23	41153	42251	43380	44541	45737	46970	48243
24	41171	42270	43399	44560	45757	46991	48265
25	41189	42288	43418	44580	45777	47012	48286
26	41207	42307	43437	44600	45797	47032	48308
27	41225	42325	43456	44619	45818	47053	48329
28	41243	42344	43475	44639	45838	47074	48351
29	41261	42362	43494	44660	45858	47095	48373

MERIDIONAL PARTS.							
L.	56	57	58	59	60	61	62
M.	Min.	Min.	Min.	Min.	Min.	Min.	Min.
30	41279	42381	43513	44678	45878	47116	48394
31	41297	42400	43533	44698	45899	47137	48416
32	41316	42418	43552	44718	45919	47158	48438
33	41334	42437	43571	44738	45939	47179	48459
34	41352	42456	43590	44757	45960	47200	48481
35	41370	42474	43609	44777	45980	47221	48503
36	41388	42493	43628	44797	46001	47242	48525
37	41406	42512	43648	44817	46021	47263	48546
38	41425	42530	43667	44836	46041	47284	48568
39	41443	42549	43686	44856	46062	47305	48590
40	41461	42568	43705	44876	46082	47326	48612
41	41479	42586	43725	44896	46103	47347	48633
42	41497	42605	43744	44916	46123	47369	48655
43	41516	42624	43763	44935	46143	47390	48677
44	41534	42643	43782	44955	46164	47411	48699
45	41552	42661	43802	44975	46184	47432	48721
46	41570	42680	43821	44995	46205	47453	48743
47	41588	42699	43840	45015	46225	47474	48764
48	41607	42718	43859	45035	46246	47495	48786
49	41625	42736	43879	45055	46266	47517	48808
50	41643	42755	43898	45075	46287	47538	48830
51	41662	42774	43917	45094	46307	47559	48852
52	41680	42793	43937	45114	46328	47580	48874
53	41698	42811	43956	45134	46348	47601	48896
54	41717	42830	43975	45154	46369	47623	48918
55	41735	42849	43995	45174	46390	47644	48940
56	41753	42868	44014	45194	46410	47665	48962
57	41772	42887	44034	45214	46431	47686	48984
58	41790	42906	44053	45234	46451	47708	49006
59	41808	42925	44072	45254	46472	47729	49028

A TABLE of

L.	63	64	65	66	67	68	69
M.	Min.	Min.	Min.	Min.	Min.	Min.	Min.
0	49050	50395	51788	53236	54740	56309	57946
1	49072	50417	51812	53260	54766	56335	57974
2	49094	50440	51836	53285	54792	56362	58002
3	49116	50463	51860	53309	54817	56389	58030
4	49138	50486	51883	53334	54843	56415	58058
5	49160	50509	51907	53359	54869	56442	58086
6	49182	50532	51931	53383	54894	56469	58114
7	49204	50555	51954	53408	54920	56490	58142
8	49226	50577	51978	53433	54946	56523	58170
9	49248	50600	52002	53457	54971	56550	58198
10	49271	50623	52026	53482	54997	56576	58226
11	49293	50646	52050	53507	55023	56603	58254
12	49315	50669	52073	53532	55049	56630	58282
13	49337	50692	52097	53556	55075	56657	58310
14	49359	50715	52121	53581	55100	56684	58339
15	49381	50738	52145	53606	55126	56711	58367
16	49404	50761	52169	53631	55152	56738	58395
17	49426	50784	52193	53656	55178	56765	58423
18	49448	50807	52217	53681	55204	56792	58452
19	49470	50830	52241	53705	55230	56819	58480
20	49493	50893	52265	53730	55256	56846	58508
21	49515	50877	52289	53755	55282	56873	58537
22	49537	50900	52313	53780	55308	56900	58565
23	49560	50923	52337	53805	55334	56928	58593
24	49582	50946	52361	53830	55360	56955	58622
25	49604	50969	52385	53855	55386	56982	58650
26	49627	50992	52409	53880	55412	57009	58679
27	49649	51015	52433	53905	55438	57036	58707
28	49671	51039	52457	53930	55464	57063	58735
29	49694	51062	52481	53955	55490	57091	58764

MERIDIONAL PARTS.

L.	63	64	65	66	67	68	69
M.	Min.	Min.	Min.	Min.	Min.	Min.	Min.
30	49716	51085	52505	53980	55516	57118	58793
31	49739	51108	52529	54005	55542	57145	58821
32	49761	51131	52553	54030	55568	57173	58850
33	49783	51155	52577	54056	55595	57200	58878
34	49806	51178	52601	54081	55621	57227	58907
35	49828	51201	52626	54106	55647	57255	58936
36	49851	51225	52650	54131	55673	57282	58964
37	49873	51248	52674	54156	55699	57310	58993
38	49896	51271	52698	54181	55726	57337	59022
39	49918	51295	52723	54207	55752	57364	59051
40	49941	51318	52747	54232	55778	57392	59079
41	49963	51341	52771	54257	55805	57419	59108
42	49986	51365	52795	54282	55831	57447	59137
43	50009	51388	52820	54308	55857	57475	59166
44	50031	51412	52844	54333	55884	57502	59195
45	50054	51435	52868	54358	55910	57530	59224
46	50076	51459	52893	54384	55937	57557	59252
47	50099	51482	52917	54409	55963	57585	59281
48	50122	51506	52942	54435	55990	57613	59310
49	50144	51529	52966	54460	56016	57640	59339
50	50167	51553	52990	54485	56043	57668	59368
51	50190	51576	53015	54511	56069	57696	59397
52	50212	51600	53039	54536	56096	57723	59426
53	50235	51623	53064	54562	56122	57751	59455
54	50258	51647	53088	54587	56149	57779	59485
55	50281	51670	53113	54613	56175	57807	59514
56	50303	51694	53137	54638	56202	57835	59543
57	50326	51718	53162	54664	56229	57862	59572
58	50349	51741	53186	54689	56255	57890	59601
59	50372	51765	53211	54715	56282	57918	59630

G gg

A TABLE of

L.	70	71	72	73	74	75	76
M.	Min.	Min.	Min.	Min.	Min.	Min.	Min.
0	59660	61457	63349	65345	67457	69703	72101
1	59689	61488	63381	65379	67494	69742	72142
2	59718	61519	63414	65413	67530	69781	72183
3	59747	61550	63446	65447	67566	69819	72225
4	59777	61580	63478	65482	67603	69858	72266
5	59806	61611	63511	65516	67639	69897	72308
6	59835	61642	63543	65550	67676	69936	72349
7	59865	61673	63576	65585	67712	69975	72391
8	59894	61704	63609	65619	67749	70014	72433
9	59924	61735	63641	65654	67785	70053	72475
10	59953	61766	63674	65688	67822	70092	72516
11	59983	61797	63706	65723	67858	70131	72558
12	60012	61828	63739	65757	67895	70170	72600
13	60042	61859	63772	65792	67932	70209	72642
14	60071	61890	63805	65826	67969	70248	72684
15	60101	61921	63837	65861	68005	70287	72726
16	60130	61952	63870	65895	68042	70327	72768
17	60160	61983	63903	65930	68079	70366	72810
18	60190	62014	63936	65965	68116	70405	72852
19	60219	62046	63969	66000	68153	70445	72894
20	60249	62077	64002	66034	68190	70484	72937
21	60279	62108	64035	66069	68227	70524	72979
22	60308	62139	64068	66104	68264	70563	73021
23	60338	62171	64101	66139	68301	70603	73064
24	60368	62202	64134	66174	68338	70642	73106
25	60398	62233	64167	66209	68376	70682	73149
26	60427	62265	64200	66244	68413	70722	73191
27	60457	62296	64233	66279	68450	70762	73234
28	60487	62327	64266	66314	68487	70801	73277
29	60517	62359	64299	66350	68525	70842	73320

MERIDIONAL PARTS.

L.	70	71	72	73	74	75	76
M.	*Min.*	*Min.*	*Min.*	*Min.*	*Min.*	*Min.*	*Min.*
30	60547	62390	64332	66385	68562	70881	73362
31	60577	62422	64366	66420	68600	70921	73405
32	60607	62453	64399	66455	68637	70961	73448
33	60637	62485	64432	66491	68675	71001	73491
34	60667	62517	64466	66526	68712	71041	73534
35	60697	62548	64499	66561	68750	71082	73577
36	60727	62580	64533	66597	68787	71122	73620
37	60757	62612	64566	66632	68825	71162	73664
38	60788	62644	64600	66668	68863	71202	73707
39	60818	62675	64633	66703	68901	71243	73750
40	60848	62707	64667	66739	68938	71283	73794
41	60878	62739	64700	66774	68976	71323	73837
42	60908	62771	64734	66810	69014	71364	73880
43	60939	62803	64768	66846	69052	71404	73924
44	60969	62835	64801	66881	69090	71445	73968
45	60999	62866	64835	66917	69128	71486	74011
46	61030	62898	64869	66953	69166	71526	74055
47	61060	62930	64903	66989	69204	71567	74099
48	61091	62962	64936	67024	69242	71608	74142
49	61121	62994	64970	67060	69281	71649	74186
50	61151	63027	65004	67096	69319	71690	64230
51	61182	63059	65038	67132	69357	71730	74274
52	61212	63091	65072	67168	69395	71771	74318
53	61243	63123	65106	67204	69434	71812	74362
54	61274	63155	65140	67240	69472	71853	74406
55	61304	63187	65174	67276	69511	71895	74450
56	61335	63220	65208	67312	69549	71936	74495
57	61365	63252	65242	67349	69588	71977	74539
58	61396	63284	65276	67385	69626	72018	74583
59	61427	63317	65310	67421	69665	72059	74628

A TABLE of

L.	77	78	79	80	81	82	83
M.	*Min.*	*Min.*	*Min.*	*Min.*	*Min.*	*Min.*	*Min.*
0	74672	77446	80457	83753	87391	91456	96059
1	74717	77494	80510	83810	87455	91527	96141
2	74761	77542	80562	83868	87519	91599	96224
3	74806	77590	80615	83926	87583	91672	96306
4	74850	77639	80668	83983	87648	91744	96389
5	74895	77687	80720	84041	87712	91816	96472
6	74940	77735	80773	84099	87777	91889	96555
7	74985	77784	80826	84158	87841	91962	96638
8	75029	77832	80879	84216	87906	92035	96722
9	75074	77881	80932	84274	87971	92108	96806
10	75119	77930	80985	84333	88036	92181	96890
11	75164	77978	81038	84391	88101	92254	96974
12	75209	78027	81092	84450	88166	92328	97058
13	75254	78076	81145	84509	88232	92402	97142
14	75300	78125	81198	84568	88297	92476	97227
15	75345	78174	81252	84626	88363	92550	97312
16	75390	78223	81306	84686	88428	92624	97397
17	75436	78272	81359	84745	88494	92699	97483
18	75481	78322	81413	84804	88560	92773	97568
19	75527	78371	81467	84863	88626	92848	97654
20	75572	78420	81521	84923	88693	92923	97740
21	75618	78470	81575	84982	88759	92998	97827
22	75663	78519	81629	85042	88826	93073	97913
23	75709	78569	81683	85102	88892	93148	98000
24	75755	78619	81737	85162	88959	93224	98086
25	75801	78668	81792	85222	89026	93300	98173
26	75847	78718	81846	85282	89093	93375	98261
27	75893	78768	81901	85342	89160	93452	98348
28	75939	78818	81955	85402	89227	93528	98436
29	75985	78868	82010	85462	89295	93604	98524

MERIDIONAL PARTS.

L.	77	78	79	80	81	82	83
M.	*Min.*	*Min.*	*Min.*	*Min.*	*Min.*	*Min.*	*Min.*
30	76031	78918	82065	85523	89362	93681	98613
31	76077	78968	82120	85584	89430	93758	98701
32	76123	79019	82175	85644	89498	93835	98790
33	76170	79069	82230	85705	89566	93912	98878
34	76216	79119	82285	85766	89634	93989	98967
35	76263	79170	82341	85827	89702	94066	99057
36	76309	79221	82396	85889	89771	94144	99146
37	76356	79271	82451	85950	89839	94221	99236
38	76402	79322	82507	86011	89908	94299	99327
39	76449	79373	82563	86073	89977	94378	99417
40	76496	79424	82618	86135	90046	94456	99508
41	76543	79475	82674	86196	90115	94534	99598
42	76590	79526	82730	86258	90184	94613	99689
43	76637	79577	82786	86320	90254	94691	99780
44	76684	79628	82842	86382	90323	94770	99872
45	76731	79680	82899	86445	90393	94849	99963
46	76778	79731	82955	86507	90463	94929	100055
47	76826	79782	83011	86569	90533	95008	100148
48	76873	79834	83068	86632	90603	95088	100240
49	76920	79885	83124	86695	90673	95168	100333
50	76968	79937	83181	86757	90744	95248	100426
51	77015	79989	83238	86820	90814	95329	100519
52	77063	80040	83294	86883	90885	95409	100613
53	77110	80092	83351	86946	90956	95489	100706
54	77158	80144	83408	87010	91027	95570	100800
55	77206	80196	83466	87073	91098	95651	100894
56	77254	80248	83523	87136	91169	95732	100989
57	77302	80300	83580	87200	91240	95814	101084
58	77350	80353	83637	87264	91312	95895	101179
59	77398	80405	83695	87327	91384	95977	101274

A TABLE of

L.	84	85	86	87	88	89
M.	Min.	Min.	Min.	Min.	Min.	Min.
0	101370	107647	115326	125223	139166	162998
1	101466	107762	115470	125414	139454	163575
2	101562	107877	115614	125607	139744	164163
3	101658	107993	115759	125800	140037	164761
4	101754	108109	115905	125995	140332	165370
5	101851	108225	116050	126191	140630	165949
6	101948	108342	116198	126389	140930	166620
7	102046	108459	116345	126586	141233	167262
8	102144	108577	116493	126786	141539	167917
9	102242	108696	116641	126986	141847	168585
10	102340	108814	116791	127188	142158	169265
11	102438	108933	116940	127391	142472	169906
12	102537	109052	117091	127595	142789	170669
13	102636	109172	117242	127800	143109	171303
14	102735	109291	117394	128007	143432	172132
15	102835	109412	117547	128215	143758	172887
16	102935	109533	117700	128425	144087	173660
17	103035	109655	117854	128635	144419	174450
18	103136	109777	118009	128847	144754	175259
19	103237	109899	118164	129060	145093	176087
20	103338	110022	118320	129274	145435	176936
21	103440	110145	118476	129489	145781	177807
22	103541	110269	118634	129706	146130	178699
23	103643	110393	118792	129925	146483	179616
24	103745	110517	118951	130144	146839	180558
25	103848	110642	119110	130366	147199	181526
26	103951	110768	119271	130588	147563	182523
27	104054	110893	119431	130812	147930	183540
28	104158	111020	119594	131038	148302	184600
29	104262	111146	119756	131265	148678	185

MERIDIONAL PARTS.

L.	84	85	86	87	88	89
M.	Min.	Min.	Min.	Min.	Min.	Min.
30	104366	111274	119920	131493	149058	185825
31	104471	111401	120084	131723	149442	187991
32	104575	111529	120249	131955	149830	189197
33	104680	111658	120415	132188	150223	190447
34	104785	111787	120582	132423	150621	191744
35	104891	111917	120749	132659	151023	193092
36	104997	112047	120917	132897	151430	194495
37	105104	112177	121086	133137	151842	195958
38	105211	112309	121256	133378	152258	197486
39	105318	112440	121427	133621	152680	199085
40	105426	112572	121599	133866	153107	200752
41	105533	112705	121771	134112	153540	202525
42	105641	112838	121944	134361	153978	204383
43	105749	112971	122118	134611	154421	206348
44	105858	113106	122293	134863	154870	208431
45	105967	113240	122469	135116	155326	210649
46	106077	113376	122646	135372	155787	213020
47	106187	113511	122824	135630	156255	215566
48	106297	113648	123002	135889	156730	218317
49	106408	113784	123182	136151	157210	221306
50	106519	113922	123363	136414	157698	224580
51	106630	114060	123544	136680	158193	228199
52	106741	114198	123727	136947	158695	232243
53	106853	114337	123910	137217	159204	236829
54	106965	114477	124095	137489	159721	242118
55	107077	114617	124280	137763	160246	248369
56	107191	114758	124467	138039	160779	256008
57	107304	114899	124653	138317	161320	265829
58	107418	115041	124842	138598	161870	279580
59	107533	115183	125031	138881	162429	303640

A

TABLE

OF THE

MILES *of* East *and* West, *anſwering to the* Degrees *of* Longitude *in the Fourth* Rumb.

Hhh

A TABLE of

Latitude.	Miles East and West.	Longitude.	Latitude.	Miles East and West.	Longitude.	Latitude.	Miles East and West.	Longitude.
D. M.		D. M.	D. M.		D. M.	D. M.		D. M.
0 0	0	0 0	4 20	260	4 20	8 40	520	8 42
0 10	10	0 10	4 30	270	4 30	8 50	530	8 52
0 20	20	0 20	4 40	280	4 40	9 0	540	9 02
0 30	30	0 30	4 50	290	4 50	9 10	550	9 12
0 40	40	0 40	5 0	300	5 0	9 20	560	9 22
0 50	50	0 50	5 10	310	5 10	9 30	570	9 32
1 0	60	1 0	5 20	320	5 20	9 40	580	9 42
1 10	70	1 10	5 30	330	5 30	9 50	590	9 52
1 20	80	1 20	5 40	340	5 40	10 0	600	10 03
1 30	90	1 30	5 50	350	5 50	10 10	610	10 13
1 40	100	1 40	6 0	360	6 0	10 20	620	10 23
1 50	110	1 50	6 10	370	6 10	10 30	630	10 33
2 0	120	2 0	6 20	380	6 20	10 40	640	10 44
2 10	130	2 10	6 30	390	6 30	10 50	650	10 54
2 20	140	2 20	6 40	400	6 40	11 0	660	11 04
2 30	150	2 30	6 50	410	6 50	11 10	670	11 14
2 40	160	2 40	7 0	420	7 0	11 20	680	11 24
2 50	170	2 50	7 10	430	7 11	11 30	690	11 34
3 0	180	3 0	7 20	440	7 21	11 40	700	11 44
3 10	190	3 10	7 30	450	7 31	11 50	710	11 55
3 20	200	3 20	7 40	460	7 41	12 0	720	12 05
3 30	210	3 30	7 50	470	7 51	12 10	730	12 15
3 40	220	3 40	8 0	480	8 2	12 20	740	12 26
3 40	230	3 50	8 10	490	8 12	12 30	750	12 36
4 0	240	4 0	8 20	500	8 22	12 40	760	12 46
4 10	250	4 10	8 30	510	8 32	12 50	770	12 56

Miles of East and West.								
Latitude.	Miles East and West	Longitude.	Latitude.	Miles East and West	Longitude.	Latitude.	Mile East and West	Longitude.
D. M.		D. M.	D. M.		D. M.	D. M.		D. M.
13 0	780	13 07	17 20	1040	17 36	21 40	1300	22 12
13 10	790	13 17	17 30	1050	17 47	21 50	1310	22 22
13 20	800	13 27	17 40	1060	17 57	22 0	1320	22 23
13 30	810	13 38	17 50	1070	18 08	22 10	1330	22 44
13 40	820	13 48	18 0	1080	18 18	22 20	1340	22 55
13 50	830	13 58	18 10	1090	18 29	22 30	1350	23 06
14 0	840	14 08	18 20	1100	18 39	22 40	1360	23 17
14 10	850	14 18	18 30	1110	18 49	22 50	1370	23 28
14 20	860	14 25	18 40	1120	19 00	23 0	1380	23 39
14 30	870	14 39	18 50	1130	19 10	23 10	1390	23 49
14 40	880	14 49	19 0	1140	19 21	23 20	1400	24 00
14 50	890	15 00	19 10	1150	19 31	23 30	1410	24 11
15 0	900	15 10	19 20	1160	19 42	23 40	1420	24 22
15 10	910	15 21	19 30	1170	19 53	23 50	1430	24 33
15 20	920	15 31	19 40	1180	20 04	24 0	1440	24 44
15 30	930	15 41	19 50	1190	20 14	24 10	1450	24 55
15 40	940	15 51	20 0	1200	20 25	24 20	1460	25 06
15 50	950	16 02	20 10	1210	20 35	24 30	1470	25 17
16 0	960	16 12	20 20	1220	20 46	24 40	1480	25 28
16 10	970	16 23	20 30	1230	20 57	24 50	1490	25 32
16 20	980	16 33	20 40	1240	21 07	25 0	1500	25 50
16 30	990	16 44	20 50	1250	21 18	25 10	1510	26 01
16 40	1000	16 55	21 0	1260	21 23	25 20	1520	26 12
16 50	1010	17 05	21 10	1270	21 39	25 30	1530	26 23
17 0	1020	17 15	21 20	1280	21 50	25 40	1540	26 34
17 10	1030	17 25	21 30	1290	21 01	25 50	1550	26 45

A TABLE of

Latitude	Miles East and West	Longitude	Latitude	Miles East and West	Longitude	Latitude	Miles East and West	Longitude
D. M.		D. M.	D. M.		D. M.	D. M.		D. M.
26 0	1560	26 56	30 20	1820	31 51	34 40	2080	36 59
26 10	1570	27 07	30 30	1830	32 03	34 50	2090	37 12
26 20	1580	27 08	30 40	1840	32 15	35 0	2100	37 24
26 30	1590	27 29	30 50	1850	32 26	35 10	2110	37 36
26 40	1600	27 40	31 0	1860	32 38	35 20	2120	37 48
26 50	1610	27 51	31 10	1870	32 49	35 30	2130	38 00
27 0	1620	28 03	31 20	1880	33 00	35 40	2140	38 13
27 10	1630	28 14	31 30	1890	33 12	35 50	2150	38 25
27 20	1640	28 25	31 40	1900	33 25	36 0	2160	38 38
27 30	1650	28 37	31 50	1910	33 37	36 10	2170	38 50
27 40	1660	28 48	32 0	1920	33 48	36 20	2180	39 03
27 50	1670	28 59	32 10	1930	34 00	36 30	2190	39 15
28 0	1680	29 11	32 20	1940	34 12	36 40	2200	39 27
28 10	1690	29 22	32 30	1950	34 24	36 50	2210	39 40
28 20	1700	29 34	32 40	1960	34 36	37 0	2220	39 53
28 30	1710	29 46	32 50	1970	34 43	37 10	2230	40 05
28 40	1720	29 57	33 0	1980	35 00	37 20	2240	40 18
28 50	1730	30 08	33 10	1990	35 12	37 30	2250	40 31
29 0	1740	30 19	33 20	2000	35 23	37 40	2260	40 43
29 10	1750	30 31	33 30	2010	35 35	37 50	2270	40 56
29 20	1760	30 43	33 40	2020	35 47	38 0	2280	41 08
29 30	1770	30 54	33 50	2030	35 59	38 10	2290	41 21
29 40	1780	31 05	34 0	2040	36 11	38 20	2300	41 33
29 50	1790	31 17	34 10	2050	36 23	38 30	2310	41 46
30 0	1800	31 28	34 20	2060	36 03	38 40	2320	42 00
30 10	1810	31 40	34 30	2070	36 47	38 50	2330	42 13

Miles of East and West.														
Latitude		Mile East and West	Longitude		Latitude		Mile East and West	Longitude		Latitude		Miles East and West	Longitude	
D.	M.	D.	D.	M.	D.	M.	D.	D.	M.	D.	M.	D.	D.	M.
39	0	234	42	26	43	20	260	48	10	47	40	286	54	21
39	10	235	42	35	43	50	261	48	25	47	50	287	54	36
39	20	236	42	52	43	40	262	48	39	48	0	288	54	52
39	30	237	43	04	43	50	263	48	53	48	10	289	55	05
39	40	238	43	17	44	0	264	49	06	48	20	290	55	22
39	50	239	43	30	44	10	265	49	20	48	30	291	55	37
40	0	240	43	43	44	20	266	49	54	48	40	292	55	51
40	10	241	43	56	44	30	267	49	48	48	50	293	56	07
40	20	242	44	09	44	40	268	50	02	49	0	294	56	22
40	30	243	44	21	44	50	269	50	16	49	10	295	56	38
40	40	244	44	34	45	0	270	50	30	49	20	296	56	52
40	50	245	44	48	45	10	271	50	43	49	30	297	57	08
41	0	246	45	02	45	20	272	50	57	49	40	298	57	23
41	10	247	45	16	45	30	273	51	12	49	50	299	57	39
41	20	248	45	29	45	40	274	51	26	50	0	300	57	54
41	30	249	45	42	45	50	275	51	40	50	10	301	58	10
41	40	250	45	55	46	0	276	51	54	50	20	302	58	26
41	50	251	46	08	46	10	277	52	10	50	30	303	58	42
42	0	252	46	22	46	20	278	52	25	50	40	314	58	58
42	10	253	46	36	46	30	279	52	35	50	50	315	59	14
42	20	254	46	49	46	40	280	52	54	51	0	306	59	30
42	30	255	47	02	46	50	281	53	08	51	10	307	59	46
42	40	256	47	16	47	0	282	53	23	51	20	308	60	01
42	50	257	47	30	47	10	283	53	37	51	30	309	60	17
42	0	258	47	43	47	20	284	53	52	51	40	310	60	33
42	10	259	47	56	47	30	285	54	06	51	50	311	60	49

A TABLE of

Latitude D.	M.	Miles East and West	Longitude D.	M.	Latitude D.	M.	Miles East and West	Longitude D.	M.	Latitude D.	M.	Miles East and West	Longitude D.	M.
52	0	3120	61	05	56	20	3380	68	29	60	40	3640	76	48
52	10	3130	61	21	56	30	3390	68	47	60	50	3650	77	08
52	20	3140	61	37	56	40	3400	69	05	61	0	3660	77	29
52	30	3150	61	54	56	50	3410	69	24	61	10	3670	77	49
52	40	3160	62	10	57	0	3420	69	42	61	20	3680	78	10
52	50	3170	62	26	57	10	3430	70	00	61	30	3690	78	31
53	0	3180	62	43	57	20	3440	70	19	61	40	3700	78	51
53	10	3190	63	00	57	30	3450	70	38	61	50	3710	79	12
53	20	3200	63	17	57	40	3460	70	58	62	0	3720	79	34
53	30	3210	63	34	57	50	3470	71	14	62	10	3730	79	55
53	40	3220	63	51	58	0	3480	71	34	62	20	3740	80	17
53	50	3230	64	08	58	10	3490	71	53	62	30	3750	80	38
54	0	3240	64	24	58	20	3500	72	12	62	40	3760	81	00
54	10	3250	64	41	58	30	3510	72	31	62	50	3770	81	22
54	20	3260	64	58	58	40	3520	72	50	63	0	3780	81	44
54	30	3270	65	15	58	50	3530	73	09	63	10	3790	82	06
54	40	3280	65	32	59	0	3540	73	28	63	20	3800	82	28
54	50	3290	65	50	59	10	3550	73	48	63	30	3810	82	51
55	0	3300	66	08	59	20	3560	74	08	63	40	3820	83	14
55	10	3310	66	26	59	30	3570	74	26	63	50	3830	83	36
55	20	3320	66	42	59	40	3580	74	46	64	0	3840	83	59
55	30	3330	67	01	59	50	3590	75	07	64	10	3850	84	22
55	40	3340	67	19	60	0	3600	75	26	64	20	3860	84	45
55	50	3350	67	36	60	10	3610	75	47	64	30	3870	85	09
56	0	3360	67	54	60	20	3620	76	08	64	40	3880	85	35
56	10	3370	68	11	60	30	3630	76	28	64	50	3890	85	55

Miles of East and West.								
Latitude.		Miles East and West.	Longitude.		Latitude.		Miles East and West.	Longitude.		Latitude.		Miles East and West.	Longitude.	

D.	M.		D.	M.	D.	M.		D.	M.	D.	M.		D.	M.
65	0	3900	86	19	66	50	4010	90	48	68	40	4120	95	38
65	10	3910	86	42	67	0	4020	91	13	68	50	4130	96	05
65	20	3920	87	06	67	10	4030	91	38	69	0	4140	96	33
65	30	3930	87	29	67	20	4040	92	04	69	10	4150	97	02
65	40	3940	87	55	67	30	4050	92	30	69	20	4160	97	30
65	50	3950	88	20	67	40	4060	92	56	69	30	4170	98	00
66	0	3960	88	44	67	50	4070	93	23	69	40	4180	98	29
66	10	3970	89	08	68	0	4080	93	50	69	50	4190	98	58
66	20	3980	89	32	68	10	4090	94	17	70	0	4200	99	26
66	30	3990	89	57	68	20	4100	94	44					
66	40	4000	90	23	68	30	4110	95	11					

A TABLE

A
TABLE

FOR

Changing the **Degrees** *and* **Minutes** *of* **East**
and **West** *into* MILES.

Iii

A *Table* for changing the Degrees and

Parallel	1 Minutes.		2 Minutes.		3 Minutes.		4 Minutes.		5 Minutes.		6 Minutes.		7 Minutes.		8 Minutes.		9 Minutes.	
	M.	c.p	M.	c.p	M.	c.p	M.	c.p	M.	c.p	M.	c.p	M.	c.p	M.	c.p	M.	c.p
0	1	0	2	0	3	0	4	0	5	0	6	0	7	0	8	0	9	0
1	1	0	2	0	3	0	4	0	5	0	6	0	7	0	8	0	9	0
2	1	0	2	0	3	0	4	0	5	0	6	0	7	0	8	0	9	0
3	1	0	2	0	3	0	4	0	5	0	6	0	7	0	8	0	9	0
4	1	0	2	0	3	0	4	0	5	0	6	0	7	0	8	0	9	0
5	1	0	2	0	3	0	4	0	5	0	6	0	7	0	8	0	9	0
6	1	0	2	0	3	0	4	0	5	0	6	0	7	0	8	0	8	9
7	1	0	2	0	3	0	4	0	5	0	6	0	6	9	7	9	8	9
8	1	0	2	0	3	0	4	0	5	0	5	9	6	9	7	9	8	9
9	1	0	2	0	3	0	4	0	4	9	5	9	6	9	7	9	8	9
10	1	0	2	0	3	0	3	9	4	9	5	9	6	9	7	9	8	8
11	1	0	2	0	2	9	3	9	4	9	5	9	6	9	7	9	8	8
12	1	0	2	0	2	9	3	9	4	9	5	9	6	8	7	8	8	8
13	1	0	1	9	2	9	3	9	4	9	5	8	6	8	7	8	8	8
14	1	0	1	9	2	9	3	9	4	9	5	8	6	8	7	8	8	7
15	1	0	1	9	2	9	3	9	4	8	5	8	6	8	7	7	8	7
16	1	0	1	9	2	9	3	8	4	8	5	8	6	7	7	7	8	7
17	1	0	1	9	2	9	3	8	4	8	5	7	6	7	7	7	8	6
18	1	0	1	9	2	9	3	8	4	8	5	7	6	7	7	6	8	6
19	0	9	1	9	2	8	3	8	4	7	5	7	6	6	7	6	8	5
20	0	9	1	9	2	8	3	8	4	7	5	6	6	6	7	5	8	5
21	0	9	1	9	2	8	3	7	4	7	5	6	6	5	7	5	8	4
22	0	9	1	9	2	8	3	7	4	6	5	6	6	5	7	4	8	3
23	0	9	1	8	2	8	3	7	4	6	5	5	6	4	7	4	8	3
24	0	9	1	8	2	7	3	7	4	6	5	5	6	4	7	3	8	2
25	0	9	1	8	2	7	3	6	4	5	5	4	6	3	7	2	8	2
26	0	9	1	8	2	7	3	6	4	5	5	4	6	3	7	2	8	1
27	0	9	1	8	2	7	3	6	4	5	5	3	6	2	7	1	8	0
28	0	9	1	8	2	6	3	5	4	4	5	3	6	2	7	0	7	9
29	0	9	1	7	2	6	3	5	4	4	5	2	6	1	7	0	7	9
30	0	9	1	7	2	6	3	5	4	3	5	2	6	0	6	9	7	8
31	0	9	1	7	2	6	3	4	4	3	5	1	6	0	6	9	7	7
32	0	8	1	7	2	5	3	4	4	2	5	1	5	9	6	8	7	6
33	0	8	1	7	2	5	3	4	4	2	5	0	5	9	6	7	7	5
34	0	8	1	7	2	5	3	3	4	1	4	9	5	8	6	6	7	5

Minutes of East and West into Miles.

Parallel	1 Minutes (M.c.p.)	2 Minutes (M.c.p.)	3 Minutes (M.c.p.)	4 Minutes (M.c.p.)	5 Minutes (M.c.p.)	6 Minutes (M. c.p.)	7 Minutes (M. c.p.)	8 Minutes (M. c.p.)	9 Minutes (M. c.p.)
35	0 8	1 6	2 5	3 3	4 1	4 9	5 7	6 5	7 4
36	0 8	1 6	2 4	3 2	4 0	4 9	5 7	6 5	7 3
37	0 8	1 6	2 4	3 2	4 0	4 8	5 6	6 4	7 2
38	0 8	1 6	2 4	3 2	3 9	4 7	5 5	6 3	7 1
39	0 8	1 6	2 3	3 1	3 9	4 7	5 4	6 2	7 0
40	0 8	1 5	2 3	3 1	3 8	4 6	5 4	6 1	6 9
41	0 8	1 5	2 2	3 0	3 8	4 5	5 3	6 0	6 8
42	0 7	1 5	2 2	3 0	3 7	4 5	5 2	5 9	6 7
43	0 7	1 5	2 2	2 9	3 7	4 4	5 1	5 9	6 6
44	0 7	1 4	2 2	2 9	3 6	4 3	5 0	5 8	6 5
45	0 7	1 4	2 1	2 8	3 5	4 2	4 9	5 6	6 4
46	0 7	1 4	2 1	2 8	3 5	4 2	4 9	5 6	6 3
47	0 7	1 4	2 0	2 7	3 4	4 1	4 8	5 5	6 1
48	0 7	1 3	2 0	2 7	3 3	4 0	4 7	5 3	6 0
49	0 7	1 3	2 0	2 6	3 3	3 9	4 6	5 2	5 9
50	0 6	1 3	1 9	2 6	3 2	3 8	4 5	5 1	5 8
51	0 6	1 2	1 9	2 5	3 1	3 8	4 4	5 0	5 7
52	0 6	1 2	1 8	2 4	3 1	3 7	4 3	5 9	5 5
53	0 6	1 2	1 8	2 4	3 0	3 6	4 2	5 8	5 4
54	0 6	1 2	1 8	2 4	2 9	3 5	4 1	5 7	5 2
55	0 6	1 1	1 7	2 3	2 9	3 4	4 0	4 6	5 1
56	0 5	1 1	1 7	2 2	2 8	3 4	3 9	4 5	5 0
57	0 5	1 1	1 6	2 2	2 7	3 3	3 8	4 4	4 9
58	0 5	1 1	1 6	2 1	2 6	3 2	3 7	4 2	4 8
59	0 5	1 0	1 5	2 1	2 6	3 0	3 6	4 1	4 6
60	0 5	1 0	1 5	2 0	2 5	3 0	3 5	4 0	4 5
61	0 5	1 0	1 5	1 9	2 4	2 9	3 4	3 9	4 4
62	0 5	0 9	1 4	1 9	2 3	2 8	3 3	3 8	4 2
63	0 5	0 9	1 4	1 8	2 3	2 7	3 2	3 6	4 1
64	0 4	0 9	1 3	1 8	2 2	2 6	3 1	3 5	3 9
65	0 4	0 8	1 3	1 7	2 1	2 5	3 0	3 4	3 8
66	0 4	0 8	1 2	1 6	2 0	2 4	2 8	3 2	3 7
67	0 4	0 8	1 2	1 6	2 0	2 3	2 7	3 1	3 5
68	0 4	0 7	1 1	1 5	1 9	2 2	2 6	3 0	3 4
69	0 4	0 7	1 1	1 4	1 8	2 1	2 5	2 9	3 2

A *Table* for changing the Degrees and

Parallel	10 Minutes	20 Minutes	30 Minutes	40 Minutes	50 Minutes	1 Degree	2 Degrees	3 Degrees	4 Degrees
	M. c. p	M. c. p	M. c. p	M. c. p	M. c. p	M. c. p	M. c. p	M. c. p	M. c. p
0	10	0 20	0 30	0 40	0 50	0 60	0 120	0 180	0 240 0
1	10	0 20	0 30	0 40	0 50	0 60	0 120	0 180	0 240 0
2	10	0 20	0 30	0 40	0 50	0 60	0 119	9 179	9 239 9
3	10	0 20	0 30	0 39	9 49	9 59	9 119	8 179	8 239 7
4	10	0 20	0 29	9 39	9 49	9 59	9 119	7 179	6 239 4
5	10	0 19	9 29	9 39	9 49	8 59	8 119	5 179	3 239 1
6	9	5 19	9 29	9 39	8 49	8 59	7 119	3 179	0 238 7
7	9	9 19	9 29	9 39	8 49	7 59	6 119	1 178	7 238 2
8	9	9 19	8 29	9 39	7 49	6 59	5 118	8 178	2 237 7
9	9	9 19	8 29	8 39	6 49	5 59	4 118	5 177	8 237 0
10	9	8 19	8 29	7 29	5 39	4 49	2 59	0 118 2 177	3 236 4
11	9	8 19	6 29	4 39	3 49	0 58	9 117	8 176	7 235 6
12	9	8 19	6 29	3 39	1 48	9 58	7 117	4 176	0 234 8
13	9	7 19	5 29	2 39	0 48	7 58	5 116	9 175	4 233 8
14	9	7 19	4 29	1 38	8 48	5 58	2 116	4 174	7 232 8
15	9	7 19	3 29	0 38	6 48	3 58	0 115	9 173	9 231 8
16	9	6 19	2 28	8 38	5 48	1 57	7 115	4 173	0 230 7
17	9	6 19	1 28	7 38	2 47	8 57	4 114	8 172	1 229 5
18	9	5 19	0 28	5 38	0 47	6 57	0 114	1 171	1 228 3
19	9	5 18	9 28	4 37	8 47	2 56	7 113	5 170	5 226 9
20	9	4 18	8 28	2 37	6 47	0 56	4 112	8 169	1 225 5
21	9	3 18	7 28	0 37	3 46	7 56	0 112	0 168	0 224 0
22	9	3 18	5 27	8 37	1 46	4 55	6 111	3 166	8 222 5
23	9	2 18	4 27	6 36	8 46	0 55	2 110	5 165	7 220 0
24	9	1 18	3 27	4 36	5 45	7 54	8 109	6 164	4 219 3
25	9	0 18	1 27	1 36	3 45	3 54	4 108	7 163	1 217 5
26	9	0 18	0 27	0 36	0 44	9 53	9 107	9 161	8 215 7
27	8	9 17	8 26	7 35	6 44	6 53	5 106	9 160	4 213 8
28	8	8 17	7 26	5 35	3 44	1 53	0 106	0 158	5 211 9
29	8	7 17	5 26	2 35	0 43	7 52	5 105	0 157	4 209 9
30	8	6 17	3 26	0 34	6 43	3 52	0 103	9 155	5 207 8
31	8	6 17	1 25	7 34	3 42	8 51	4 102	9 154	3 205 7
32	8	5 17	0 25	4 33	5 42	4 50	9 101	8 152	6 203 5
33	8	4 16	8 25	1 33	5 41	9 50	3 100	6 151	0 201 3
34	8	3 16	6 24	9 33	1 41	5 49	7 99	5 149	2 199 0

Minutes of East and West into Miles.

Parallel	10 Minutes		20 Minutes		30 Minutes		40 Minutes		50 Minutes		1 Degree		2 Degrees		3 Degrees		4 Degrees	
	M.	c.p.	M.	c.p.	M.	c.p.	M.	c.p.	M.	c.p.	M.	c.p.	M.	c.p.	M.	c.p.	M.	c.p.
35	8	2	16	4	24	5	32	8	41	0	49	1	98	2	147	4	196	6
36	8	1	16	2	24	3	32	4	40	5	48	5	97	0	145	6	194	2
37	8	0	16	0	24	0	31	·	39	9	47	9	95	8	143	8	191	7
38	7	9	15	8	23	6	31	5	39	3	47	2	94	5	141	8	189	1
39	7	8	15	5	23	3	31	1	38	1	46	6	93	2	139	9	186	5
40	7	7	15	3	23	0	30	6	38	3	46	0	91	9	137	9	183	8
41	7	5	15	1	22	6	30	1	37	7	45	3	90	6	135	8	181	1
42	7	4	14	9	22	3	29	7	37	2	44	6	89	2	133	8	178	4
43	7	3	14	6	21	9	29	2	36	6	43	9	87	7	131	6	175	5
44	7	2	14	4	21	6	28	8	36	0	43	2	86	3	129	5	172	6
45	7	1	14	1	21	2	28	3	35	5	42	4	84	8	127	2	169	7
46	6	9	13	9	20	8	27	8	34	7	41	6	83	4	125	0	166	7
47	6	8	13	6	20	4	27	2	34	1	40	9	81	8	122	8	163	7
48	6	7	13	5	20	0	26	6	33	4	40	0	79	9	119	9	159	9
49	6	6	13	1	19	7	26	3	32	3	39	4	78	7	118	1	157	5
50	6	4	12	6	19	3	25	7	32	1	38	6	77	7	115	7	154	3
51	6	3	12	9	18	9	25	2	31	5	37	8	75	5	113	3	151	0
52	6	2	12	3	18	5	24	6	30	8	36	9	73	9	110	8	147	8
53	6	0	12	0	18	1	24	1	30	1	36	1	72	2	108	3	144	4
54	5	9	11	8	17	6	23	5	29	4	35	3	70	5	105	8	141	1
55	5	7	11	5	17	2	22	9	28	7	34	4	68	8	103	2	137	6
56	5	6	11	2	16	8	22	4	28	0	33	6	67	1	100	7	134	2
57	5	4	11	9	16	3	21	8	27	2	32	7	65	5	98	3	131	0
58	5	3	10	6	15	9	21	2	26	5	31	8	63	6	95	4	127	2
59	5	2	10	3	15	5	20	6	25	8	30	9	61	8	92	7	123	6
60	5	0	10	0	15	0	20	0	25	0	30	0	60	0	90	0	120	0
61	4	8	9	7	14	5	19	4	24	2	29	1	58	2	87	3	116	4
62	4	7	9	4	14	1	18	8	23	4	28	2	56	3	84	5	112	7
63	4	5	9	1	13	6	18	2	22	7	27	2	54	5	81	6	109	0
64	4	5	8	8	13	2	17	5	22	6	26	3	52	6	78	9	105	2
65	4	2	8	5	12	7	16	9	21	1	25	4	50	7	76	1	101	4
66	4	1	8	1	12	2	16	3	20	3	24	4	48	8	73	2	97	6
67	3	9	7	8	11	7	15	6	19	5	23	4	46	9	70	3	93	8
68	3	7	7	5	11	2	15	0	18	7	22	5	45	0	67	4	89	9
69	3	6	7	2	10	8	14	3	17	3	21	5	43	1	64	6	86	1
70	3	3	6	7	10	0	13	4	16	7	20	0	40	1	60	1	80	1

A *Table* for changing the Degrees and

Parallel	5 Degrees.		6 Degrees.		7 Degrees.		8 Degrees.		9 Degrees.		10 Degrees.		20 Degrees.	
	M.	c.p.	M.	c.p.	M.	c.p.	M.	c.p.	M.	c.p.	M.	c.p.	M.	c.p.
0	300	0	360	0	420	0	480	0	540	0	600	0	1200	0
1	300	0	359	9	419	9	479	9	539	9	599	9	1199	8
2	299	8	359	8	419	7	479	7	539	7	599	6	1199	3
3	299	6	359	5	419	4	479	3	539	3	599	2	1198	4
4	299	3	359	1	419	0	478	8	538	7	598	5	1197	6
5	298	9	358	6	418	4	478	2	537	9	597	7	1195	4
6	298	4	358	0	417	7	477	4	537	0	596	7	1193	4
7	297	8	357	3	416	9	476	4	536	0	595	5	1191	0
8	297	0	356	5	415	9	475	3	534	7	594	2	1188	3
9	296	3	355	6	414	8	474	1	533	4	592	6	1185	2
10	295	4	354	5	413	6	472	7	531	8	590	9	1181	8
11	294	5	353	4	412	3	471	2	529	0	588	9	1177	9
12	293	4	352	1	410	8	469	5	528	2	586	9	1173	8
13	292	3	350	7	409	2	467	7	526	2	584	6	1169	2
14	291	1	349	3	407	5	465	7	524	0	582	2	1164	4
15	289	8	347	7	405	7	463	6	521	6	579	6	1159	1
16	288	4	346	1	403	7	461	4	519	0	576	7	1153	5
17	286	9	344	3	401	6	459	0	516	4	573	8	1147	6
18	285	7	342	4	399	4	456	5	513	6	570	6	1141	3
19	283	6	340	4	397	1	453	8	510	6	567	3	1134	6
20	281	9	338	3	394	7	451	1	507	4	563	8	1127	6
21	280	1	336	1	392	1	448	1	504	1	560	1	1120	3
22	278	2	333	8	389	4	445	0	500	7	556 1/4		1112	6
23	275	1	331	4	386	6	441	8	497	1	552	3	1104	6
24	274	1	328	9	384	0	438	5	493	3	548	1	1096	2
25	272	0	326	3	380	6	435	0	489	4	543	8	1087	6
26	269	6	323	6	377	6	431	5	485	3	539	3	1078	5
27	267	3	320	7	374	2	427	7	481	1	534	6	1069	2
28	264	9	317	9	370	8	423	8	476	8	529	8	1059	6
29	262	3	314	9	367	3	419	8	472	3	524	8	1049	6
30	259	8	311	7	363	7	415	7	467	7	519	6	1039	2
31	257	1	308	6	360	0	411	4	462	4	514	3	1028	6
32	254	4	305	3	356	1	407	1	457	9	508	8	1017	6
33	251	6	301	9	352	2	402	6	452	9	503	2	1006	4
34	248	7	307	5	348	1	307	9	447	7	497	4	094	8

	Minutes of East and West into Miles.												
Parallel.	5 Degrees.		6 Degrees.		7 Degrees.		8 Degrees.		9 Degrees.		10 Degrees.		20 Degrees.	
	M.	c.p.	M.	c.p.	M.	c.p.	M.	c.p.	M.	c.p.	M.	c.p.	M.	c.p.
35	245	7	294	8	344	0	393	2	442	3	491	5	983	0
36	242	7	291	2	339	8	388	3	436	9	485	4	970	3
37	239	6	281	5	335	4	383	3	431	2	479	1	958	2
38	236	4	283	7	331	0	378	4	425	5	472	8	945	6
39	233	1	279	8	326	4	373	0	419	7	466	3	932	6
40	229	8	275	8	321	7	367	7	413	6	459	6	919	2
41	226	4	271	7	317	0	362	5	407	5	452	8	905	6
42	222	9	267	5	312	1	356	7	401	3	445	9	891	7
43	219	4	263	3	307	2	351	0	394	9	438	8	877	6
44	215	8	259	0	302	1	345	2	388	4	431	6	863	2
45	212	1	254	5	296	9	339	4	381	8	424	2	848	4
46	208	4	250	0	291	7	333	4	375	1	416	8	833	6
47	204	6	245	5	286	4	327	3	368	2	409	1	818	4
48	199	8	239	8	279	8	319	7	359	7	399	8	799	3
49	196	7	236	2	275	5	314	9	354	2	393	3	787	2
50	192	8	231	4	270	0	308	5	347	1	385	7	771	4
51	188	8	226	6	264	3	302	1	349	8	377	6	755	2
52	184	7	221	6	258	6	295	5	332	5	369	4	738	8
53	180	5	216	6	252	8	288	9	325	0	361	0	722	2
54	176	3	211	6	246	9	282	1	317	4	352	7	705	4
55	172	1	206	5	240	9	275	3	309	7	344	1	688	2
56	167	8	201	3	234	9	268	4	302	0	335	5	671	0
57	163	8	196	9	229	4	262	2	295	0	326	8	653	6
58	159	0	190	8	222	6	254	4	286	2	318	0	636	0
59	154	5	185	4	216	3	247	2	278	1	309	0	618	0
60	150	0	180	0	210	0	240	0	270	0	300	0	600	2
61	145	4	174	5	203	9	232	7	261	8	290	9	581	8
62	140	8	169	0	197	2	225	3	253	5	281	7	563	4
63	136	2	163	4	190	6	217	9	245	2	272	4	544	8
64	131	5	157	8	184	1	210	4	236	7	263	0	526	1
65	126	8	152	1	177	5	202	8	228	2	253	6	507	2
66	122	0	146	4	170	8	195	2	219	6	244	0	488	0
67	117	2	140	7	164	1	187	5	211	0	234	4	468	8
68	112	4	134	8	157	3	179	2	202	3	224	8	449	6
69	107	6	129	1	150	6	172	1	193	6	215	0	430	0
70	100	1	120	2	140	2	160	2	180	2	200	3	400	6

A
TABLE
FOR

Reducing MILES *of* Eaſt *and* Weſt *into* Degrees *of* Longitude.

Kkk

	1 Mile.			2 Miles.			3 Miles.			4 Miles.			5 Miles.		
Parallel.	D.	M.	S.	D.	M.	S.	D.	M.	S.	D.	M.	S.	D.	M.	S.
0	0	1	0	0	2	0	0	3	0	0	4	0	0	5	0
1	0	1	0	0	2	0	0	3	0	0	4	0	0	5	0
2	0	1	0	0	2	0	0	3	0	0	4	0	0	5	0
3	0	1	0	0	2	0	0	3	0	0	4	0	0	5	0
4	0	1	0	0	2	0	0	3	0	0	4	1	0	5	1
5	0	1	0	0	2	0	0	3	1	0	4	1	0	5	1
6	0	1	0	0	2	1	0	3	1	0	4	1	0	5	2
7	0	1	0	0	2	1	0	3	1	0	4	2	0	5	2
7	0	1	1	0	2	1	0	3	2	0	4	2	0	5	3
9	0	1	1	0	2	1	0	3	2	0	4	3	0	5	5
10	0	1	1	0	2	2	0	3	3	0	4	4	0	5	5
11	0	1	1	0	2	2	0	3	3	0	4	4	0	5	6
12	0	1	2	0	2	3	0	3	4	0	4	5	0	5	7
13	0	1	2	0	2	3	0	3	5	0	4	6	0	5	8
14	0	1	2	0	2	4	0	3	5	0	4	7	0	5	9
15	0	1	2	0	2	4	0	3	6	0	4	8	0	5	11
16	0	1	2	0	2	5	0	3	7	0	4	10	0	5	12
17	0	1	3	0	2	6	0	3	8	0	4	11	0	5	14
18	0	1	3	0	2	6	0	3	9	0	4	12	0	5	15
19	0	1	3	0	2	7	0	3	10	0	4	14	0	5	17
20	0	1	4	0	2	8	0	3	12	0	4	15	0	5	18
21	0	1	4	0	2	9	0	3	13	0	4	17	0	5	21
22	0	1	5	0	2	9	0	3	14	0	4	19	0	5	24
23	0	1	5	0	2	10	0	3	16	0	4	21	0	5	26
24	0	1	6	0	2	11	0	3	17	0	4	23	0	5	28
25	0	1	6	0	2	12	0	3	19	0	4	25	0	5	31
26	0	1	7	0	2	13	0	3	20	0	4	27	0	5	34
27	0	1	7	0	2	15	0	3	22	0	4	29	0	5	37
28	0	1	8	0	2	16	0	3	24	0	4	31	0	5	39
29	0	1	9	0	2	17	0	3	26	0	4	34	0	5	43
30	0	1	9	0	2	19	0	3	28	0	4	37	0	5	46
31	0	1	10	0	2	20	0	3	30	0	4	40	0	5	50
32	0	1	11	0	2	21	0	3	32	0	4	43	0	5	54
33	0	1	12	0	2	23	0	3	35	0	4	46	0	5	58
34	0	1	12	0	2	25	0	3	37	0	4	50	0	6	02

A Table for reducing Miles of East

Parallel.	6 Miles.			7 Miles.			8 Miles.			9 Miles.		
	D.	M.	S.	D.	M.	S.	D.	M.	S.	D.	M.	S.
0	0	6	0	0	7	0	0	8	0	0	9	0
1	0	6	0	0	7	0	0	8	0	0	9	0
2	0	6	0	0	7	0	0	8	0	0	9	1
3	0	6	0	0	7	1	0	8	1	0	9	1
4	0	6	1	0	7	1	0	8	1	0	9	1
5	0	6	1	0	7	2	0	8	2	0	9	2
6	0	6	2	0	7	2	0	8	3	0	9	3
7	0	6	3	0	7	3	0	8	4	0	9	4
8	0	6	3	0	7	4	0	8	4	0	9	5
9	0	6	4	0	7	5	0	8	6	0	9	6
10	0	6	6	0	7	7	0	8	7	0	9	8
11	0	6	7	0	7	8	0	8	9	0	9	10
12	0	6	8	0	7	9	0	8	11	0	9	12
13	0	6	9	0	7	11	0	8	12	0	9	14
14	0	6	11	0	7	13	0	8	15	0	9	16
15	0	6	13	0	7	15	0	8	17	0	9	19
16	0	6	14	0	7	17	0	8	19	0	9	22
17	0	6	16	0	7	19	0	8	22	0	9	25
18	0	6	18	0	7	21	0	8	25	0	9	28
19	0	6	20	0	7	24	0	8	28	0	9	31
20	0	6	22	0	7	25	0	8	29	0	9	34
21	0	6	26	0	7	30	0	8	34	0	9	38
22	0	6	28	0	7	33	0	8	38	0	9	42
23	0	6	31	0	7	37	0	8	41	0	9	47
24	0	6	33	0	7	39	0	8	44	0	9	50
25	0	6	37	0	7	43	0	8	50	0	9	56
26	0	6	40	0	7	47	0	8	54	0	10	01
27	0	6	42	0	7	51	0	9	59	0	10	06
28	0	6	47	0	7	55	0	9	03	0	10	11
29	0	6	52	0	8	00	0	9	09	0	10	17
30	0	6	56	0	8	05	0	9	14	0	10	24
31	0	7	00	0	8	10	0	9	20	0	10	30
32	0	7	04	0	8	15	0	9	26	0	10	37
33	0	7	09	0	8	21	0	9	32	0	10	44
34	0	7	14	0	8	27	0	9	39	0	10	51

A *Table* for reducing Miles of East

Parallel	1 Mile.			2 Miles.			3 Miles.			4 Miles.			5 Miles.		
	D.	M.	S.	D.	M.	S.	D.	M.	S.	D.	M.	S.	D.	M.	S.
35	0	1	13	0	2	26	0	3	40	0	4	53	0	6	6
36	0	1	14	0	2	28	0	3	42	0	4	57	0	6	11
37	0	1	15	0	2	30	0	3	45	0	5	01	0	6	16
38	0	1	16	0	2	32	0	3	48	0	5	05	0	6	21
39	0	1	17	0	2	34	0	3	52	0	5	09	0	6	26
40	0	1	18	0	2	37	0	3	55	0	5	13	0	6	32
41	0	1	19	0	2	39	0	3	58	0	5	18	0	6	37
42	0	1	21	0	2	41	0	4	02	0	5	23	0	6	44
43	0	1	22	0	2	44	0	4	06	0	5	28	0	6	50
44	0	1	23	0	2	47	0	4	10	0	5	34	0	6	57
45	0	1	25	0	2	50	0	4	15	0	5	39	0	7	4
46	0	1	26	0	2	53	0	4	19	0	5	45	0	7	12
47	0	1	28	0	2	56	0	4	24	0	5	52	0	7	20
48	0	1	30	0	2	59	0	4	29	0	5	59	0	7	28
49	0	1	31	0	3	03	0	4	34	0	6	06	0	7	37
50	0	1	33	0	3	07	0	4	40	0	6	13	0	7	47
51	0	1	35	0	3	11	0	4	46	0	6	21	0	7	57
52	0	1	37	0	3	14	0	4	52	0	6	30	0	8	07
53	0	1	40	0	3	19	0	4	59	0	6	38	0	8	18
54	0	1	42	0	3	24	0	5	06	0	6	48	0	8	30
55	0	1	45	0	3	29	0	5	14	0	6	58	0	8	43
56	0	1	47	0	3	35	0	5	22	0	7	09	0	8	56
57	0	1	50	0	3	40	0	5	30	0	7	21	0	9	11
58	0	1	53	0	3	46	0	5	40	0	7	33	0	9	26
59	0	1	56	0	3	53	0	5	49	0	7	46	0	9	42
60	0	2	00	0	4	00	0	6	00	0	8	00	0	10	00
61	0	2	04	0	4	07	0	6	11	0	8	15	0	10	19
62	0	2	08	0	4	16	0	6	23	0	8	31	0	10	39
63	0	2	12	0	4	24	0	6	36	0	8	49	0	11	01
64	0	2	17	0	4	34	0	6	51	0	9	07	0	11	24
65	0	2	22	0	4	44	0	7	06	0	9	28	0	11	50
66	0	2	28	0	4	45	0	7	23	0	9	50	0	12	18
67	0	2	33	0	5	06	0	7	39	0	10	12	0	12	45
68	0	2	40	0	5	27	0	8	14	0	11	01	0	13	48
69	0	2	47	0	5	34	0	8	21	0	11	08	0	13	55
70	0	2	55	0	5	50	0	8	45	0	11	40	0	14	35

	and West into Degrees of Longitude.											
Parallel.	6 *Miles.*			7 *Miles.*			8 *Miles.*			9 *Miles.*		
	D.	M.	S.	D.	M.	S.	D.	M.	S.	D.	M.	S.
35	0	7	19	0	8	33	0	9	46	0	10	59
36	0	7	25	0	8	39	0	9	53	0	11	08
37	0	7	31	0	8	46	0	10	01	0	11	16
38	0	7	37	0	8	53	0	10	09	0	11	25
39	0	7	43	0	9	00	0	10	18	0	11	35
40	0	7	50	0	9	08	0	10	27	0	11	45
41	0	7	57	0	9	16	0	10	36	0	11	55
42	0	8	04	0	9	25	0	10	46	0	12	07
43	0	8	12	0	9	34	0	10	56	0	12	20
44	0	8	21	0	9	44	0	11	07	0	12	31
45	0	8	29	0	9	54	0	11	19	0	12	44
46	0	8	38	0	10	05	0	11	31	0	12	57
47	0	8	48	0	10	16	0	11	44	0	13	12
48	0	8	58	0	10	28	0	11	57	0	13	27
49	0	9	08	0	10	40	0	11	12	0	13	43
50	0	9	20	0	10	53	0	12	27	0	14	00
51	0	9	22	0	10	57	0	12	33	0	14	08
52	0	9	44	0	11	22	0	13	00	0	14	37
53	0	9	58	0	11	38	0	13	18	0	14	57
54	0	10	12	0	11	55	0	13	36	0	15	19
55	0	10	28	0	12	12	0	13	57	0	15	41
56	0	10	44	0	12	31	0	14	18	0	16	06
57	0	11	01	0	12	51	0	14	41	0	16	31
58	0	11	19	0	13	13	0	15	06	0	16	59
59	0	11	39	0	13	35	0	15	32	0	17	28
60	0	12	00	0	14	00	0	16	00	0	18	00
61	0	12	22	0	14	26	0	16	30	0	18	33
62	0	12	47	0	14	55	0	17	02	0	19	10
63	0	13	13	0	15	25	0	17	37	0	19	49
64	0	13	41	0	15	58	0	18	15	0	20	32
65	0	14	12	0	16	34	0	18	55	0	21	18
66	0	14	45	0	17	12	0	19	40	0	22	08
67	0	15	18	0	17	51	0	20	24	0	22	57
68	0	16	25	0	19	22	0	22	09	0	23	56
69	0	16	42	0	19	29	0	22	16	0	25	03
70	0	17	30	0	20	25	0	23	20	0	26	13

	10 *Miles.*			20 *Miles.*			30 *Miles.*			40 *Miles.*			50 *Miles.*		
A *Table* for reducing Miles of East															
Parallel.	D.	M.	S.	D.	M.	S.	D.	M.	S.	D.	M.	S.	D.	M.	S.
0	0	10	0	0	20	0	0	30	0	0	40	0	0	50	0
1	0	10	0	0	20	0	0	30	0	0	40	0	0	50	0
2	0	10	1	0	20	1	0	30	1	0	40	1	0	50	1
3	0	10	1	0	20	2	0	30	2	0	40	3	0	50	4
4	0	10	1	0	20	3	0	30	4	0	40	6	0	50	7
5	0	10	2	0	20	5	0	30	7	0	40	9	0	50	11
6	0	10	3	0	20	6	0	30	9	0	40	13	0	50	16
7	0	10	4	0	20	9	0	30	13	0	40	18	0	50	22
8	0	10	6	0	20	1	0	30	17	0	40	23	0	50	28
9	0	10	7	0	20	5	0	30	22	0	40	30	0	50	37
10	0	10	9	0	20	18	0	30	28	0	40	37	0	50	46
11	0	10	11	0	20	22	0	30	34	0	49	45	0	50	56
12	0	10	13	0	20	27	0	30	40	0	40	53	0	51	06
13	0	10	16	0	20	32	0	30	47	0	41	03	0	51	19
14	0	10	18	0	20	37	0	30	55	0	41	13	0	51	31
15	0	10	21	0	20	42	0	31	09	0	41	24	0	51	46
16	0	10	24	0	20	49	0	31	13	0	41	37	0	52	01
17	0	10	27	0	20	55	0	31	21	0	41	49	0	52	16
18	0	10	31	0	21	02	0	31	33	0	42	04	0	52	34
19	0	10	34	0	21	09	0	31	44	0	42	18	0	52	53
20	0	10	38	0	21	17	0	31	55	0	42	34	0	53	12
21	0	10	43	0	21	25	0	32	08	0	42	51	0	53	33
22	0	10	47	0	21	34	0	32	21	0	43	09	0	53	56
23	0	10	52	0	21	44	0	32	35	0	43	27	0	54	19
24	0	10	57	0	21	54	0	32	50	0	43	47	0	54	44
25	0	11	02	0	22	04	0	33	06	0	44	08	0	55	10
26	0	11	07	0	22	15	0	33	22	0	44	30	0	55	37
27	0	11	13	0	22	27	0	33	40	0	44	53	0	56	07
28	0	11	19	0	22	37	0	33	56	0	45	16	0	56	35
29	0	11	26	0	22	52	0	34	18	0	45	44	0	57	10
30	0	11	33	0	23	06	0	34	38	0	46	11	0	57	44
31	0	11	40	0	23	19	0	34	58	0	46	38	0	58	18
32	0	11	47	0	23	35	0	35	22	0	47	10	0	58	57
33	0	11	55	0	23	50	0	35	46	0	47	41	0	59	37
34	0	12	04	0	24	08	0	36	11	0	48	15	1	00	19

	and West into Degrees of Longitude.														
Parallel.	60 *Miles.*			70 *Miles.*			80 *Miles.*			90 *Miles.*			100 *Miles.*		
	D.	M.	S.	D.	M.	S.	D.	M.	S.	D.	M.	S.	D.	M.	S.
0	1	0	0	1	10	00	1	20	00	1	30	00	1	40	00
1	1	0	1	1	10	01	1	20	01	1	30	00	1	40	01
2	1	0	2	1	10	03	1	20	03	1	30	03	1	40	04
3	1	0	5	1	10	06	1	20	07	1	30	07	1	40	08
4	1	0	9	1	10	10	1	20	12	1	30	13	1	40	14
5	1	0	14	1	10	16	1	20	18	1	30	21	1	40	23
6	1	0	19	1	10	21	1	20	24	1	30	27	1	40	31
7	1	0	27	1	10	31	1	20	35	1	30	39	1	40	45
8	1	0	34	1	10	40	1	20	46	1	30	52	1	40	59
9	1	0	45	1	10	52	1	20	59	1	31	07	1	41	15
10	1	0	55	1	11	05	1	21	14	1	31	23	1	41	34
11	1	1	07	1	11	19	1	21	30	1	31	41	1	41	52
12	1	1	20	1	11	33	1	21	46	1	31	60	1	42	13
13	1	1	35	1	11	50	1	22	06	1	32	22	1	42	38
14	1	1	50	1	12	09	1	22	27	1	32	45	1	43	04
15	1	2	07	1	12	28	1	22	49	1	33	10	1	43	32
16	1	2	26	1	12	50	1	23	14	1	33	39	1	44	03
17	1	2	43	1	13	11	1	23	38	1	34	06	1	44	34
18	1	3	05	1	13	36	1	24	07	1	34	38	1	45	08
19	1	3	27	1	14	01	1	24	37	1	35	11	1	45	46
20	1	3	51	1	14	29	1	25	08	1	35	46	1	46	20
21	1	4	16	1	14	59	1	25	41	1	36	24	1	47	06
22	1	4	43	1	15	30	1	26	17	1	37	04	1	47	52
23	1	5	11	1	16	03	1	26	55	1	37	46	1	48	38
24	1	5	41	1	16	37	1	27	34	1	38	31	1	49	27
25	1	6	12	1	17	14	1	28	18	1	39	20	1	50	22
26	1	6	44	1	17	51	1	28	58	1	40	05	1	51	13
27	1	7	20	1	18	33	1	29	47	1	41	00	1	52	13
28	1	7	55	1	19	14	1	30	34	1	41	53	1	53	15
29	1	8	36	1	20	02	1	31	28	1	42	54	1	54	20
30	1	9	17	1	20	49	1	32	23	1	44	55	1	55	28
31	1	9	58	1	21	37	1	33	18	1	44	57	1	56	38
32	1	10	45	1	22	32	1	34	20	1	46	07	1	57	55
33	1	11	42	1	23	37	1	35	33	1	47	28	1	59	13
34	1	12	23	1	24	26	1	36	31	1	48	34	2	00	38

A *Table* for reducing Miles of East															
Parallel	10 *Miles.*			20 *Miles.*			30 *Miles.*			40 *Miles.*			50 *Miles.*		
	D.	M.	S.	D.	M.	S.	D.	M.	S.	D.	M.	S.	D.	M.	S.
35	0	12	12	0	24	25	0	36	37	0	48	50	1	1	02
36	0	12	22	0	24	43	0	37	05	0	49	27	1	1	48
37	0	12	31	0	25	03	0	37	34	0	50	05	1	2	37
38	0	12	41	0	25	23	0	38	04	0	50	45	1	3	27
39	0	12	52	0	25	44	0	38	36	0	51	28	1	4	20
40	0	13	03	0	26	06	0	39	09	0	52	13	1	5	16
41	0	13	15	0	26	30	0	39	45	0	53	00	1	6	15
42	0	13	27	0	26	55	0	40	22	0	53	49	1	7	17
43	0	13	42	0	27	25	0	41	17	0	54	49	1	8	32
44	0	13	54	0	27	48	0	41	42	0	55	37	1	9	30
45	0	14	08	0	28	16	0	42	24	0	56	33	1	10	41
46	0	14	24	0	28	47	0	43	11	0	57	39	1	11	58
47	0	14	40	0	29	20	0	43	59	0	58	34	1	13	19
48	0	14	57	0	29	53	0	44	50	0	59	47	1	14	43
49	0	15	14	0	30	29	0	45	33	1	0	57	1	16	11
50	0	15	33	0	31	07	0	46	40	1	2	14	1	17	47
51	0	15	53	0	31	47	0	47	40	1	3	33	1	19	27
52	0	16	14	0	32	29	0	48	43	1	4	58	1	21	12
53	0	16	37	0	33	14	0	49	51	1	6	28	1	23	05
54	0	17	01	0	34	02	0	51	02	1	8	03	1	25	04
55	0	17	26	0	34	52	0	52	18	1	9	48	1	27	10
56	0	17	53	0	35	46	0	53	39	1	11	32	1	29	25
57	0	18	21	0	36	43	0	55	05	1	13	27	1	31	48
58	0	18	52	0	37	44	0	56	36	1	15	29	1	34	21
59	0	19	24	0	38	50	0	58	14	1	17	39	1	37	04
60	0	20	00	0	40	00	1	0	00	1	20	00	1	40	00
61	0	20	37	0	41	15	1	1	52	1	22	30	1	43	07
62	0	21	18	0	42	36	1	3	54	1	25	12	1	46	30
63	0	22	12	0	44	03	1	6	05	1	28	07	1	50	08
64	0	22	48	0	45	37	1	8	16	1	31	15	1	54	03
65	0	23	40	0	47	09	1	10	49	1	34	29	1	58	08
66	0	24	35	0	49	10	1	13	45	1	38	21	2	02	56
67	0	25	35	0	51	10	1	16	45	1	42	20	2	07	55
68	0	26	42	0	53	24	1	20	06	1	46	48	2	13	30
69	0	27	54	0	55	45	1	23	42	1	51	36	2	19	30
70	0	29	14	0	59	28	1	27	42	1	56	56	2	26	10

	and West into Degrees of Longitude.														
Parallel.	60 *Miles.*			70 *Miles.*			80 *Miles.*			90 *Miles.*			100 *Miles.*		
	D.	M.	S.	D.	M.	S.	D.	M.	S.	D.	M.	S.	D.	M.	S.
35	1	13	14	1	25	26	1	37	38	1	49	50	2	2	05
36	1	14	10	1	26	32	1	38	53	1	51	15	2	3	36
37	1	13	08	1	27	39	1	40	11	1	52	41	2	5	13
38	1	16	08	1	28	49	1	41	31	1	54	12	2	6	53
39	1	17	12	1	30	04	1	42	56	1	55	48	2	8	40
40	1	18	19	1	31	22	1	44	25	1	57	28	2	10	32
41	1	19	30	1	32	45	1	46	00	1	59	15	2	12	30
42	1	20	44	1	34	11	1	47	38	2	01	06	2	14	33
43	1	22	14	1	35	56	1	49	39	2	03	21	2	17	03
44	1	23	25	1	37	19	1	51	13	2	05	07	2	19	02
45	1	24	50	1	38	58	1	53	07	2	07	19	2	21	25
46	1	26	22	1	40	45	1	55	09	2	10	32	2	23	55
47	1	27	59	1	42	38	1	57	18	2	11	58	2	26	38
48	1	29	40	1	44	36	1	59	33	2	14	30	2	29	27
49	1	31	26	1	46	40	2	01	54	2	17	08	2	32	25
50	1	33	21	1	48	54	2	04	28	2	20	01	2	35	35
51	1	35	20	1	51	13	2	07	07	2	23	00	2	38	53
52	1	37	27	1	53	41	2	09	55	2	26	09	2	42	20
53	1	39	42	1	56	19	2	12	56	2	29	33	2	46	15
54	1	42	05	1	59	06	2	16	07	2	33	07	2	50	08
55	1	44	36	2	02	02	2	19	28	2	36	54	2	54	20
56	1	47	18	2	05	11	2	23	04	2	41	57	2	58	50
57	1	50	10	2	08	32	2	26	53	2	45	15	2	03	36
58	1	53	13	2	12	05	2	30	57	2	49	49	3	08	43
59	1	56	29	2	15	54	2	35	19	2	54	43	3	14	08
60	2	00	00	2	20	00	2	40	00	3	00	00	3	20	00
61	2	03	45	2	24	22	2	45	40	3	05	37	3	26	15
62	2	07	48	2	29	06	2	50	24	3	11	42	3	33	00
63	2	12	10	2	34	12	2	56	13	3	18	15	3	40	16
64	2	16	52	2	39	41	3	02	29	3	25	18	3	48	06
65	2	21	47	2	45	27	3	09	06	3	32	46	3	56	37
66	2	27	21	2	51	56	3	16	31	3	41	07	4	05	41
67	2	33	30	2	59	05	3	24	40	3	50	15	4	15	50
68	2	40	12	2	06	54	3	33	36	4	00	18	4	27	00
69	2	47	14	2	15	18	3	43	12	4	11	16	4	39	30
70	2	55	24	2	24	38	3	53	52	4	23	06	4	52	20

A Table for reducing Miles of East

Parallel	200 Miles.			300 Miles.			400 Miles.			500 Miles.			600 Miles.		
	D.	M.	S.	D.	M.	S.	D.	M.	S.	D.	M.	S.	D.	M.	S.
0	3	20	00	5	0	00	6	40	00	8	20	00	10	0	00
1	3	20	02	5	0	03	6	40	04	8	20	04	10	0	05
2	3	20	07	5	0	11	6	40	14	8	20	18	10	0	22
3	3	20	17	5	0	25	6	40	33	8	20	45	10	0	53
4	3	20	29	5	0	44	6	40	59	8	20	13	10	1	27
5	3	20	46	5	1	09	6	41	32	8	21	55	10	2	18
6	3	21	03	5	1	35	6	42	07	8	22	39	10	3	11
7	3	21	30	5	2	15	6	43	00	8	23	45	10	4	30
8	3	21	58	5	2	57	6	43	56	8	24	55	10	5	54
9	3	22	29	5	3	44	6	44	59	8	26	14	10	7	29
10	3	23	06	5	4	39	6	46	12	8	27	45	10	9	18
11	3	23	44	5	5	36	6	47	28	8	29	20	10	11	12
12	3	24	26	5	6	49	6	49	02	8	31	15	10	13	28
13	3	25	16	5	7	54	6	50	32	8	33	10	10	15	48
14	3	26	08	5	9	12	6	52	16	8	35	20	10	18	24
15	3	27	04	5	10	36	6	54	08	8	37	40	10	21	12
16	3	28	06	5	12	09	6	56	12	8	40	15	10	24	18
17	3	29	08	5	13	42	6	58	26	8	43	00	10	27	34
18	3	30	18	5	15	27	7	00	36	8	45	45	10	30	56
19	3	31	32	5	17	18	7	03	04	8	48	50	10	34	36
20	3	32	40	5	19	00	7	05	20	8	51	40	10	38	00
21	3	34	12	5	21	18	7	08	24	8	55	30	10	42	36
22	3	35	44	5	23	36	7	11	28	8	59	20	10	47	12
23	3	37	16	5	27	54	7	16	32	9	05	10	10	53	48
24	3	38	54	5	28	21	7	17	48	9	07	15	10	56	42
25	3	40	44	5	31	06	7	21	28	9	11	50	11	02	12
26	3	42	26	5	33	39	7	24	52	9	16	05	11	07	18
27	3	44	26	5	36	39	7	28	52	9	21	05	11	13	18
28	3	46	30	5	39	45	7	33	00	9	26	15	11	19	30
29	3	48	40	5	43	00	7	37	20	9	31	40	11	26	00
30	3	50	56	5	46	24	7	41	52	9	37	20	11	32	48
31	3	53	16	5	49	54	7	45	32	9	42	10	11	38	48
32	3	55	50	5	53	45	7	51	40	9	49	35	11	47	20
33	3	58	26	5	57	39	7	56	52	9	56	05	11	55	18
34	4	01	16	6	01	54	8	02	32	10	03	10	12	03	48

Parallel.	700 Miles.			800 Miles.			900 Miles.			1000 Miles.		
	D.	M.	S.	D.	M.	S.	D.	M.	S.	D.	M.	S.
0	11	40	00	13	20	00	15	0	00	16	40	00
1	11	40	06	13	20	07	15	0	09	16	40	10
2	11	40	26	13	20	30	15	0	34	16	40	38
3	11	41	00	13	21	08	15	1	17	16	41	25
4	11	41	41	13	21	55	15	2	19	16	42	23
5	11	42	41	13	23	04	15	3	27	16	43	48
6	11	43	43	13	24	15	15	4	47	16	45	18
7	11	45	15	13	26	00	15	6	45	16	47	30
8	11	46	53	13	27	52	15	8	51	16	49	50
9	11	48	44	13	29	59	15	11	14	16	52	27
10	11	50	51	13	32	24	15	13	57	16	55	37
11	11	53	04	13	34	56	15	16	48	16	58	42
12	11	55	41	13	37	54	15	20	07	17	02	10
13	11	57	26	13	40	04	15	22	42	17	04	18
14	12	01	28	13	44	32	15	27	36	17	10	36
15	12	04	44	13	48	16	15	31	48	17	15	17
16	12	08	21	13	52	24	15	36	27	17	20	31
17	12	12	08	13	56	42	15	41	16	17	25	41
18	12	16	05	14	01	14	15	46	23	17	31	27
19	12	20	22	14	06	08	15	51	05	17	37	37
20	12	24	20	14	10	44	15	57	00	17	43	20
21	12	29	42	14	16	46	16	03	54	17	50	00
22	12	35	04	14	22	54	16	10	48	17	58	31
23	12	42	26	14	31	04	16	19	42	18	08	20
24	12	46	09	14	35	32	16	25	30	18	14	30
25	12	52	34	14	42	56	16	33	18	18	23	40
26	12	58	38	14	49	44	16	40	57	18	32	10
27	13	05	31	14	57	44	16	49	57	18	42	13
28	13	12	45	15	06	00	16	59	15	18	52	30
29	13	20	20	15	14	40	17	09	00	19	03	20
30	13	28	16	15	23	44	17	19	12	19	14	43
31	13	35	26	15	32	04	17	29	42	19	26	23
32	13	45	15	15	43	10	17	41	05	19	39	10
33	13	54	31	15	53	44	17	52	57	19	52	13
34	14	04	26	16	05	04	18	05	42	20	06	23

	A Table for reducing Miles of East													
Parallel	200 Miles.			300 Miles.			400 Miles.			500 Miles.			600 Miles.		
	D.	M.	S.	D.	M.	S.	D.	M.	S.	D.	M.	S.	D.	M.	S.
35	4	04	10	6	06	15	8	08	20	10	10	25	12	12	30
36	4	07	14	6	10	51	8	14	28	10	18	05	12	21	42
37	4	10	26	6	15	39	8	20	52	10	26	05	12	31	18
38	4	13	46	6	20	39	8	27	32	10	34	25	12	41	18
39	4	17	20	6	26	00	8	34	40	10	43	20	12	52	00
40	4	21	4	6	31	36	8	42	08	10	52	40	13	03	12
41	4	25	0	6	37	30	8	50	00	11	02	30	13	15	00
42	4	29	8	6	43	42	8	58	16	11	12	50	13	27	24
43	4	34	6	6	51	09	9	08	12	11	25	15	13	42	18
44	4	38	4	6	57	06	9	16	08	11	35	10	13	54	12
45	4	42	50	7	04	15	9	25	40	11	47	05	14	08	30
46	4	47	52	7	11	48	9	35	44	11	59	40	14	23	36
47	4	53	16	7	19	54	9	46	32	12	13	10	14	39	48
48	4	58	54	7	28	21	9	57	48	12	27	15	14	56	42
49	5	04	50	7	37	15	10	09	40	12	42	05	14	14	30
50	5	11	10	7	46	45	10	22	20	12	57	55	15	33	20
51	5	17	46	7	56	39	10	35	22	13	14	15	15	53	08
52	5	24	50	8	07	15	10	49	40	13	32	05	16	14	30
53	5	32	20	8	18	30	11	04	40	13	50	50	16	37	00
54	5	40	16	8	30	24	11	20	32	14	10	40	17	00	48
55	5	48	40	8	43	30	11	37	20	14	31	55	17	26	00
56	5	57	40	8	56	30	11	55	20	14	54	15	17	53	00
57	6	07	14	9	10	91	12	14	28	15	18	05	18	21	42
58	6	17	26	9	26	09	12	34	52	15	43	35	18	52	18
59	6	28	16	9	42	25	12	56	33	16	10	40	19	24	50
60	6	40	00	10	00	00	13	20	00	16	40	00	20	00	00
61	6	52	30	10	18	45	13	45	00	17	11	15	20	37	30
62	7	06	00	10	39	00	14	12	00	17	48	00	21	18	00
63	7	20	32	11	00	48	14	41	04	18	21	20	22	01	36
64	7	36	12	11	24	18	15	12	24	19	00	30	22	48	36
65	7	53	13	11	49	48	15	46	24	19	43	00	23	39	36
66	8	11	22	12	17	03	16	22	44	20	28	25	24	34	06
67	8	31	40	12	47	30	17	03	20	21	19	10	25	35	00
68	8	54	00	13	21	00	17	48	00	22	15	00	26	42	00
69	9	18	00	13	57	00	18	36	00	23	15	00	27	54	00
70	9	44	40	14	37	00	19	29	20	24	21	00	29	14	00

	and West into Degrees of Longitude.										
Parallel.	700 *Miles.*			800 *Miles.*			900 *Miles.*			1000 *Miles.*		
	D.	M.	S.	D.	M.	S.	D.	M.	S.	D.	M.	S.
35	14	14	35	16	16	40	18	18	45	20	20	50
36	14	25	19	16	28	56	18	32	33	20	36	10
37	14	36	31	16	41	44	18	46	57	20	53	53
38	14	48	11	16	55	04	19	01	57	21	08	30
39	15	00	40	17	09	20	19	08	00	21	27	40
40	15	13	44	17	24	16	19	34	48	21	45	26
41	15	27	30	17	40	00	19	52	30	22	05	00
42	15	41	58	17	56	22	20	10	56	23	25	30
43	15	59	21	18	16	24	20	33	27	22	50	13
44	16	13	14	18	32	16	20	51	18	23	10	20
45	16	29	55	18	51	20	21	12	45	23	34	10
46	16	47	32	19	11	28	21	35	24	23	59	10
47	17	06	26	19	33	04	21	59	42	24	26	10
48	17	26	09	19	55	36	22	25	03	24	54	26
49	17	46	55	20	19	20	22	51	45	25	24	10
50	18	08	55	20	44	30	23	20	05	25	55	50
51	18	32	01	21	10	54	23	49	47	26	28	53
52	18	57	15	21	39	30	24	21	55	27	04	20
53	19	23	10	22	09	20	24	55	30	27	41	40
54	19	50	56	22	41	04	25	31	12	28	21	20
55	20	20	20	23	14	40	26	09	00	29	03	20
56	20	51	50	23	50	40	26	49	30	29	43	20
57	21	25	19	24	28	56	27	32	23	30	36	00
58	22	01	01	25	09	44	28	18	27	31	27	10
59	22	38	58	25	53	06	29	07	15	32	21	23
60	23	20	00	26	40	00	30	00	00	33	20	00
61	24	03	45	27	30	00	30	56	15	34	22	30
62	24	51	00	28	24	00	31	57	00	35	30	00
63	25	41	52	29	29	08	33	02	24	36	42	40
64	26	36	42	30	24	48	34	12	54	38	06	00
65	27	36	12	31	32	58	35	29	24	39	26	00
66	28	39	47	32	45	28	36	51	09	40	56	50
67	29	50	50	34	06	40	38	22	30	42	38	20
68	31	09	00	35	36	00	40	03	00	44	30	00
69	32	33	00	37	12	00	41	51	00	46	30	00
70	34	06	24	38	57	40	43	51	00	48	43	20

LOXODROMIQUES,

O R

TRAVESE-TABLES

Of MILES, *with the Difference of*
Longitudes *and* Latitudes.

| Loxodromiques , or Traverse-Tables of Miles, | | | | | | | | |

Lati-tude.	1 Rumb, 11°15'.		2 Rumb, 22°30'.		3 Rumb, 33°45'.		4 Rumb, 45°00'.	
	Longit.	Dist. in	Longit.	Dist. in	Longit.	Dist. in	Longit.	Dist. in
D. M.	D. M.	Miles.	D. M.	Miles.	D. M.	Miles.	D. M.	Miles.
0 0	0 0	0	0 0	0	0 0	0	0 0	0
0 10	0 2	10	0 4	11	0 7	12	0 10	14
0 20	0 4	20	0 8	22	0 13	25	0 20	28
0 30	0 6	31	0 12	32	0 20	36	0 30	42
0 40	0 8	41	0 16	43	0 27	48	0 40	57
0 50	0 10	51	0 20	54	0 33	60	0 50	71
1 0	0 12	61	0 24	65	0 40	72	1 0	85
1 10	0 14	71	0 29	76	0 47	84	1 10	99
1 20	0 16	82	0 33	87	0 53	96	1 20	113
1 30	0 18	92	0 37	197	1 00	100	1 30	127
1 40	0 20	102	0 41	108	1 07	110	1 40	141
1 50	0 22	112	0 45	119	1 13	122	1 50	156
2 0	0 24	122	0 49	130	1 20	144	2 0	170
2 10	0 26	133	0 53	141	1 27	156	2 00	184
2 20	0 28	143	0 57	152	1 33	168	2 10	198
2 30	0 30	153	1 02	162	1 40	180	2 20	212
2 40	0 32	163	1 06	173	1 47	192	2 30	226
2 50	0 34	173	1 10	184	1 53	204	2 40	240
3 0	0 36	184	1 15	194	2 00	216	3 0	255
3 10	0 38	194	1 19	206	2 07	229	3 10	269
3 20	0 40	204	1 23	216	2 13	241	3 20	283
3 30	0 42	214	1 27	227	2 20	253	3 30	297
3 40	0 44	224	1 32	238	2 27	265	3 40	311
3 50	0 46	234	1 36	249	2 33	277	3 50	325
4 0	0 48	245	1 39	260	2 40	289	4 0	339
4 10	0 50	255	1 44	271	2 47	301	4 10	344
4 20	0 52	265	1 48	281	2 54	313	4 20	358
4 30	0 54	275	1 52	292	3 01	325	4 30	372
4 40	0 56	285	1 56	303	3 07	337	4 40	386
4 50	0 58	296	2 00	314	3 14	349	4 50	390
5 0	1 0	306	2 04	325	3 20	361	5 0	424
5 10	1 2	316	2 08	336	3 27	373	5 10	438
5 20	1 4	326	2 12	346	3 34	385	5 20	453
5 30	1 6	336	2 16	357	3 40	397	5 30	467
5 40	1 8	347	2 20	368	3 47	409	5 40	481
5 50	1 10	357	2 24	378	3 54	421	5 50	495

with the Difference of Longitudes and Latitudes.

Latitude.		5 Rumb, 56° 15'.			6 Rumb, 67° 30'.			7 Rumb, 78° 45'.		
		Longitude.		Dist. in Miles.	Longitude.		Dist. in Miles.	Longitude.		Dist. in Miles.
D.	M.	D.	M.		D.	M.		D.	M.	
0	0	0	0	0	0	0	0	0	0	0
0	10	0	15	18	0	24	26	0	51	51
0	20	0	30	36	0	48	52	1	38	103
0	30	0	45	54	1	12	78	2	29	154
0	40	1	00	72	1	36	105	3	20	205
0	50	1	15	90	2	00	131	4	11	256
1	0	1	30	108	2	24	157	5	02	308
1	10	1	45	126	2	49	183	5	53	359
1	20	2	00	144	3	12	299	6	44	410
1	30	2	15	162	3	38	235	7	35	461
1	40	2	30	180	4	02	261	8	22	513
1	50	2	45	198	4	26	287	9	13	564
2	0	3	00	216	4	50	314	10	04	615
2	10	3	15	234	5	14	340	10	55	666
2	20	3	30	252	5	38	366	11	42	718
2	30	3	45	270	6	02	392	12	33	769
2	40	4	00	288	6	25	418	13	24	820
2	50	4	15	309	6	50	444	14	16	871
3	0	4	30	324	7	14	470	15	07	923
3	10	4	45	342	7	38	496	15	54	974
3	20	5	00	360	8	02	523	16	49	1025
3	30	5	15	378	8	26	549	17	36	1076
3	40	5	30	396	8	52	575	18	27	1128
3	50	5	45	414	9	15	601	19	18	1179
4	0	6	00	432	9	39	627	20	09	1230
4	10	6	15	450	10	04	653	21	00	1281
4	20	6	30	468	10	29	679	21	48	1333
4	30	6	45	486	10	53	706	22	39	1384
4	40	7	00	504	11	17	732	23	30	1435
4	50	7	15	522	11	40	758	24	21	1487
5	0	7	30	540	12	05	784	25	08	1538
5	10	7	45	558	12	29	810	26	03	1589
5	20	8	00	576	12	53	836	26	51	1640
5	30	8	15	594	13	17	862	27	42	1642
5	40	8	30	612	13	42	888	28	33	1743
5	50	8	45	630	14	06	915	29	25	1704

M m m

Loxodromiques, or Traverse-Tables of Miles,

Lati-tude.		1 Rumb, 11°15'. Longit.		Dist. in Miles.	2 Rumb, 22°30'. Longit.		Dist. in Miles.	3 Rumb, 33°45'. Longit.		Dist. in Miles.	4 Rumb, 45°00'. Longit.		Dist. in Miles.
D.	M.	D.	M.	Miles.	D.	M.	Miles.	D.	M.	Miles.	D.	M.	Miles.
6	0	1	12	367	2	28	390	4	01	433	6	0	509
6	10	1	14	377	2	33	401	4	07	445	6	10	523
6	20	1	16	387	2	37	411	4	14	457	6	20	537
6	30	1	18	398	2	41	422	4	21	469	6	30	552
6	40	1	20	408	2	45	433	4	27	481	6	40	536
6	50	1	22	418	2	50	443	4	34	493	6	50	580
7	0	1	24	428	2	54	455	4	41	505	7	0	594
7	10	1	26	438	2	58	465	4	47	517	7	11	608
7	20	1	28	449	3	02	476	4	54	529	7	20	622
7	30	1	30	459	3	06	487	5	01	541	7	31	636
7	40	1	32	469	3	11	498	5	08	553	7	41	651
7	50	1	34	479	3	15	509	5	14	565	7	51	665
8	0	1	36	480	3	19	520	5	21	577	8	2	679
8	10	1	38	500	3	24	530	5	28	589	8	12	693
8	20	1	40	510	3	28	541	5	35	601	8	22	707
8	30	1	42	520	3	32	552	5	41	613	8	32	721
8	40	1	44	530	3	36	563	5	48	625	8	42	735
8	50	1	46	540	3	40	574	5	55	637	8	52	749
9	0	1	48	558	3	44	584	6	02	649	9	1	764
9	10	1	50	561	3	49	595	6	08	661	9	12	778
9	20	1	52	571	3	53	606	6	15	673	9	22	792
9	30	1	54	581	3	57	617	6	22	686	9	33	806
9	40	1	56	591	4	01	628	6	25	698	9	42	820
9	50	1	58	602	4	05	638	6	35	709	9	52	834
10	0	2	0	612	4	10	649	6	42	722	10	3	849
10	10	2	2	622	4	14	660	6	45	734	10	13	863
10	20	2	4	632	4	18	671	6	56	746	10	23	877
10	30	2	6	642	4	22	682	7	01	758	10	33	891
10	40	2	8	653	4	26	693	7	05	770	10	44	905
10	50	2	10	663	4	30	704	7	16	782	10	54	919
11	0	2	12	673	4	34	714	7	23	794	11	4	933
11	10	2	14	683	4	38	725	7	30	806	11	14	948
11	20	2	16	693	4	42	732	7	36	818	11	24	956
11	30	2	18	704	4	48	747	7	44	830	11	34	966
11	40	2	20	714	4	54	758	7	51	842	11	44	980
11	50	2	22	724	4	50	768	7	58	854	11	55	994

with the Difference of Longitudes and Latitudes.								
Latitude.	5 *Rumb,* 56° 15′.			6 *Rumb,* 67° 30′.			7 *Rumb,* 78° 45′.		
	Longitude.		Diff. in	Longitude.		Diff. in	Longitude.		Diff. in
D. M.	D.	M.	Miles.	D.	M.	Miles.	D.	M.	Miles.
6 0	9	0	648	14	30	941	30	14	1845
6 10	9	15	666	14	55	969	31	03	1897
6 20	9	30	684	15	19	993	31	54	1948
6 30	9	45	702	15	44	1019	32	46	2999
6 40	10	00	720	16	08	1045	33	37	2050
6 50	10	15	738	16	33	1071	34	25	2102
7 0	10	30	756	16	57	1098	35	18	2153
7 10	10	45	774	17	21	1124	36	07	2204
7 20	11	00	792	17	45	1150	37	00	2255
7 30	11	16	810	18	10	1176	38	00	2307
7 40	11	32	828	18	34	1202	38	42	2368
7 50	11	46	846	18	58	1228	39	29	2409
8 0	12	.01	864	19	22	1254	40	21	2460
8 10	12	16	882	19	47	1280	40.	12	2512
8 20	12	31	900	20	11	1307	42	04	2563
8 30	12	47	918	20	35	1333	42	51	2614
8 40	13	02	936	21	00	1359	43	43	2665
8 50	13	17	954	21	24	1385	44	34	2717
9 0	13	32	972	21	50	1411	45	26	2768
9 10	13	47	990	22	14	1437	46	18	2819
9 20	14	02	1008	22	39	1463	47	09	2870
9 30	14	16	1026	23	03	1489	47	57	2922
9 40	14	32	1044	23	27	1516	48	40	2973
9 50	14	48	1062	23	52	1542	49	41	3024
10 0	15	03	1080	24	16	1568	50	32	3075
10 10	15	17	1098	24	41	1594	51	24	3127
10 20	15	33	1116	25	05	1620	52	16	3178
10 30	15	47	1134	25	30	1646	52	04	3229
10 40	16	03	1152	25	54	1662	53	56	3271
10 50	16	18	1170	26	18	1689	54	48	3322
11 0	16	34	1188	26	43	1725	55	36	3373
11 10	16	49	1206	27	08	1751	56	28	3424
11 20	17	05	1224	27	32	1777	57	20	3476
11 30	17	19	1242	27	58	1803	58	12	3527
11 40	17	35	1260	28	22	1829	59	41	3578
11 50	17	50	1278	28	47	1855	59	56	3636

Loxodromiques, or Traverse-Tables of Miles,

Lati-tude.		1 Rumb, 11°15'.			2 Rumb, 22°30'.			3 Rumb, 33°45'.			4 Rumb, 45°00'.		
		Longit.		Dist. in	Longit.		Dist. in	Longit.		Dist. in	Longit.		Dist. in
D.	M.	D.	M.	Miles.	D.	M.	Miles.	D.	M	Miles.	D.	M.	Miles.
12	0	2	24	734	5	2	779	8	4	856	12	5	1018
12	10	2	26	744	5	5	790	8	11	878	12	15	1032
12	20	2	28	755	5	9	801	8	18	890	12	26	1047
12	30	2	30	765	5	13	812	8	25	902	12	35	1051
12	40	2	32	775	5	17	823	8	32	914	12	46	1075
12	50	2	34	785	5	21	833	8	35	926	12	56	1089
13	0	2	36	795	5	26	844	8	46	938	13	6	1103
13	10	2	38	805	5	30	855	8	52	950	13	17	1117
13	20	2	40	816	5	34	866	8	55	962	13	28	1131
13	30	2	43	826	5	38	877	9	06	974	13	38	1146
13	40	2	45	836	5	43	887	9	14	986	13	48	1160
13	50	2	47	846	5	47	898	9	21	998	13	58	1174
14	0	2	49	856	5	51	909	9	28	1010	14	8	1188
14	10	2	51	867	5	55	920	9	34	1022	14	18	1202
14	20	2	53	877	6	00	931	9	41	1034	14	29	1216
14	30	2	55	887	6	04	942	9	47	1046	14	30	1230
14	40	2	57	897	6	08	952	9	54	1058	14	49	1244
14	50	2	59	907	6	13	963	10	01	1070	14	59	1259
15	0	3	1	918	6	17	974	10	08	1082	15	10	1272
15	10	3	3	928	6	21	985	10	15	1094	15	21	1281
15	20	3	6	938	6	26	996	10	22	1106	15	31	1301
15	30	3	8	948	6	30	1007	10	30	1118	15	41	1315
15	40	3	10	958	6	34	1017	10	37	1131	15	51	1329
15	50	3	12	969	6	39	1028	10	44	1143	16	1	1343
16	0	3	14	979	6	43	1035	10	50	1155	16	12	1358
16	10	3	16	989	6	47	1050	10	56	1167	16	23	1372
16	20	3	18	999	6	51	1061	11	03	1179	16	33	1386
16	30	3	20	1009	6	55	1072	11	16	1191	16	44	1400
16	40	3	22	1029	6	59	1082	11	17	1203	16	55	1414
16	50	3	24	1039	7	04	1093	11	24	1215	17	5	1428
17	0	3	26	1040	7	09	1104	11	31	1227	17	15	1442
17	10	3	28	1050	7	13	1115	11	35	1235	17	25	1457
17	20	3	30	1060	7	17	1126	11	46	1251	17	36	1471
17	30	3	32	1070	7	21	1136	11	53	1263	17	47	1485
17	40	3	34	1081	7	26	1147	12	00	1275	17	57	1499
17	50	3	36	1091	7	30	1158	12	07	1287	18	8	1513

with the Difference of Longitudes and Latitudes.

Latitude		5 Rumb, 56° 15'.			6 Rumb, 67° 30'.			7 Rumb, 78° 45'.		
		Longitude.		Dift. in	Longitude.		Dift. in	Longitude.		Dift. in
D.	M.	D.	M.	Miles.	D.	M.	Miles.	D.	M.	Miles.
12	0	18	05	1256	29	12	1881	60	48	3691
12	10	18	20	1314	29	36	1908	61	36	3742
12	20	18	36	1332	30	00	1934	62	28	3793
12	30	18	51	1350	30	25	1960	63	20	3844
12	40	19	07	1368	30	50	1986	64	12	3896
12	50	19	22	1386	31	15	2012	65	05	3947
13	0	19	38	1404	31	39	2038	65	55	3998
13	10	19	52	1422	32	04	2064	66	46	4049
13	20	20	08	1440	32	29	2090	67	38	4101
13	30	20	34	1458	32	53	2117	68	30	4152
13	40	20	39	1475	33	18	2143	69	23	4203
13	50	20	54	1494	33	43	2169	70	15	4252
14	0	21	10	1512	34	10	2195	71	08	4306
14	10	21	25	1530	34	34	2221	71	57	4357
14	20	21	40	1548	34	59	2247	72	49	4408
14	30	21	55	1566	35	24	2273	73	42	4459
14	40	22	11	1584	35	49	2300	74	34	4511
14	50	22	26	1592	36	13	2326	75	23	4562
15	0	22	43	1620	36	38	2352	76	16	4613
15	10	22	57	1638	37	03	2378	77	09	4664
15	20	23	14	1656	37	28	2404	78	02	4716
15	30	23	28	1674	37	52	2430	78	55	4767
15	40	23	44	1692	38	18	2456	79	47	4818
15	50	24	04	1710	38	42	2482	80	38	4870
16	0	24	16	1728	39	08	2509	81	29	4921
16	10	24	31	1746	39	35	2535	82	28	4972
16	20	24	47	1764	39	58	2561	83	16	5023
16	30	25	02	1782	40	25	2587	84	09	5075
16	40	25	18	1800	40	50	2613	84	58	5126
16	50	25	34	1818	41	15	2639	85	55	5177
17	0	25	49	1836	41	40	2665	86	44	5228
17	10	26	05	1854	42	05	2691	87	38	5280
17	20	26	21	1872	42	30	2718	88	51	5331
17	30	26	36	1890	42	56	2744	89	25	5382
17	40	26	52	1908	43	21	2770	90	18	5433
17	50	27	08	1926	43	46	2796	91	07	5485

Loxodromiques, or Traverse-Tables of Miles,

Lati-tude.		1 Rumb, 11°15'.			2 Rumb, 22°30'.			3 Rumb, 33°45'.			4 Rumb, 45°00'.		
		Longit.		Diſt. in	Longit.		Diſt. in	Longit.		Diſt. in	Longit.		Diſt. in
D.	M.	D.	M.	Miles.	D.	M.	Miles.	D.	M.	Miles.	D.	M.	Miles.
18	0	3	38	1101	7	34	1169	12	14	1290	18	19	1527
18	10	3	40	1111	7	40	1180	12	21	1318	18	29	1541
18	20	3	42	1122	7	44	1191	12	28	1323	18	39	1556
18	30	3	44	1132	7	48	1201	12	35	1335	18	49	1570
18	40	3	46	1142	7	52	1212	12	42	1347	19	00	1584
18	50	3	49	1152	7	57	1223	12	49	1359	19	10	1598
19	0	3	51	1162	8	01	1234	12	56	1371	19	21	1612
19	10	3	53	1173	8	05	1245	13	03	1383	19	31	1626
19	20	3	55	1183	8	10	1256	13	11	1395	19	42	1640
19	30	3	57	1193	8	14	1266	13	18	1407	19	53	1655
19	40	3	59	1203	8	19	1277	13	25	1419	20	04	1669
19	50	4	01	1213	8	23	1288	13	32	1431	20	14	1683
20	0	4	04	1224	8	28	1299	13	39	1443	20	25	1697
20	10	4	07	1234	8	32	1310	13	46	1455	20	35	1711
20	20	4	09	1244	8	36	1321	13	53	1467	20	46	1725
20	30	4	11	1254	8	41	1331	14	02	1479	20	57	1739
20	40	4	13	1264	8	45	1342	14	08	1491	21	07	1744
20	50	4	15	1274	8	49	1353	14	15	1503	21	18	1758
21	0	4	17	1285	8	54	1364	14	22	1515	21	28	1782
21	10	4	19	1295	8	59	1375	14	29	1527	21	39	1796
21	20	4	21	1305	9	03	1385	14	36	1539	21	50	1811
21	30	4	23	1315	9	07	1396	14	44	1551	22	01	1825
21	40	4	25	1325	9	12	1407	14	51	1563	22	12	1839
21	50	4	27	1336	9	16	1418	14	58	1575	22	22	1854
22	0	4	29	1346	9	20	1429	15	05	1588	22	33	1867
22	10	4	31	1356	9	25	1440	15	12	1600	22	44	1871
22	20	4	33	1366	9	30	1450	15	19	1612	22	55	1885
22	30	4	35	1376	9	34	1461	15	26	1624	23	06	1899
22	40	4	37	1387	9	38	1472	15	32	1636	23	17	1903
22	50	4	40	1397	9	43	1483	15	41	1648	23	28	1917
23	0	4	42	1407	9	48	1494	15	48	1660	23	39	1952
23	10	4	44	1417	9	52	1505	15	55	1672	23	45	1966
23	20	4	46	1427	9	57	1515	16	03	1684	24	00	1980
23	30	4	48	1438	10	02	1526	16	10	1656	24	11	1993
23	40	4	50	1448	10	07	1537	16	17	1708	24	22	2008
23	50	4	52	1458	10	11	1548	16	24	1720	24	33	2022

	with the Difference of Longitudes and Latitudes.							
Latitude.	5 Rumb, 56° 15'.			6 Rumb, 67° 30'.			7 Rumb, 78° 45'.	
	Longitude.		Dist. in	Longitude.		Dist. in	Longitude.		Dist. in
D. M.	D.	M.	Miles.	D.	M.	Miles.	D.	M.	Miles.
18 0	27	24	1944	44	11	2822	92	01	5536
18 10	27	39	1962	44	37	2848	92	55	5587
18 20	27	55	1980	44	02	2874	93	44	5638
18 30	28	11	1998	45	27	2901	94	40	5690
18 40	28	27	2016	45	55	2927	95	32	5741
18 50	28	42	2034	46	18	2953	95	56	5792
19 0	28	58	2052	46	45	2979	97	29	5843
19 10	29	14	2070	47	11	3005	98	14	5895
19 20	29	30	2088	47	36	3031	99	04	5946
19 30	29	46	2106	48	02	3057	99	58	5997
19 40	30	02	2124	48	27	3083	100	52	6048
19 50	30	18	2142	48	53	3110	101	46	6100
20 0	30	34	2160	49	18	3136	102	40	6151
20 10	30	50	2178	49	44	3162	103	30	6202
20 20	31	06	2196	50	09	3188	104	25	6254
20 30	31	21	2214	50	35	3216	105	19	6305
20 40	31	37	2232	51	01	3230	106	14	6356
20 50	31	53	2250	51	26	3266	107	08	6407
21 0	32	09	2268	51	52	3292	108	03	6458
21 10	32	25	2286	52	20	3319	108	53	6510
21 20	32	42	2304	52	43	3345	109	48	6561
21 30	32	57	2322	53	11	3371	110	43	6612
21 40	33	14	2340	53	37	3397	111	38	6664
21 50	33	30	2358	54	03	3423	112	32	6715
22 0	33	46	2376	54	29	3449	113	25	6746
22 10	34	02	2394	54	55	3475	114	20	6797
22 20	94	18	2412	55	21	3502	115	13	6849
22 30	34	35	2430	55	47	3528	116	09	6900
22 40	34	51	2448	56	13	3554	117	04	6951
22 50	35	07	2466	56	39	3580	117	59	7002
23 0	35	23	2484	57	05	3606	118	50	7075
23 10	35	40	2502	57	31	3632	119	45	7126
23 20	35	56	2520	57	57	3658	120	41	7177
23 30	36	12	2538	58	23	3684	121	37	7228
23 40	36	28	2556	58	52	3711	122	32	7280
23 50	36	45	2574	59	18	3736	123	24	7331

Loxodromiques, or Traverse-Tables of Miles,

Latitude	1 Rumb, 11°15'.		2 Rumb, 22°30'.		3 Rumb, 33°45'.		4 Rumb, 45°00'.	
	Longit.	Dist. in	Longit.	Dist. in	Longit.	Dist. in	Longit.	Dist. in
D. M.	D. M.	Miles.	D. M.	Miles.	D. M.	Miles.	D. M.	Miles.
24 0	4 55	1468	10 15	1559	16 31	1732	24 44	2036
24 10	4 57	1478	10 19	1570	16 39	1744	24 55	2041
24 20	4 59	1489	10 24	1580	16 46	1756	25 06	2055
24 30	5 02	1499	10 29	1591	16 53	1768	25 17	2069
24 40	5 04	1509	10 33	1602	17 01	1780	25 28	2083
24 50	5 06	1519	10 37	1613	17 08	1792	25 32	2097
25 0	5 08	1530	10 42	1624	17 16	1804	25 50	2121
25 10	5 10	1540	10 46	1634	17 23	1816	26 01	2135
25 20	5 12	1550	10 51	1645	17 30	1828	26 12	2150
25 30	5 14	1560	10 56	1656	17 38	1840	26 23	2164
25 40	5 16	1570	11 01	1667	17 45	1852	26 34	2178
25 50	5 18	1580	11 05	1678	17 52	1864	26 45	2192
26 0	5 21	1591	11 09	1688	18 00	1876	26 56	2206
26 10	5 23	1601	11 14	1699	18 07	1888	27 07	2220
26 20	5 26	1611	11 19	1710	18 15	1900	27 08	2234
26 30	5 28	1621	11 23	1721	18 22	1912	27 29	2249
26 40	5 30	1631	11 28	1732	18 30	1924	27 40	2263
26 50	5 32	1642	11 32	1743	18 37	1936	27 51	2277
27 0	5 34	1652	11 37	1753	18 45	1948	28 03	2291
27 10	5 36	1662	11 42	1764	18 52	1960	28 14	2305
27 20	5 39	1672	11 47	1775	19 00	1972	28 25	2319
27 30	5 41	1682	11 51	1786	19 07	1984	28 37	2333
27 40	5 43	1693	11 56	1797	19 15	1996	28 49	2348
27 50	5 46	1703	12 01	1808	19 22	2008	28 59	2362
28 0	5 48	1713	12 06	1818	19 30	2021	29 11	2376
28 10	5 50	1723	12 10	1829	19 37	2033	29 22	2390
28 20	5 52	1733	12 15	1840	19 45	2045	29 34	2404
28 30	5 55	1744	12 20	1851	19 56	2057	29 46	2418
28 40	5 57	1754	12 25	1862	20 00	2069	29 57	2432
28 50	6 00	1764	12 30	1872	20 08	2081	30 08	2447
29 0	6 02	1774	12 34	1883	20 16	2093	30 19	2461
29 10	6 04	1784	12 35	1894	20 23	2105	30 31	2475
29 20	6 07	1795	12 43	1905	20 30	2117	30 43	2489
29 30	6 09	1805	12 48	1916	20 38	2125	30 54	2503
29 40	6 11	1815	12 52	1927	20 46	2141	31 05	2517
29 50	6 13	1825	12 57	1937	20 54	2153	31 17	2531

with the Difference of Longitudes and Latitudes.									
Latitude.		*5 Rumb,* 56° 15'.			*6 Rumb,* 67° 30'.			*7 Rumb,* 78° 45'.	
		Longitude.		Dift. in	Longitude.		Dift. in	Longitude.		Dift. in
D.	M.	D.	M.	Miles.	D.	M.	Miles.	D.	M.	Miles.
24	0	37	01	2592	59	44	3763	124	24	7381
24	10	37	17	2610	60	11	3789	125	15	7432
24	20	37	34	2628	60	37	3815	126	11	7484
24	30	37	50	2646	61	03	3841	127	07	7535
24	40	38	07	2664	61	30	3867	128	00	7586
24	50	38	22	2682	61	56	3894	128	57	7537
25	0	38	40	2200	62	22	3920	129	52	7689
25	10	38	56	2718	62	49	3946	130	48	7740
25	20	39	13	2736	63	15	3972	131	44	7791
25	30	39	29	2754	63	42	3998	132	56	7842
25	40	39	46	2772	64	09	4024	133	33	7894
25	50	40	02	2790	64	34	4050	134	29	7945
26	0	40	19	2808	65	02	4076	135	26	7996
26	10	40	36	2826	65	31	4103	136	23	8048
26	20	40	53	2844	65	56	4129	137	20	8098
26	30	41	09	2862	66	25	4155	138	17	8150
26	40	41	26	2880	66	30	4181	139	09	8201
26	50	41	42	2898	67	19	4207	140	07	8253
27	0	42	00	2916	67	45	4233	141	04	8304
27	10	42	17	2934	68	12	4259	142	01	8355
27	20	42	33	2952	68	39	4285	142	56	8406
27	30	42	50	2970	69	06	4312	143	51	8458
27	40	43	07	2988	69	33	4338	144	49	8500
27	50	43	24	2906	70	00	4344	145	46	8560
28	0	43	14	3024	70	27	4390	146	44	8611
28	10	43	58	3042	70	55	4416	147	33	8663
28	20	44	15	3060	71	22	4442	148	35	8714
28	30	44	32	3078	71	49	4468	149	33	8765
28	40	44	49	3096	72	19	4495	150	31	8816
28	50	45	06	3114	72	44	4521	151	29	8868
29	0	45	23	3132	73	14	4547	152	28	8919
29	10	45	40	3150	73	41	4573	153	24	8970
29	20	45	57	3168	74	09	4599	154	23	9021
29	30	46	15	3186	74	37	4625	155	19	9073
29	40	46	22	3204	75	04	4651	156	17	9124
29	50	46	49	3222	75	32	4677	157	16	9175

N n n

Loxodromiques, or Traverse-Tables of Miles,

Lati-tude.		1 Rumb, 11°15'.			2 Rumb, 22°30'.			3 Rumb, 33°45'.			4 Rumb, 45°00'.	
		Longit.		Dist.in	Longit.		Dist.in	Longit.		Dist.in	Longit.		Dist.in
D.	M.	D.	M.	Miles.	D.	M.	Miles.	D.	M.	Miles.	D.	M.	Miles.
30	0	6	16	1835	13	02	1948	21	01	2165	31	28	2546
30	10	6	18	1845	13	07	1959	21	09	2177	31	40	2560
30	20	6	20	1856	13	11	1970	21	17	2189	31	51	2574
30	30	6	23	1866	13	16	1981	21	24	2201	32	03	2588
30	40	6	25	1876	13	21	1992	21	32	2213	32	15	2602
30	50	6	27	1886	13	26	2002	21	40	2225	32	29	2616
31	0	6	29	1896	13	30	2013	21	47	2237	32	38	2630
31	10	6	31	1907	13	35	2024	21	55	2249	32	49	2645
31	20	6	34	1917	13	40	2035	22	03	2261	33	00	2659
31	30	6	36	1927	13	45	2046	22	11	2273	33	12	2673
31	40	6	39	1937	13	50	2056	22	19	2285	33	25	2687
31	50	6	41	1947	13	55	2067	22	27	2297	33	37	2701
32	0	6	43	1958	14	01	2078	22	34	2309	33	48	2715
32	10	6	46	1968	14	06	2089	22	42	2321	34	00	2730
32	20	6	48	1978	14	10	2100	22	50	2333	34	12	2744
32	30	6	50	1988	14	15	2111	22	58	2345	34	24	2758
32	40	6	52	1998	14	20	2121	23	06	2357	34	36	2772
32	50	6	55	2009	14	25	2132	23	14	2369	34	48	2786
33	0	6	57	2019	14	30	2143	23	21	2381	35	00	2800
33	10	7	00	2029	14	35	2154	23	30	2393	35	12	2814
33	20	7	03	2039	14	40	2165	23	38	2405	35	23	2828
33	30	7	05	2049	14	45	2176	23	46	2417	35	35	2843
33	40	7	08	2060	14	49	2186	23	54	2429	35	47	2857
33	50	7	10	2070	14	54	2197	24	02	2441	35	59	2871
34	0	7	13	2080	14	59	2208	24	10	2454	36	11	2885
34	10	7	15	2090	15	04	2219	24	18	2466	36	23	2890
34	20	7	18	2000	15	09	2230	24	26	2478	36	35	2903
34	30	7	20	2111	15	14	2240	24	34	2490	36	47	2917
34	40	7	22	2121	15	20	2251	24	42	2502	36	56	2932
34	50	7	24	2131	15	25	2262	24	50	2514	37	12	2946
35	0	7	26	2141	15	29	2273	24	55	2526	37	24	2970
35	10	7	29	2151	15	34	2284	25	07	2538	37	36	2984
35	20	7	31	2162	15	39	2295	25	15	2550	37	48	3098
35	30	7	34	2172	15	44	2305	25	23	2562	38	00	3012
35	40	7	36	2182	15	50	2316	25	31	2574	38	13	3026
35	50	7	38	2192	15	55	2327	25	40	2586	38	25	3041

with the Difference of Longitudes and Latitudes.

Latitude		5 Rumb, 56° 15'.			6 Rumb, 67° 30'.			7 Rumb, 78° 45'.		
		Longitude.		Dist. in	Longitude.		Dist. in	Longitude.		Dist. in
D.	M.	D.	M.	Miles.	D.	M.	Miles.	D.	M.	Miles.
30	0	47	06	3240	76	00	4704	158	15	9226
30	10	47	24	3258	76	27	4730	159	14	9278
30	20	47	41	3276	76	55	4756	160	08	9329
30	30	47	58	3294	77	23	4782	161	08	9380
30	40	48	16	3312	77	51	4808	162	07	9431
30	50	48	33	3330	78	19	4834	163	06	9483
31	0	48	50	3348	78	47	4860	164	03	9534
31	10	49	08	3366	79	17	4886	165	03	9585
31	20	49	25	3384	79	43	4913	166	01	9636
31	30	49	43	3402	80	13	4939	167	00	9688
31	40	50	01	3420	80	41	4965	168	00	9739
31	50	50	18	3438	81	10	4991	169	00	9790
32	0	50	36	3456	81	38	5017	170	00	9842
32	10	50	53	3474	82	06	5043	170	55	9893
32	20	51	11	3492	82	35	5069	171	55	9944
32	30	51	29	3510	83	31	5095	172	56	9995
32	40	51	47	3528	83	32	5122	173	56	10047
32	50	52	04	3546	84	01	5148	174	53	10098
33	0	52	22	3564	84	29	5174	175	54	10149
33	10	52	40	3582	84	58	5200	176	55	10200
33	20	52	58	3600	85	26	5226	177	56	10252
33	30	53	16	3618	85	55	5252	178	57	10303
33	40	53	34	3636	86	26	5278	179	58	10354
33	50	53	50	3654	86	52	5305	180	57	10405
34	0	54	10	3672	87	24	5331	181	56	10457
34	10	54	28	3690	87	52	5357	182	57	10508
34	20	54	46	3708	88	02	5383	183	59	10559
34	30	55	04	3726	88	50	5409	184	59	10610
34	40	55	22	3744	89	20	5435	185	58	10662
34	50	55	41	3762	89	49	5461	187	00	10713
35	0	55	59	3780	90	19	5488	188	02	10764
35	10	56	17	3798	90	48	5514	189	05	10815
35	20	56	35	3816	91	18	5540	190	04	10867
35	30	56	54	3834	91	48	5566	191	06	10918
35	40	57	12	3852	92	17	5592	192	09	10969
35	50	57	30	3870	92	46	5618	193	11	11020

Loxodromiques , or Traverse-Tables of Miles ,								
Lati-tude.	1 *Rumb,* 11°15'.		2 *Rumb,* 22°30'.		3 *Rumb,* 33°45'.		4 *Rumb,* 45°00'.	
	Longit.	Dist. in	Longit.	Dist. in	Longit.	Dist. in	Longit.	Dist. in
D. M.	D. M.	Miles.	D. M.	Miles.	D. M.	Miles.	D. M.	Miles.
36 0	7 41	2202	16 00	2338	25 49	2598	38 38	3055
36 10	7 43	2212	16 05	2349	25 57	2610	38 50	3069
36 20	7 46	2223	16 11	2360	26 05	2622	39 03	3083
36 30	7 48	2233	16 16	2370	26 13	2634	39 15	3097
36 40	7 50	2243	16 21	2381	26 22	2646	39 27	3111
36 50	7 53	2253	16 26	2392	26 30	2658	39 40	3125
37 0	7 56	2263	16 32	2403	26 38	2670	39 53	3140
37 10	7 59	2274	16 37	2414	26 47	2681	40 05	3154
37 20	8 02	2284	16 42	2425	26 55	2693	40 18	3168
37 30	8 04	2294	16 47	2435	27 03	2705	40 31	3182
37 40	8 06	2304	16 52	2446	27 12	2717	40 43	3196
37 50	8 09	2314	16 57	2457	27 20	2729	40 56	3210
38 0	8 11	2325	17 02	2468	27 28	2742	41 08	3224
38 10	8 13	2335	17 08	2479	27 37	2754	41 21	3239
38 20	8 16	2345	17 13	2490	27 45	2766	41 33	3259
38 30	8 19	2355	17 18	2500	27 54	2778	41 46	3262
38 40	8 21	2366	17 23	2511	28 03	2790	42 00	3281
38 50	8 24	2376	17 28	2522	28 11	2802	42 13	3296
39 0	8 26	2386	17 32	2533	28 2c	2814	42 26	3309
39 10	8 28	2396	17 19	2544	28 29	2826	42 39	3323
39 20	8 31	2406	17 44	2554	28 37	2838	42 52	3338
39 30	8 34	2417	17 49	2565	28 46	2850	43 04	3352
39 40	8 37	2427	17 55	2576	28 55	2862	43 17	3366
39 50	8 39	2437	18 00	2587	29 04	2874	43 30	3380
40 0	8 42	2447	18 06	2598	29 12	2886	43 43	3394
40 10	8 44	2457	18 12	2609	29 21	2898	43 56	3408
40 20	8 47	2467	18 20	2619	29 30	2911	44 09	3422
40 30	8 49	2478	18 22	2630	29 38	2923	44 21	3437
40 40	8 52	2488	18 28	2641	29 47	2935	44 34	3451
40 50	8 55	2498	18 34	2652	29 56	2947	44 48	3465
41 0	8 57	2508	18 40	2663	30 05	2959	45 02	3479
41 10	9 00	2519	18 45	2674	30 14	2971	45 16	3493
41 20	9 03	2529	18 50	2684	30 23	2983	45 29	3501
41 30	9 06	2539	18 56	2695	30 32	2995	45 42	3521
41 40	9 08	2549	19 01	2706	30 40	3006	45 55	3535
41 50	9 12	2569	19 06	2717	30 50	3019	46 08	3550

with the Difference of Longitudes and Latitudes.								
Latitude.		5 Rumb, 56° 15'.			6 Rumb, 67° 30'.		7 Rumb, 78° 45'.	
		Longitude.	Dift. in Miles.	Longitude.	Dift. in Miles.		Longitude.	Dift. in Miles.
D.	M.	D.	M.		D.	M.		D.	M.		D.	M.	

D.	M.	D.	M.	Dift. in Miles.	D.	M.	Dift. in Miles.	D.	M.	Dift. in Miles.
36	0	57	49	3888	93	16	5644	194	14	11072
36	10	58	08	3906	93	48	5670	195	17	11123
36	20	58	26	3924	94	16	5697	196	18	11174
36	30	58	45	3942	94	48	5723	197	20	11226
36	40	59	03	3960	95	18	5741	198	22	11277
36	50	59	22	3978	95	48	5775	199	26	11328
37	0	59	41	3996	96	18	5801	200	29	11379
37	10	60	00	4014	96	48	5827	291	31	11421
37	20	60	18	4032	97	18	5853	202	33	11472
37	30	60	37	4050	97	48	5879	203	37	11423
37	40	60	56	4068	98	19	5906	204	41	11574
37	50	61	15	4086	98	49	5932	205	46	11626
38	0	61	34	4104	99	19	5958	206	50	11687
38	10	61	52	4122	99	49	5984	207	54	11738
38	20	62	12	4140	100	20	6010	208	55	11789
38	30	62	31	4158	100	51	6036	210	06	11841
38	40	62	50	4176	101	24	6062	211	05	11892
38	50	63	09	4194	101	52	6089	212	10	11943
39	0	63	29	4212	102	26	6115	213	15	11994
39	10	63	48	4230	102	56	6141	214	20	12046
39	20	64	07	4248	103	28	6167	215	24	12097
39	30	64	27	4266	103	58	6193	216	29	12148
39	40	64	46	4284	104	30	6219	217	35	12199
39	50	65	06	4302	105	02	6245	218	41	12251
40	0	65	25	4320	105	32	6271	219	42	12302
40	10	65	45	4338	106	04	6298	220	49	12353
40	20	66	04	4356	106	35	6324	221	56	12404
40	30	66	24	4374	107	06	6350	223	03	12456
40	40	66	44	4392	107	39	6378	224	10	12507
40	50	67	04	4410	108	10	6402	225	18	12558
41	0	67	23	4428	108	42	6428	226	23	12609
41	10	67	43	4436	109	16	6454	227	27	12661
41	20	68	03	4454	109	46	6480	228	35	12712
41	30	68	23	4472	110	20	6507	229	43	12763
41	40	68	43	4490	110	52	6533	230	46	12815
41	50	69	03	4418	111	25	6559	231	58	12866

Loxodromiques , or Traverse-Tables of Miles,											
Lati-tude.	1 Rumb, 11° 15'.			2 Rumb, 22° 30'.			3 Rumb, 33° 45'.			4 Rumb, 45° 00'.		
	Longit.		Diſt.in	Longit.		Diſt.in	Longit.		Diſt.in	Longit.		Diſt.in
Θ. M.	D.	M.	Miles.	D.	M.	Miles.	D.	M.	Miles.	D.	M.	Miles.
42 0	9	13	2569	19	12	2728	31	00	3031	46	22	3564
42 10	9	16	2580	19	18	2739	31	08	3043	46	36	3578
42 20	9	18	2590	19	23	2749	31	17	3055	46	49	3592
42 30	9	21	2600	19	29	2760	31	26	3067	47	02	3606
42 40	9	24	2610	19	35	2771	31	35	3079	47	16	3620
42 50	9	27	2620	19	40	2782	31	44	3091	47	30	3634
43 0	9	30	2631	19	46	2793	31	53	3103	47	43	3649
43 10	9	32	2641	19	51	2803	32	02	3115	47	56	3663
43 20	9	35	2651	19	57	2814	32	11	3127	48	10	3677
43 30	9	37	2661	20	03	2825	32	20	3139	48	25	3691
43 40	9	40	2671	20	06	2836	32	29	3151	48	39	3715
43 50	9	43	2682	20	13	2847	32	38	3163	48	53	3729
44 0	9	46	2692	20	19	2858	32	47	3175	49	06	3733
44 10	9	48	2702	20	26	2868	32	57	3187	49	20	3748
44 20	9	51	2712	20	32	2879	33	07	3199	49	34	3762
44 30	9	54	2722	20	38	2890	33	16	3211	49	48	3776
44 40	9	57	2733	20	44	2901	33	25	3223	50	02	3790
44 50	10	00	2743	20	50	2912	33	34	3235	50	16	3804
45 0	10	03	2753	20	56	2923	33	48	3247	50	30	3818
45 10	10	06	2763	21	01	2933	33	53	3259	50	43	3832
45 20	10	08	2773	21	07	2944	34	02	3271	50	57	3847
45 30	10	11	2784	21	16	2955	34	12	3283	51	12	3861
45 40	10	14	2794	21	19	2966	34	23	3295	51	26	3875
45 50	10	17	6804	21	25	2977	34	33	3307	51	40	3889
46 0	10	19	2814	21	31	2988	34	42	3319	51	54	3903
46 10	10	22	2824	21	33	2998	34	52	3331	52	10	3917
46 20	10	25	2835	21	42	3009	35	01	3343	52	25	3931
46 30	10	28	2845	21	48	3020	35	11	3356	52	39	3946
46 40	10	31	2855	21	54	3031	35	21	3368	52	54	3960
46 50	10	34	2865	22	00	3041	35	30	3380	53	28	3974
47 0	10	37	2875	22	06	3052	35	40	3392	53	23	3988
47 10	10	40	2886	22	12	3063	35	50	3404	53	37	4002
47 20	10	43	2896	22	18	3074	36	00	3416	53	52	4016
47 30	10	45	2906	22	24	3085	36	10	3428	54	06	4030
47 40	10	48	2916	22	30	3096	36	20	3440	54	21	4045
47 50	10	51	2926	22	36	3106	36	30	3442	54	36	4059

with the Difference of Longitudes and Latitudes.

Latitude		5 Rumb, 56° 15'.			6 Rumb, 67° 30'.			7 Rumb, 78° 45'.		
		Longitude.		Diſt. in Miles.	Longitude.		Diſt. in Miles.	Longitude.		Diſt. in Miles.
D.	M.	D.	M.		D.	M.		D.	M.	
42	0	69	23	4536	111	56	6585	233	03	12917
42	10	69	43	4554	112	29	6611	234	12	12968
42	20	70	04	4572	113	01	6637	235	21	13020
42	30	70	24	4590	113	34	6663	236	30	13061
42	40	70	44	4608	114	06	6690	237	34	13112
42	50	71	04	4626	114	39	6716	238	43	13163
43	0	71	25	4644	115	12	6742	239	53	13215
43	10	71	45	4662	115	45	6768	241	02	13266
43	20	72	06	4680	116	19	6794	242	12	13317
43	30	72	27	4698	116	51	6820	243	20	13368
43	40	72	47	4716	117	27	6846	244	34	13420
43	50	73	08	4734	117	58	6872	245	39	13471
44	0	73	29	4752	118	34	6899	246	50	13532
44	10	73	50	4770	119	06	6925	248	01	13583
44	20	74	11	4788	119	40	6951	249	12	13635
44	30	74	31	4806	120	13	6977	250	23	13686
44	40	74	52	4824	120	48	7003	251	29	13737
44	50	75	12	4842	121	21	7029	252	41	13788
45	0	75	35	4860	121	56	7055	253	34	13840
45	10	75	56	4878	122	30	7082	255	04	13891
45	20	76	17	4896	123	04	7108	256	11	13942
45	30	76	38	4914	123	38	7134	257	24	13993
45	40	77	00	4932	124	13	7160	258	38	14045
45	50	77	21	4950	124	47	7186	259	50	14096
46	0	77	43	4968	125	22	7212	261	04	14147
46	10	78	04	4986	125	56	7238	262	12	14198
46	20	78	26	5004	126	31	7264	263	28	14250
46	30	78	47	5022	127	08	7291	264	39	14301
46	40	79	09	5040	127	43	7317	265	53	14352
46	50	79	30	5058	128	18	7343	267	08	14403
47	0	79	53	5076	128	53	7369	268	23	14455
47	10	80	15	5094	129	28	7395	269	23	14506
47	20	80	37	5112	130	04	7321	270	47	14557
47	30	80	59	5130	130	40	7347	272	02	14609
47	40	81	22	5148	131	13	7373	273	18	14660
47	50	81	43	5166	131	52	7400	274	34	14711

Loxodromiques, or Traverse-Tables of Miles,												
Lati-tude.	1 *Rumb*, 11°15'.			2 *Rumb*, 22°30'.			3 *Rumb*, 33°45'.			4 *Rumb*, 45°00'.		
	Longit.		*Dist.in*	*Longit.*		*Dist.in*	*Longit.*		*Dist.in*	*Longit.*		*Dist.in*
D. M.	D.	M.	*Miles.*	D.	M.	*Miles.*	D.	M.	*Miles.*	D.	M.	*Miles.*
48 0	10	54	2936	22	41	3117	36	40	3464	55	52	4073
48 10	10	57	2947	22	49	3128	36	50	3476	55	05	4087
48 20	11	00	2957	22	56	3139	37	00	3488	55	22	4101
48 30	11	04	2967	23	03	3150	37	10	3500	55	37	4115
48 40	11	07	2977	23	10	3161	37	20	3512	55	51	4129
48 50	11	10	2987	23	16	3171	37	30	3524	56	07	4144
49 0	11	13	2998	23	22	3182	37	40	3536	56	22	4158
49 10	11	16	3008	23	27	3193	37	50	3548	56	38	4172
49 20	11	19	3018	23	33	3204	38	00	3560	56	52	4186
49 30	11	22	3028	23	40	3215	38	10	3572	57	08	4200
49 40	11	25	3038	23	46	3225	38	20	3584	57	23	4214
49 50	11	28	3049	23	52	3236	38	30	3596	57	39	4228
50 0	11	31	3059	23	59	3247	38	42	3608	57	54	4243
50 10	11	34	3069	24	05	3258	38	52	3620	58	10	4255
50 20	11	37	3079	24	12	3269	39	03	3632	58	26	4271
50 30	11	40	3089	24	19	3280	39	13	3644	58	42	4285
50 40	11	44	3100	24	26	3290	39	24	3656	58	58	4299
50 50	11	47	3110	24	32	3301	39	34	3668	59	14	4313
51 0	11	50	3120	24	38	3312	39	45	3680	59	30	4327
51 10	11	53	3130	24	45	3323	39	55	3692	59	46	4342
51 20	11	56	3140	24	52	3334	40	06	3704	60	01	4356
51 30	12	00	3150	24	58	3345	40	16	3716	60	17	4370
51 40	12	03	3161	25	05	3355	40	27	3728	60	33	4384
51 50	12	06	3171	25	12	3366	40	38	3741	60	49	4398
52 0	12	09	3181	25	18	3377	40	49	3753	61	05	4412
52 10	12	12	3191	25	25	3388	41	00	3765	61	21	4426
52 20	12	15	3202	25	32	3399	41	10	3777	61	37	4441
52 30	12	19	3212	25	39	3410	41	22	3789	61	54	4455
52 40	12	22	3222	25	45	3420	41	32	3801	62	10	4469
52 50	12	26	3232	25	52	3431	41	43	3813	62	26	4483
53 0	12	29	3242	25	56	3442	41	55	3825	62	43	4497
53 10	12	32	3253	26	05	3453	42	06	3837	63	00	4511
53 20	12	36	3263	26	13	3464	42	17	3849	63	17	4525
53 30	12	39	3273	26	20	3474	42	28	3861	63	34	4540
53 40	12	42	3283	26	27	3485	42	40	3873	63	51	4554
53 50	12	45	3293	26	34	3496	42	51	3885	64	08	4568

with the Difference of Longitudes and Latitudes.							
Latitude.		5 Rumb, 56° 15'.			6 Rumb, 67° 30'.			7 Rumb, 78° 45'.		
		Longitude.		Dist. in Miles.	Longitude.		Dist. in Miles.	Longitude.		Dist. in Miles.
D.	M.	D.	M.		D.	M.		D.	M.	
48	0	82	06	5184	132	28	7426	275	50	14752
48	10	82	28	5202	133	03	7452	277	00	14814
48	20	82	51	5220	133	38	7478	278	17	14865
48	30	83	14	5238	134	16	7504	279	34	14916
48	40	83	36	5256	134	54	7530	280	51	14967
48	50	83	58	5274	135	28	7556	282	08	15019
49	0	84	22	5292	136	07	7683	283	23	15070
49	10	84	44	5310	136	42	7709	284	38	15121
49	20	85	07	5328	137	20	7735	285	56	15172
49	30	85	30	5346	137	57	7761	287	24	15224
49	40	85	54	5364	138	34	7787	288	33	15275
49	50	86	16	5382	139	12	7813	289	46	15320
50	0	86	44	5400	139	49	7839	291	05	15377
50	10	87	03	5418	140	26	7865	292	24	15429
50	20	87	27	5436	141	04	7892	295	44	15480
50	30	87	50	5454	141	41	7918	296	04	15531
50	40	88	14	5472	142	20	7944	296	24	15582
50	50	88	39	5480	142	58	7970	297	45	15634
51	0	89	01	5508	143	36	7996	299	00	15685
51	10	89	25	5526	144	17	8022	300	21	15736
51	20	89	49	5544	144	52	8048	301	43	15787
51	30	90	13	5562	145	34	8074	303	04	15839
51	40	90	37	5580	146	12	8101	304	20	15890
51	50	91	00	5598	146	51	8127	305	42	15941
52	0	91	25	5616	147	30	8153	307	05	15992
52	10	91	50	5634	148	09	8179	308	28	16044
52	20	92	14	5652	148	48	8205	309	52	16095
52	30	92	39	5670	149	28	8231	311	12	16146
52	40	93	03	5688	150	07	8257	312	36	16198
52	50	93	28	5706	150	47	8284	313	57	16249
53	0	93	53	5724	151	27	8310	315	22	16300
53	10	94	18	5742	152	07	8336	316	46	16351
53	20	94	43	5760	152	47	8362	318	12	16403
53	30	95	03	5778	153	27	8388	319	35	16454
53	40	95	33	5796	154	11	8414	320	56	16505
53	50	95	58	5814	154	48	8440	322	23	16556

Loxodromiques, or Traverse-Tables of Miles,

Latitude		1 Rumb, 11°15'.			2 Rumb, 22°30'.			3 Rumb, 33°45'.			4 Rumb, 45°00'.		
		Longit.		Dist. in	Longit.		Dist. in	Longit.		Dist. in	Longit.		Dist. in
D.	M.	D.	M.	Miles.	D.	M.	Miles.	D.	M.	Miles.	D.	M.	Miles.
54	0	12	49	3303	26	41	3507	43	03	3897	64	24	4582
54	10	12	52	3314	26	48	3518	43	14	3909	64	41	4596
54	20	12	56	3324	26	58	3529	43	25	3921	64	58	4610
54	30	12	59	3334	27	02	3539	43	37	3933	65	15	4624
54	40	13	02	3344	27	09	3550	43	48	3945	65	32	4639
54	50	13	05	3355	27	16	3561	44	00	3957	65	50	4653
55	0	13	09	3365	27	28	3572	44	12	3969	66	08	4667
55	10	13	12	3375	27	31	3583	44	23	3981	66	26	4681
55	20	13	16	3385	27	38	3594	44	35	3993	66	42	4695
55	30	13	20	3395	27	45	3604	44	46	4005	67	01	4709
55	40	13	23	3405	27	53	3615	44	58	4017	67	19	4723
55	50	13	26	3416	28	00	3626	45	10	4029	67	36	4738
56	0	13	30	3426	28	07	3637	45	22	4041	67	54	4752
56	10	13	33	3436	28	15	3648	45	33	4053	68	11	4766
56	20	13	37	3446	28	22	3658	45	45	4065	68	29	4780
56	30	13	41	3456	28	30	3669	45	58	4077	68	47	4794
56	40	13	44	3467	28	37	3680	46	10	4089	69	05	4808
56	50	13	48	3477	28	45	3691	46	22	4101	69	24	4822
57	0	13	52	3487	28	52	3702	46	35	4113	69	42	4837
57	10	13	56	3497	29	00	3713	46	47	4125	70	00	4851
57	20	14	00	3507	29	08	3723	46	55	4137	70	20	4865
57	30	14	03	3518	29	15	3734	47	11	4150	70	38	4879
57	40	14	07	3528	29	23	3745	47	24	4162	70	58	4893
57	50	14	11	3538	29	31	3756	47	37	4174	71	14	4907
58	0	14	15	3548	29	39	3767	47	50	4186	71	34	4921
58	10	14	18	3558	29	46	3778	48	02	4198	71	53	4936
58	20	14	22	3569	29	54	3788	48	15	4210	72	12	4950
58	30	14	26	3579	30	02	3799	48	27	4222	72	31	4963
58	40	14	30	3589	30	10	3810	48	40	4234	72	50	4978
58	50	14	34	3599	30	19	3821	48	53	4246	73	09	4992
59	0	14	37	3609	30	27	3832	49	06	4258	73	28	5006
59	10	14	41	3620	30	35	3843	49	19	4270	73	48	5020
59	20	14	44	3630	30	43	3853	49	32	4282	74	08	5035
59	30	14	48	3640	30	51	3864	49	45	4294	74	26	5049
59	40	14	52	3650	30	59	3875	49	55	4306	74	46	5062
59	50	14	56	3660	31	08	3886	50	01	4318	75	07	5077

with the Difference of Longitudes and Latitudes.									
Latitude.	5 *Rumb,* 56° 15'.			6 *Rumb,* 67° 30'.			7 *Rumb,* 78° 45'.		
	Longitude.		Dist. in	Longitude.		Dist. in	Longitude.		Dist. in
D. M.	D.	M.	Miles.	D.	M.	Miles.	D.	M.	Miles.
54 0	96	24	5832	155	32	7467	323	49	16608
54 10	96	50	5850	156	13	8493	325	16	16659
54 20	97	15	5868	156	54	8519	326	40	16710
54 30	97	40	5886	157	36	8545	328	04	16761
54 40	98	07	5904	158	17	8571	329	33	16813
54 50	98	32	5922	158	58	8597	331	01	16864
55 0	98	59	5940	159	40	8623	332	30	16915
55 10	99	24	5958	160	22	8649	334	00	16960
55 20	99	51	5976	161	05	8675	335	21	17018
55 30	100	17	5994	161	47	8702	336	51	17069
55 40	100	46	6012	162	29	8728	338	21	17120
55 50	101	10	6030	163	12	8754	339	52	17171
56 0	101	37	6048	163	55	8780	341	23	17223
56 10	102	04	6066	164	40	8806	342	47	17274
56 20	102	31	6084	165	21	8832	344	19	17325
56 30	102	58	6102	166	08	8858	345	51	17376
56 40	103	25	6120	166	52	8885	347	24	17428
56 50	103	52	6138	167	36	8911	348	57	17479
57 0	104	20	6156	168	20	8937	350	27	17530
57 10	104	46	6174	169	04	8963	352	00	17581
57 20	105	15	6192	169	48	8989	353	31	17633
57 30	105	43	6210	170	33	9015	355	06	17684
57 40	106	11	6228	171	18	9041	356	41	17735
57 50	106	39	6246	172	02	9067	358	16	17787
58 0	107	07	6264	172	47	9094	359	45	17838
58 10	107	35	6282	173	32	9120	361	21	17889
58 20	108	04	6300	174	18	9146	362	59	17940
58 30	108	32	6318	175	04	9172	364	36	17992
58 40	109	01	6336	175	54	9198	366	14	18043
58 50	109	30	6354	176	37	9224	367	52	18094
59 0	109	59	6372	177	27	9250	369	29	18145
59 10	110	28	6390	178	13	9277	371	23	18197
59 20	110	57	6408	179	00	9303	372	43	18248
59 30	111	27	6426	179	47	9329	374	23	18299
59 40	111	56	6444	180	35	9355	375	58	18350
59 50	112	26	6462	181	22	9381	377	37	18402

Loxodromiques, or Traverse-Tables of Miles,												
Lati-tude.	1 Rumb, 1°15'.			2 Rumb, 22°30'.			3 Rumb, 33°45'.			4 Rumb, 45°00'.		
	Longit.		Dist. in	Longit.		Dist. in	Longit.		Dist. in	Longit.		Dist. in
D. M.	D.	M.	Miles.	D.	M.	Miles.	D.	M.	Miles.	D.	M.	Miles.
60 0	15	00	3671	31	16	3897	50	25	4330	75	26	5091
60 10	15	04	3691	31	24	3907	50	38	4342	75	47	5105
60 20	15	08	3701	31	32	3918	50	52	4354	76	08	5119
60 30	15	12	3711	31	41	3929	51	06	4366	76	28	5134
60 40	15	16	3721	31	48	3940	51	20	4378	76	48	5148
60 50	15	20	3722	31	57	3951	51	33	4390	77	08	5162
61 0	15	25	3732	32	06	3962	51	46	4402	77	29	5176
61 10	15	29	3742	32	15	3972	52	00	4414	77	49	5190
61 20	15	33	3752	32	23	3983	52	14	4426	78	10	5204
61 30	15	37	3762	32	31	3994	52	28	4438	78	31	5218
61 40	15	41	3773	32	40	4005	52	42	4450	78	51	5233
61 50	15	45	3783	32	49	4016	52	56	4462	79	12	5247
62 0	15	49	3793	32	58	4027	53	10	4474	79	34	5260
62 10	15	54	3803	33	07	4037	53	24	4486	79	55	5275
62 20	15	58	3813	33	16	4048	53	38	4498	80	17	5289
62 30	16	03	3824	33	25	4059	53	53	4510	80	38	5303
62 40	16	07	3834	33	34	4070	54	27	4532	81	00	5317
62 50	16	12	3844	33	43	4080	54	21	4554	81	22	5332
63 0	16	16	3854	33	52	4091	54	38	4546	81	44	5346
63 10	16	20	3864	34	01	4102	54	52	4558	82	06	5360
63 20	16	24	3875	34	10	4113	55	06	4570	82	28	5374
63 30	16	29	3895	34	19	4124	55	22	4582	82	51	5388
63 40	16	34	3905	34	28	4135	55	37	4594	83	14	5402
63 50	16	38	3910	34	38	4145	55	52	4616	83	36	5416
64 0	16	43	3915	34	48	4156	56	07	4618	83	59	5431
64 10	16	47	3925	34	58	4167	56	22	4630	84	22	5445
64 20	16	52	3936	35	00	4178	56	38	4642	84	45	5459
64 30	16	56	3946	35	16	4189	56	53	4654	85	09	5473
64 40	17	01	3956	35	26	4200	57	08	4666	85	31	5487
64 50	17	05	3966	35	36	4210	57	24	4679	85	55	5501
65 0	17	10	3976	35	46	4221	57	40	4690	86	19	5515
65 10	17	15	3987	35	55	4232	57	56	4701	86	42	5530
65 20	17	20	3997	36	05	4243	58	12	4715	87	06	5544
65 30	17	25	4007	36	15	4254	58	28	4727	87	29	5558
65 40	17	20	4017	36	25	4267	58	44	4739	87	55	5572
65 50	17	34	4027	36	36	4275	59	01	4750	88	20	5586

with the Difference of Longitudes and Latitudes.

Latitude.		5 *Rumb,* 56° 15'.			6 *Rumb,* 67° 30'.			7 *Rumb,* 78° 45'.		
		Longitude.		Dist. in	Longitude.		Dist. in	Longitude.		Dist. in
D.	M.	D.	M.	Miles.	D.	M.	Miles.	D.	M.	Miles.
60	0	112	56	6480	182	11	9407	379	19	18453
60	10	113	25	6498	182	59	9433	381	01	18504
60	20	113	56	6506	183	48	9459	382	44	18555
60	30	114	26	6524	184	36	9486	384	27	18607
60	40	114	57	6542	185	26	9512	386	10	18658
60	50	115	27	6560	186	13	9538	387	50	18709
61	0	115	58	6588	187	04	9564	389	32	18760
61	10	116	29	6606	187	57	9590	391	18	18812
61	20	116	59	6624	188	43	9616	393	04	18863
61	30	117	31	6642	188	38	9642	394	50	18914
61	40	118	03	6660	190	28	9668	396	29	18965
61	50	118	34	6678	191	19	9695	398	17	19017
62	0	119	06	6696	192	10	9721	400	05	19068
62	10	119	37	6714	193	01	9747	401	54	19119
62	20	120	10	6732	193	52	9773	403	43	19170
62	30	120	43	6750	194	44	9799	405	28	19222
62	40	121	15	6768	195	37	9825	407	16	19273
62	50	121	44	6786	196	30	9851	409	08	19324
63	0	122	21	6804	197	23	9878	411	00	19376
63	10	122	53	6822	198	16	9904	412	52	19427
63	20	123	27	6840	199	09	9930	414	36	19478
63	30	124	00	6858	200	03	9956	416	31	19529
63	40	124	34	6876	201	00	9982	418	25	19581
63	50	125	08	6894	201	51	10008	420	20	19632
64	0	125	42	6912	202	49	10034	422	16	19683
64	10	126	16	6930	203	43	10060	424	08	19734
64	20	126	51	6948	204	38	10087	426	01	19787
64	30	127	26	6966	205	34	10113	428	00	19837
64	40	128	00	6984	206	31	10139	430	00	19888
64	50	128	35	7002	207	27	10165	431	58	19939
65	0	129	11	7020	208	24	10191	433	58	19991
65	10	129	46	7038	209	20	10217	435	50	20041
65	20	130	22	7056	210	18	10243	437	52	20093
65	30	130	58	7074	211	16	10269	439	54	20144
65	40	131	34	7092	212	15	10296	441	57	20195
65	50	132	10	7110	213	12	10322	444	02	20247

Loxodromiques, or Traverse-Tables of Miles,								
Lati-tude.	1 Rumb, 11°15'.		2 Rumb, 22°30'.		3 Rumb, 33°45'.		4 Rumb, 45°00'.	
	Longit.	Dist.in	Longit.	Dist.in	Longit.	Dist.in	Longit.	Dist.in
D. M.	D. M.	Miles.	D. M.	Miles.	D. M.	Miles.	D. M.	Miles.
66 0	17 35	4038	36 46	4286	59 17	4763	88 44	5600
66 10	17 43	4048	36 56	4297	59 33	4775	89 08	5614
66 20	17 48	4058	37 06	4308	59 50	4787	89 32	5629
66 30	17 53	4068	37 16	4319	60 06	4799	89 57	5643
66 40	17 58	4078	37 26	4330	60 23	4811	90 23	5657
66 50	18 03	4089	37 38	4349	60 40	4823	90 48	5671
67 0	18 08	4099	37 48	4351	60 58	4835	91 13	5685
67 10	18 13	4109	37 58	4362	61 15	4847	91 38	5699
67 20	18 18	4119	38 09	4373	61 32	4859	92 04	5713
67 30	18 24	4129	38 20	4384	61 49	4861	92 30	5728
67 40	18 30	4140	38 30	4394	62 07	4883	92 56	5732
67 50	18 33	4150	38 40	4405	62 24	4895	93 23	5740
68 0	18 40	4160	38 52	4416	62 41	4907	93 50	5770
68 10	18 45	4170	39 04	4427	62 59	4919	94 17	5784
68 20	18 51	4180	39 15	4438	63 17	4931	94 44	5798
68 30	18 56	4190	39 26	4448	63 35	4943	95 11	5812
68 40	19 02	4201	39 38	4459	63 54	4955	95 38	5826
68 50	19 07	4211	39 59	4470	64 12	4967	96 05	5841
69 0	19 12	4221	40 00	4481	64 31	4989	96 33	5855
69 10	19 17	4231	40 12	4492	64 50	5001	97 02	5869
69 20	19 24	4241	40 24	4503	65 08	5013	97 30	5883
69 30	19 29	4252	40 35	4514	65 27	5025	98 00	5897
69 40	19 35	4262	40 45	4524	65 46	5037	98 29	5911
69 50	19 40	4272	40 58	4535	66 05	5049	98 58	5925
70 0	19 45	4282	41 10	4546	66 25	5051	99 26	5940
70 10	19 50	4292	42 26	4557	68 27	5063	99 54	5954
70 20	19 55	4302	43 45	4568	70 33	5075	100 24	5968
70 30	20 00	4312	45 07	4578	72 47	5087	100 54	5982
70 40	20 05	4322	46 31	4589	75 07	5099	101 24	5996
70 50	20 10	4332		4600		5111	101 54	6010

with the Difference of Longitudes and Latitudes.									
Latitude.		5 *Rumb*, 56° 15'.			6 *Rumb*, 67° 30'.			7 *Rumb*, 78° 45'.	
		Longitude.		*Diſt. in*	*Longitude*		*Diſt. in*	*Longitude.*		*Diſt. in*
D.	M.	D.	M.	*Miles.*	D.	M.	*Miles.*	D.	M.	*Miles.*
66	0	132	47	7128	214	11	10348	445	58	20298
66	10	133	24	7146	215	10	10374	448	03	20349
66	20	134	01	7164	216	13	10400	450	10	20400
66	30	134	38	7182	217	15	10426	452	18	20451
66	40	135	16	7200	218	17	10452	454	26	20503
66	50	135	54	7218	219	18	10479	456	28	20554
67	0	136	32	7236	229	19	10567	458	40	20605
67	10	137	09	7254	221	20	10531	460	46	20656
67	20	137	49	7272	222	21	10557	462	58	20708
67	30	138	28	7290	223	23	10583	465	11	20759
67	40	139	08	7308	224	26	10609	467	24	20810
67	50	139	47	7326	225	29	10635	469	28	20862
68	0	140	27	7344	226	33	10661	471	44	20913
68	10	141	07	7362	227	36	10688	474	01	20964
68	20	141	47	7380	228	41	10714	476	18	21015
68	30	142	29	7398	229	45	10740	478	37	21067
68	40	143	09	7416	230	49	10766	480	57	21118
68	50	143	49	7434	232	41	10792	483	12	21169
69	0	144	32	7452	233	13	10818	485	28	21220
69	10	145	14	7470	234	21	10844	487	51	21272
69	20	145	56	7488	235	29	10870	490	15	21323
69	30	146	39	7506	236	37	10897	492	40	21374
69	40	147	22	7524	237	46	10923	494	59	21425
69	50	148	06	7542	238	54	10949	497	24	21477
70	0	148	48	7560	240	04	10975	499	51	21528
70	10	149	33	7578	241	13	11001	502	21	21579
70	20	150	17	7596	242	23	11027	504	52	21630
70	30	151	01	7614	242	34	11053	507	20	21682
70	40	151	47	7632	244	48	11080	509	47	21733
70	50	152	31	7650	244	00	11106	512	21	21784

A
TABLE

OF THE

Latitudes *and* Longitudes *of the principal* Ports, Harbours, Capes *and* Iflands, *in moft of the known parts of the World: Beginning from the Meridian of* Pico Teneriffa. *Collected from the beft Charts,* Defcriptions *, and Obfervations of feveral able and experienced* Navigators *of our own and other* Nations.

A *Table* of Latitude and Longitude.

Places Names.	Latitude.			Longitude	
The Sea-Coaft of Greenland.	D.	M.		D.	M.
Hacluits Headland	79	50	N	26	55
Fair Foreland	79	15	N	24	50
Black Point	78	32	N	25	10
Point Look-out	76	25	N	32	00
Cape Blanco	78	25	N	38	00
Point Negro	77	10	N	42	00
Hopelefs Ifles	77	00	N	42	30
Iflands in the North Sea.					
Hope Ifland	76	13	N	41	50
Cherry Ifland	74	34	N	34	10
South Point of Trinity Ifland	71	00	N	07	55
Youngs Foreland in Trinity Ifland	71	23	N	10	20
Sea-Coaft of Nova-Zembla.					
Orange Ifland	78	25	N	91	35
Ice Point	77	45	N	90	50
Admiralties Ifland	75	50	N	73	55
Langenefs	74	55	N	68	50
Crofs-Point	72	25	N	68	05
Fretum Burrough	70	40	N	75	00
The River *Obij* in the Tartarian Sea	69	12	N	80	49
Mauritias Ifle	71	24	N	72	10
Sea-Coaft in the White Sea.					
Archangel	63	22	N	55	28
Swelgenofe	69	10	N	61	38
Cape Candenofe	69	35	N	58	02
Orlogenofe	66	55	N	54	56
Sea-Coaft of Lapland *and* Norway.					
Fox Nofe	64	12	N	37	07
Cape Grace	65	17	N	39	30
Cape Gallant	67	11	N	39	32

A *Table* of Latitude and Longitude.				
Places Names.	Latitude.		Longitude	
Sea-Coaſt of Lapland *and* Norway.	D.	M.	D.	M.
Cape Race	65	49 N	40	04
Iſland Kilduym	68	54 N	38	05
North Cape	71	22 N	32	35
Roſs-Iſles	67	01 N	25	06
Catſneſs	61	54 N	18	42
Bomel	59	32 N	19	38
Naze of Norway, ...	58	00 N	21	02
The Sea-Coaſt in the Sound.				
The Nyding	57	00 N	25	40
Cape Cole	56	46 N	26	12
Scarlet Iſland	56	02 N	26	38
Falſterborne	56	53 N	26	25
Abbo	61	08 N	34	30
Wyburrough	61	16 N	40	18
Dagaret	59	44 N	34	31
Dormamel	56	55 N	34	31
Gotland	58	20 N	31	05
Horroſound	58	48 N	32	58
Gothſound	59	15 N	32	29
Earth Holme	56	10 N	28	14
Burnt Holme	56	00 N	28	16
Elſenore	56	40 N	25	57
The Sea-Coaſt of Flanders, *from the* Scaw *to* Callice.				
The Scaw	57	52 N	24	27
Bovenberg	56	20 N	23	56
Holy Land	54	30 N	22	14
The Texel	53	20 N	20	56
The Brill	52	08 N	19	44
Callice	51	13 N	17	52

A *Table* of Latitude and Longitude.

Places Names.	Latitude.			Longitude	
	D.	M.		D.	M.
The Sea-Coaft of Ifland.					
Langenefs	67	20	N	03	45
Grinfe	67	00	N	352	05
Maze	68	15	N	351	10
Andifer	66	30	N	345	00
Snowhill	65	40	N	344	40
Alera Point	64	08	N	344	20
Weftmonia Ifles	63	35	N	349	00
Merchants Foreland	63	36	N	358	40
Whales Back	65	18	N	003	00
Iflands near the Coaft of Scotland.					
St. Kilda	58	02	N	05	56
Skie Ifland	57	40	N	10	08
Lewis Ifland	58	30	N	08	00
Fair Iflands	61	43	N	09	00
Shotland	60	22	N	14	30
Fair ifle	59	30	N	14	20
Ifles of Orkney	58	50	N	13	25
Sea-Coaft of Scotland, England, *and* Ireland.					
Catenefs	58	37	N	13	24
Buchannefs	58	00	N	14	32
St. Abbs Head	56	25	N	14	12
Tinmouth	55	08	N	15	00
Flambrough Head	54	08	N	16	26
The Sporne	53	45	N	16	58
Wintertonnefs	52	52	N	18	00
Orfordnefs	52	20	N	18	02
The North Foreland	51	32	N	17	40
The South Foreland	51	22	N	17	42
Dongenefs	51	09	N	17	14
Ifle of Wight	50	24	N	14	47
Portland	50	20	N	13	24
The Start	50	27	N	12	
The Lizard	50	00	N		

A *Table* of Latitude and Longitude.				
Places-Names.	Latitude.		Longitude	
	D.	M.	D.	M.
Sea-Coast of Scotland, England, *and* Ireland.				
Iſlands of Silly	50	07	N	09	47
Londy Iſle	51	22	N	11	57
St. David's Head	51	54	N	11	18
Bradſey Iſle	52	40	N	1	39
Holy Head	54	44	N	11	44
Iſle of Man	54	25	N	11	45
Coſwel-Point	54	36	N	11	19
Fair Foreland	55	35	N	10	16
Aron Iſle	53	21	N	8	30
Black Rock	53	52	N	6	00
Sline Head	53	16	N	6	00
Blaſques	52	15	N	5	20
Cape Clear	51	15	N	6	28
Old Head	51	40	N	7	32
Hearn Point	52	05	N	10	04
Sea-Coast of France, Spain, *and* Portugal.				
Sain Head	50	04	N	16	50
Cape Hage	50	04	N	15	05
Caskets	50	07	N	14	30
Garoſey	49	43	N	14	20
Jerſey	49	30	N	14	46
Uſhant	48	40	N	11	16
Orleon	45	58	N	11	00
Cape de Machicaⱥ	44	22	N	14	20
Cape Piras	44	04	N	11	10
Cape Ortegal	44	08	N	9	16
Cap Finiſterre	43	10	N	6	58
The Rock of Liſbon	38	52	N	6	43
Cape St. Vincent	37	00	N	7	20
Cape St. Maria	37	05	N	8	42
The Straits of Gilbralter	36	00	N	10	40

A *Table* of Latitude and Longitude.					
Places Names.	Latitude.			Longitude	
The Sea-Coasts on the Main Continent *in the* Straits.	D.	M.		D.	M.
Cape de Gata . . . . . . ; . .	36	47	N	16	08
Cape St. Martin	38	46	N	18	57
Cape Daga Frito	41	41	N	21	49
Cape Larei	42	58	N	24	28
Cape Melle	43	51	N	26	21
Terracina	41	26	N	31	26
Cape Sparteventura	37	46	N	36	06
Cape Collom	38	50	N	37	30
Cape St. Maria	39	52	N	38	16
Angelo	41	31	N	36	28
Ancana	43	25	N	32	40
Cape Cefta	43	27	N	34	55
Ragufa	42	29	N	37	36
Cape Linga	40	19	N	38	50
Cape Matopan	36	28	N	42	00
Cape St. Angelo	37	15	N	42	56
Cape Doro	39	14	N	45	12
Cape Barbarnou	37	58	N	46	19
Cape Saradoni	35	35	N	48	46
Cape Pollopollo	34	54	N	55	34
Cape de Becur	32	40	N	50	48
Cape Roatini	32	18	N	43	32
Cape Rozato	32	58	N	40	28
Cape Bona	37	05	N	30	12
Collo	37	09	N	24	52
Tunis	36	40	N	19	46
Cape Falcon	36	08	N	17	28
Cape Tres Forcas . . . , . . .	35	40	N	15	24
Tangier	35	36	N	11	35
Iflands within the Straits.					
Alboran	37	52	N	15	18
Formentara	38	44	N	19	38
Ivica	39	05	N	19	50

A *Table* of Latitude and Longitude.				
Places Names.	Latitude.		Longitude	

Iflands within the Straits.	D.	M.		D.	M.
Majorca	39	38	N	21	20
Minorca	39	55	N	22	30
Gallatta	37	57	N	27	50
Cape Pulo in Sardinia	38	56	N	27	36
Cape Corſo in Corſica	42	51	N.	27	32
Lilbo	42	51	N	29	00
Palmorolla	40	50	N	31	12
Uſtica	38	46	N	32	48
Maritimo	37	52	N	30	54
Pantalaria	36	53	N.	31	00
Limpadoſa	35	58	N	31	52
Limoſa	36	25	N	32	05
Malta	36	00	N	33	12
Cape Paſſaro in Sicilia	37	10	N	34	52
Meſſina	38	07	N	35	08
Liſſa	43	00	N	35	22
Trinity	41	50	N	35	14
Palagoſa	42	17	N	35	50
Auguſta	42	36	N	36	12
Mallida	42	37	N	37	04
Corfu	39	25	N	39	18
Cephalonia	38	28	N	40	29
Zant	37	37	N	40	40
Weſt end of Candia	35	15	N	43	00
Eaſt end of Candia	35	04	N	46	28
Rhodes	35	40	N	48	00
Weſt end of Cyprus	34	22	N	51	34
Eaſt end of Cyprus	34	48	N	54	35
· *Sea-Coaſt of* Barbary *and* Guinny.					
Cape Spartel	35	38	N	11	35
Cape Cantin	32	27	N	7	35
Cape Bojador	26	55	N	2	24
Cape Blanco	20	32	N	358	56
Cape Verde	14	36	N	358	50

A *Table* of Latitude and Longitude.				
Places Names.	Latitude.		Longitude	
	D.	M.	D.	M.
Sea-Coast of Barbary *and* Guinny.				
Sirre Leone	08	00 N	03	32
The South side of St. Anne	06	40 N	01	11
Cape de Palmas	04	10 N	10	00
Cape Tres Punctas	04	06 N	16	00
Cape Formosa	04	10 N	24	15
The North Point of Fernando	03	25 N	27	25
Island Principas	01	50 N	28	20
Island Chochos	00	40 N	03	32
Island St. Thoma	00	10 N	27	30
Cape Lupus	01	00 S	27	40
Cape Negro	16	00 S	30	50
Cape Sacos	29	40 S	36	20
Cape Bona Esperanza	34	24 S	38	10
Western Islands.				
The West side of Corva	40	00 N	345	30
The West side of Flores	39	40 N	345	30
The Road before Fyal	38	50 N	347	47
The West end of Pico	38	40 N	348	18
The West end of St. George	39	00 N	348	30
The West end of Tercera	30	00 N	349	10
The East end of St. Michaels	38	00 N	351	40
The East end of St. Maries	37	00 N	351	30
The Canary *Islands.*				
The North Part of Ferro	27	40 N	358	25
The East side of Palme	28	36 N	358	40
Gomera	28	10 N	359	15
Pico Tenerisa	28	20 N	00	00
The East end of Madera	32	32 N	00	10
The East end of Port Sancto	33	00 N	01	00
The North-east Point of Canaria	28	10 N	01	20
The North-east Point of Forteventura	28	20 N	02	50
The East part of Lancerotta	28	30 N	03	20

A *Table* of Latitude and Longitude.				
Places Names.	**Latitude.**		**Longitude**	
Cape de Verde Islands.	D.	M.	D.	M.
The West side of Antonio	17	00 N	350	00
The East Point of St. Vincent	17	40 N	350	08
The East side of St. Lucia	16	50 N	351	40
Isle Brava	14	40 N	351	08
Isle Fogo	14	00 N	351	30
The East side of St. Jago	15	00 N	352	30
East side of de Mopo Isle	15	00 N	353	04
East side of the Isle of Sal	16	00 N	353	04
East side of Bona Vista	16	00 N	353	04
St. Matthews	01	40 S	11	32
Island Anabona	01	22 S	26	20
Ascension	08	00 S	04	30
St. Helena	16	00 S	13	50
St. Helena Nova	16	03 S	24	48
Island Degialica	37	56 S	12	00
Island Desistian	36	57 S	12	42
Sea-Coast of the Main Continent in East-India.				
Cape Anquilhas	35	00 S	39	30
Cape Corintes	23	30 S	56	00
Cape de Guada	15	17 S	59	56
Cape de Guardasin	11	40 N	74	15
Cape de Rasalgate	22	07 N	84	10
Surat	21	04 N	96	20
Goa	15	40 N	97	00
Cape Comerin	07	52 N	99	12
The South-west Point of Ceylon	06	00 N	101	56
River Bengale	22	00 N	110	20
River de Care	10	45 N	119	10
Johr	01	25 N	125	06
Siam	14	52 N	122	45

A *Table* of Latitude and Longitude.				
Places Names.	Latitude.		Longitude	
Sea-Coast on the Main Continent in East-India.	D.	M.	D.	M.
Cambodia	12	42 N	126	00
Vifchers Point	20	18 N	131	00
The Point of Cavallos	25	16 N	140	58
Cape Somber	28	07 N	142	50
Ninghai	36	40 N	142	26
Corea	36	05 N	146	00
Iflands in the Eaft-India.				
Romeyros	28	19 S	89	40
John de Lisbon	25	24 S	75	52
Diego Roize	20	05 S	85	05
St. Branda	17	13 S	87	50
Dolgatias	15	20 S	82	15
Moroflas	20	10 S	78	36
Domafcaicaes	20	50 S	74	00
St. Apollonia	20	50 S	74	00
South end of St. Lawrence	25	37 S	68	00
North end of St. Lawrence	11	03 S	73	00
Baflos de India	22	10 S	60	40
John de Nova	17	20 S	63	15
Mayotta	12	40 S	66	30
Comore	11	20 S	65	10
De Natal	08	20 S	68	05
Cofmolede	09	40 S	72	08
John de Nova	09	00 S	79	46
De Almiranta	03	57 S	76	00
Domefcaicubas	03	21 S	80	26
St. Hermanas	03	02 S	84	00
Diego Gratiofa	08	03 S	92	20
De Gamo	02	40 S	99	05
Adu	05	39 S	99	00
Apoluria	09	20 S	100	40
Ifland Pracel	10	23 N	93	14
Gubile	8	53 N	93	30
Molique	9	05 N	94	25

A *Table* of Latitude and Longitude.				
Places Names.	Latitude.		Longitude	
Iſlands in the Eaſt-India.	D.	M.	D.	M.
Andaro	11	30 N	95	38
The North-weſt Point of Sumatra . .	5	30 N	116	00
South-eaſt end of Sumatra	5	52 S	125	40
Bantam	6	15 S	126	30
Batavia	6	10 S	127	05
Combava	8	36 S	138	00
Flores	8	50 S	140	48
Timor	9	52 S	145	53
Ceram	3	26 S	148	20
Amboina	3	52 S	147	25
South end of Celebes	5	45 S	139	30
North Point of Celebes	2	16 N	144	6
The middle of Gilolo	0	00 N	147	20
Bachian	0	03 N	146	10
Machian	0	14 N	146	6
Motir	0	25 N	146	8
Pottobackers	0	32 N	146	10
Tidore	0	41 N	146	16
Miferra	0	43 N	146	14
Ternate	0	47 N	146	12
St. Johannes	4	10 N	138	50
South Point of Burneo	4	16 S	135	4
North Point of Burneo	7	40 N	134	35
Weſt end of Mindano	6	50 N	141	25
Anamba	2	38 N	126	42
Natura	3	36 N	127	45
St. Juan	8	16 N	146	20
Tandaia	12	00 N	145	05
Masbat	11	50 N	143	35
Sebu	9	55 N	143	10
Pandi	11	05 N	142	00
Mindora	12	50 N	140	28
Paragoa	9	40 N	136	30
South end of Lucon	12	42 N	143	45
North end of Lucon	18	42 N	141	56
The Middle of Aynam	19	00 N	131	00

A *Table* of Latitude and Longitude.

Places Names.	Latitude			Longitude	
	D.	M.		D.	M.
Iflands in the East-India.					
South end of Formofa	21	.20	N	142	05
North end of Formofa	28	10	N	143	10
Firando Ifle	33	00	N	137	15
Weft end of Japan	34	00	N	150	05
North end of Cikoko	34	05	N	150	10
Tonfa	33	25	N	153	00
North Point of Japan	40	05	N	163	20
Cape Eroen in Jefo	42	50	N	183	20
Cape Patience in Jefo	49	00	N	186	00
The Coaft of America *in the* South-fea.					
The Straits of Anian	57	10	N	251	56
Cape Blanco	42	00	N	245	10
Sir Francis Drake's Bay	38	16	N	246	30
Ifland Peraros	30	08	N	253	50
Cape St. Lucas	23	10	N	266	10
Cape Corintes	20	28	N	270	56
Aquatulco	16	45	N	283	50
Gulf of Salina	10	00	N	288	20
Cape St. Maria	07	08	N	293	06
Cape Corintas	05	26	N	295	20
Cape de Francifco	01	25	N	292	20
Cape de Paffao	00	00	N	291	35
Payta	04	30	S	293	00
Truxilla	08	05	S	295	00
Villa la Nafca	14	10	S	297	50
Arica	18	35	S	2c2	00
Ifland Ferando	33	47	S	292	20
Baldivia	40	00	S	297	35
P. St. Cyprian	43	16	S	296	38
Weft Entrance of Magellan	53	00	S	296	42
Cape Horn	57	54	S	303	00

A *Table* of Latitude and Longitude.				
Places Names.	Latitude.		Longitude	
Iflands in the Great South-fea.	D.	M.	D.	M.
Honder Ifland	14	00	S	237	30
Water Iflands	14	50	S	229	00
Iflands Tiburones	12	00	S	218	05
St. Pedro	22	10	S	229	10
Prince William's Iflands	18	14	S	204	50
Iflands of Good Hope	17	12	S	195	40
States Land	38	00	S	192	00
Green Iflands	04	00	S	172	00
Salteadores Ifle	06	40	N	172	05
Miracomo	06	24	N	175	40
Iflands de Ladrones	10	00	N	170	00
Nadadores	04	22	N	186	55
Barbadoes Ifles	07	00	N	195	05
St. Peter's Ifle	11	14	N	205	00
Hermanes Ifle	15	10	N	181	05
Sea-Coaft on the Main Continent of *America.*				
Lemairs Strait	55	00	S	310	16
Cape Pennas	53	45	S	307	20
Eaft Entrance of Magelan	52	20	S	305	00
Cape Blanco	47	30	S	309	45
Cape St. Antonio	36	38	S	325	00
Cape St. Maria	35	00	S	325	40
Cape Frio	22	52	S	338	38
Baja de toda Santos	13	00	S	341	50
Cape St. Auguftine	08	40	S	345	40
Black Point	03	10	S	343	00
River Caflepore	04	00	N	328	05
Suranam	05	05	N	323	40
Cape Three Points	11	18	N	314	20
Cape de Coquibocao	12	42	N	305	24
Carthagene	10	25	N	298	25
Cape de Gratias	15	32	N	291	55
Cape de Catoche	21	23	N	287	00
Cape Rexo	22	40	N	275	26
Cape Blanco	26	55	N	274	36

A *Table* of Latitude and Longitude.					
Places Names.	Latitude.			Longitude	
Sea-Coast on the Main Continent of *America.*	D.	M.		D.	M.
Cape Efcondido	29	52	N	285	40
La Florida	25	51	N	272	16
Cape Fair	34	02	N	298	25
Cape Henry	37	00	N	300	40
Cape Charles	37	48	N	300	54
Cape May	39	55	N	302	00
The Eaft end of Long-Ifland	40	45	N	303	17
Cape Codd	42	20	N	308	40
Cape Ann	42	45	N	308	16
Cape Furcu	44	00	N	314	10
Eaft end of the Ifle of Sables	43	40	N	323	00
Cape Britain	45	30	N	323	50
Cape Raze	46	36	N	328	30
Conception Bay	48	22	N	327	50
Bay of Bulls	47	27	N	328	05
Cape Bona Vifta	49	19	N	328	36
Pingwins Ifle	50	02	N	328	40
Cape Gate	52	00	N	325	32
Bell Ifle	52	25	N	325	30
Iflands in the Weft-India.					
Bermudas	32	18	N	310	50
Bahama	27	57	N	395	20
North-eaft Point of Lucaioneque . . .	27	52	N	398	20
Signateo	26	18	N	300	00
Guatro	25	47	N	301	00
Guamina	25	15	N	301	40
Tiango	24	33	N	302	20
Majagana	23	05	N	303	00
Caicos	22	05	N	304	05
Ihagua	21	19	N	301	40
Yamata	22	32	N	301	20
Samana	24	20	N	395	45
Yamia	24	30	N	301	05
Anguilla	18	48	N	313	35

A *Table* of Latitude and Longitude.				
Places Names.	Latitude.		Longitude	
	D. M.		D. M.	
Iflands in the Weft-India.				
St. Chriftophers	17 30	N	313	30
South end of Barbada	17 36	N	316	00
Antego	16 32	N	315	10
Gadalupa	16 00	N	314	40
Marigallata . . . . x	15 41	N	315	25
Dominica	15 00	N	314	50
Martineco	14 30	N	316	36
St. Lucia	13 13	N	315	20
Barbadoes	13 12	N	319	40
St. Vincent	12 50	N	313	45
Granada	12 10	N	314	40
Tobago	11 12	N	317	00
Point de Gallaia	10 45	N	316	55
Mevis	17 00	N	314	00
Monferat	16 20	N	314	20
Margaretta	11 28	N	312	25
Tortogas	11 30	N	313	30
Dolkilla	12 19	N	310	30
Bonayre	12 32	N	308	35
Qniffa	12 25	N	307	30
Eaft end of Hifpaniola	18 47	N	308	10
Weft end of Hifpaniola	18 25	N	300	10
Port Royal in Jamaica	18 15	N	397	10
Eaft end of Cuba	20 27	N	201	20
Weft end of Cuba	22 00	N	388	26
Camnamis	19 41	N	294	45
Great Caiman	19 21	N	293	30
Santa Villa	17 28	N	294	00
Mofquito	14 50	N	294	12
Guanabo	16 33	N	287	14
Guanabimo	16 10	N	286	20
Cozumal	19 25	N	287	10
Lafalleiranes	22 00	N	284	40

A *Table* of Latitude and Longitude.

Places Names.	Latitude.			Longitude	
	D.	M.		D.	M.
The Northern Parts of America.					
Cape Camas	53	40	N	226	52
Refolution Ifles	61	00	N	309	00
The Kings Foreland	61	50	N	295	30
Queen Ann's Foreland	63	52	N	293	40
Cape Charles	62	55	N	291	42
North end of Mansfield's Ifle	62	40	N	284	39
Prince Rupert's River	51	00	N	289	12
Cape Monmouth	54	40	N	283	00
Cape Henrietta	56	16	N	279	10
Port Nelfon	58	32	N	267	50
Cape Southampton	62	30	N	279	25
Seahorfe Point	64	46	N	282	20
Sir Dudley Digg's Cape	75	10	N	298	00
Sanderfon's Tower	68	00	N	314	08
Cape Walfingham	65	42	N	311	20
Cape Comfort	62	21	N	321	20
Cape Defolation	61	20	N	325	05
Cape Farewel	59	45	N	329	02

The

The Use of the foregoing

TABLES.

The Use of Table I. page 2.

T. THE first is a *Table* shewing the *Dominical Letter*, whose Use is already taught *page 3.*

The Use of Table II. page 3.

T. THE second is a *Table* of *Moveable Feasts* and *Terms*, whose Use is so easie by the Directions on the head of each Column, that it needs no Example.

The Use of Table III. page 4.

T. THE third is a *Table* or *Calender* containing several things, amongst which is the *Suns Declination.*

S. Pray then explain it to me, that I may understand well its Use, and what each Column signifie ?

T. Take notice then, that in each Page there are Eleven Columns, that the first sheweth the *Days of the Month*, the second *Days of the Week*, expressed by the *Week-Day Letters*, as in the Year 1685. in the Month of *October*, you will find A for *Thursday*, B for *Friday*, C for *Saturday*, and D for *Sunday*, which is the *Dominical Letter* for that Year. (As you see in the first *Table* of this Book, *page* 3.) The third Column sheweth the *Fixed Feasts, and Remarkable Days and Things*, the Time that the *Sun Rises* and *Sets*, and the *Southing of several Stars at Midnight*, as in *November*, you will find against the 27th. Day *Orion's left Foot*, which shews, that the said Star comes to the Meridian the 27th. Day at Midnight.   The fourth Column sheweth the *Place* of the Sun for *Leap-year*, as against the 4th. of *October*, you will find the Sun to be in 21° 55′ of *Libra*.   The fifth Column sheweth the *Declination* of the Sun in the *Leap-year*, as against the 4th. of *October*, you will find the Suns Declination 8° 33′ South, and after the same manner the other six Columns are to be used; as against the said 4th. Day of *October*, you will find in the sixth Column, in the *First-year* after *Leap-year*, that the Suns place is 21° 39′

in

in *Libra*; and in the seventh Column, the *Suns Declination* to be 8° 28′;
and in the eighth Column, in the *Second-year* after Leap-year, the *Suns
Place* 21° 24′ in *Libra*; and in the ninth Column, the *Declination* of the
Sun is 8° 22′; and in the tenth Column, in the *Third-year* after *Leap-
year* the *Suns Place* is 21° 10′ in *Libra*; and in the eleventh Column, the
Suns Declination is 8° 17′.

The Use of Table *IV. page* 16.

T. **T**HE fourth is a *Table* shewing what Time *Aldebaran*, or the *Bulls
Eye*, comes to the Meridian throughout the Year, whose use is so
well known by the *Month* on the head of the *Table*, and the *Days* of it
on the Left-hand Column, that it needs no Example.

The Use of Table *V. page* 17.

T. **T**HE fifth is a *Table* shewing what Time some of the chief Stars
comes upon the Meridian before or after the *Bulls Eye*, whose
use requires no other Direction, than what the said Table sheweth, for
the *H* signifie the hours, and *M* the Minutes.

The Use of Table *VI. page* 18.

T. **T**HE sixth is a *Table* of the *Right Ascensions* and *Declinations* of
the chiefest and most known Stars in the Firmament, with their
Magnitude, Latitude, Longitude, and distance from the Pole.

S. What is the chiefest use of this *Table*?

T. Its chiefest use is for the *Declination* of the Stars, which you will
find in the fifth Column; and for to know what Time a Star will come
to the Meridian after another.

S. How shall I know what Time a Star will come to the Meridian
after another?

T. You may easily know it by the Stars *Right Ascension*, (which you
will find in the last Column of the same *Table*) thus: Subtract the *Right
Ascension* of the Star already upon the Meridian, from the *Right Ascension*
of the given Star, and add the remainder to the hour that you observe
the first Star on the Meridian, and the Sum will shew you what time the
given Star will come upon the Meridian.

Example 1.

The Bull's Eye (Aldebaran) being on the *Meridian at Nine a Clock at Night*;
I would know what Time the Lions Heart *will come to it*?

Therefore,

Therefore, I look for the *Right Afcenfion* of the *Bulls Eye*, which in the laft Column I find to be 4 Hours 18 Minutes, which I fubtract from the *Right Afcenfion* of the *Lions Heart*, 9 Hours 52 Minutes, and there remaineth 5 Hours 34 Minutes; which being added to 9 Hours, the Time that the *Bulls Eye* was on the Meridian, comes 14 Hours 34 Minutes in the Afternoon; but becaufe we do not ufe to Reckon our Hours beyond 12, I cut off the 12 Hours, (from 12 a Clock to Midnight) and there remaineth 2 Hours 34 Minutes, which fhews that the *Lions Heart* will come upon the Meridian at 2 a Clock 34 Minutes in the Morning.

Example 2.

Admit that at 10 a Clock at Night I fee the Eagles Heart *on the Meridian, (or South) and would know what Time the* Little Dogs Thigh Procion *will come to it ?*

First, I look in the precedent *Table* for the *Right Afcenfion* of the *Eagles Heart*, which in the laft Column I find to be 19 Hours 35 Min. and that of the *Little Dogs Thigh* 7 Hours 23 Minutes; but becaufe I cannot fubtract 19 Hours 33 Minutes, from 7 Hours 23 Minutes of the *Little Dog*, I add 24 Hours to it, and there will come 31 Hours 23 Minutes, from which I fubtract the 19 Hours 35 Minutes, and there will remain 11 Hours 48 Minutes, which being added to 10 Hours, the Time that the *Eag'les Heart* was on the Meridian, comes 21 Hours 48 Minutes, from which I fubtract the 12 Hours to Midnight, and there remaineth 9 Hours 48 Minutes in the Morning, the Time that the *Little Dogs Thigh* will come to the South or Meridian; by which I know that I cannot make any Obfervation at it, becaufe being then Broadday that Star will not be feen.

S. Is there no way to find what Time a Star will be on the Meridian, without comparing it to another Star ?

T. Yes, but not by this *Table* alone, for in that cafe, befides the *Right Afcenfion* of the Star, you muft alfo know the *Right Afcenfion* of the Sun, which you will find in *Table* V. *page* 20. of the Fifth Book.

How to find the Time that any Star comes upon the Meridian, and by it the hours of the Night.

S. How fhall I find when a Star will be South or on the Meridian, without comparing it to another Star ?

T. To find when a Star will be on the Meridian, firft look for the *Right Afcenfion* of the propofed Star, (as in the precedent Examples) and alfo for the *Right Afcenfion* of the Sun in its proper *Table*, page 20. and from the *Right Afcenfion* of the Star, fubtract the *Right Afcenfion* of the Sun, and the remainder will shew you the hours in the Afternoon, when the Star will be on the Meridian; and if it exceed 12, subtract 12 hours therefrom, and the remainder will shew you the Hours and Minutes of the Star, coming upon the Meridian after Midnight: But take notice, that when the *Right Ascension* of the Star is less than the *Right Ascension* of the Sun, you must add 24 Hours thereto, and then subtracting from it the *Right Ascension* of the Sun, the remainder shall shew you as before the Hour in the Afternoon, or in the Morning (if it passeth 12.)

Example 1.

The 29th. of March, *I would know when the* Virgins Spike *comes upon the* Meridian?

Therefore, I look for the *Right Ascension* of that Star, and find it to be 13 Hours 9 Minutes: I look also for the *Right Ascension* of the Sun in its proper *Table*, and right against the 29 of *March*, I find it to be 1 Hour 10 Minutes, which being subtracted from the *Right Ascension* of the Star, 13 Hours 9 Minutes, there remaineth 11 Hours 59 Minutes, which is the time that the *Virgin Spike* comes to the Meridian in the Afternoon. But you are to take notice, that the *Table* of the *Suns Right Ascension* is Calculated but for Noon, and that it doth increase about 4 Minutes each Day; and therefore to be more exact, you ought to proportion that difference, by allowing for every 6 Hours in the Afternoon 1 Minute; by which Rule you will find that the proposed Star will be on the Meridian at 11 Hours 57 Minutes in the Afternoon, because of the 2 Minutes, which must be subtracted for the increase of the *Suns Right Ascension* in 12 Hours, (one Minute making no difference.).

Example 2.

The 18th. of October, *I would know when the great* Dog (Sirius) *comes upon the* Meridian?

Therefore, I look as before for the *Right Ascension* of that Star, which I find to be 6 Hours 31 Minutes; I look also for the *Suns Right Ascension* on the 18th. of *October*, and find it to be 14 Hours 12 Minutes: Now because the *Right Ascension* of the Star is lesser than the *Right Ascension* of the Sun, I add 24 Hours to the Stars *Right Ascension*, and there comes 30 Hours 31 Minutes, from which I subtract the *Suns Right Ascension* 14 Hours 12 Minutes, and the remainder is 16 Hours 19 Min. from which I subtract the 12 Hours, from Noon to Midnight, and there remaineth 4 Hours 19 Minutes, which is the Time that the proposed Star (*Sirius*) comes to the Meridian in the Morning. But to be
more

more exact, you ought (as in the precedent Example) to subtract 3 Minutes for the increase of the *Suns Right Ascension* in almost 18 Hours, and so 4 Hours 16 Minutes will be the Time that the Great *Dog Sirius* will be South, or on the Meridian.

S. I understand now very well how to find what Time any Star comes upon the Meridian, but how shall I know by it what Time of the Night it is?

T. You may know it by the same Rules by which I did find, the the Time of the Star coming upon the Meridian, there being no more in it, then to subtract as before the *Right Ascension* of the Sun, from the *Right Ascension* of the Star, and the remainder will shew you what Time of the Night it is.

Example 1.

The Night on the 19th. of November, *the brightest Star in the* Pleiades *being South, (or on the Meridian) I would know what Hour of the Night it is?*

First, I look for the *Right Ascension* of the proposed Star, and find it to be 3 Hours 29 Minutes; I look also for the *Right Ascension* of the Sun (in its proper Table) on the 19th. of *November*, and find it to be 16 Hours 23 Minutes, and therefore must add 24 Hours to the *Right Ascension* of the *Pleiades*, comes 27 Hours 29 Minutes, from which I subtract the *Right Ascension* of the Sun 16 Hours 23 Minutes, and there remaineth 11 Hours 6 Minutes, from which I subtract 2 Minutes more for the increase of the *Suns Right Ascension* in 12 Hours, comes 11 Hours 4 Minutes for the true time of the Night, as was required.

Example 2.

The Night on the 16th. of October, *the Star in the Left Foot of* Orion (Regel) *being South, or upon the Meridian; I desire to know what time of the Night it is?*

	Hours.	Min.
Regel *Right Ascension*	4	59 $\frac{1}{4}$
Add	24	00
Comes	28	59 $\frac{1}{4}$
The Suns *Right Ascension*, Subtract	14	04
Remaineth	14	55 $\frac{1}{4}$
The 12 Hours from Noon to Midnight, Subt:	12	00
Time in the Morning	02	55 $\frac{1}{4}$
Two Minutes for the increase of the Suns *Right Ascension*, (in 12 Hours) Subtract	00	02
The Time of the Night required, is	02	53 $\frac{1}{4}$ in Mor.

How.

How to find the Hour of the Night without the Sun or Stars Right Ascension.

S. Is there no way to find the Hour of the Night without the *Tables* of the Sun and Stars *Right Ascension*?

T. Yes, there is, but you must be used to it before you can tell readily what Time of the Night it is, but then being very easie I will shew you this way, which is thus: Having observed such a Night in the year, that such a Star was at the Horizon, Meridian, or any other point or part of the Heaven at such Hour of the Night, if a Fortnight after you see it in the same place or point, you may conclude that it is an Hour later than the first time you saw it, for it will come every Day later to it 4 Minutes, which is 2 Hours in a Month, 4 Hours in two Months, and soforth; this being known, you may easily find out the Hour of the Night only by subtracting 4 Minutes for every Day past, since you first did observe the Star on the Meridian, or any other point in the Heavens.

Example 1.

The 13th. of December, having observed that the Left Shoulder of Orion was South (or on the Meridian.) at 11 a Clock at Night, and 20 Days after seeing the same Star on the Meridian; I would know what time of the Night it is?

Therefore, I Multiply the 20 Days past by 4 Minutes, (that the Star comes later every Day to the Meridian) and there comes 80 Minutes, which being divided by 60 (because 60 Minutes makes an Hour) comes 1 Hour 20 Minutes, which I subtract from 11 Hours, (the Time of my first Observation) and the remains 9 Hours 40 Minutes is the Time of the Night required. The same is to be done of any other Star, observed either above or under the Pole Star, or any other Point about it.

How to find the Hour of the Night by the shadow of the Moon.

S. How is the Hour of the Night known by the shadow of the Moon?

T. It is easily known by a Sun Dial, observing by it the Hour as at the Sun, then adding to those Hours the Time that the Moon comes to the South, (that Day) the sum will shew you the Hour of the Night, (but if it be above 12, subtract 12 therefrom, and the remainder will be the Hour required.)

S. How shall I know when the Moon comes to the South?

T. The way to know it, is to Multiply the Moons Age by 4, and to divide the Product by 5, and the Quotient will give the Hours that the Moon comes to the South, and if there remaineth any thing in

your

your Division, you must Multiply it by 12, and the Product will be the Minutes, which must be added to the Hours that the Moon comes to the South; for if One remaineth it is worth 12 Minutes; if Two, 24 Minutes; if Three, 36 Minutes; and if Four, 48 Minutes; by which Rule, you will find that the Moon will be South at 9 a Clock 36 Minutes past, when she is 12 Days Old: This being understood, you may easily find the Hour of the Night by the shadow of the Moon.

Example.

The Moon being 14 Days Old, I find by the precedent Rule that she comes to the South at 11 of the Clock and 12 Minutes past, and by my Sun-Dial it is half an Hour past 3, the Hour of the Night is required?

Therefore, I add the 3 Hours 30 Minutes that the shadow of the Moon sheweth (upon my Dial) to 11 Hours 12 Minutes, the Time of the Moon's Southing, comes 14 Hours 42 Minutes, from which I subtract 12 Hours, remaineth 2 Hours 42 Minutes, for the Time of the Night required.

The Use of Table VII. page 20.

T. THE seventh is a *Table* of the *Suns Right Ascension*, whose use is already taught with that of the *Stars Right Ascensions*.

The Use of Table VIII. page 21.

T. THE eighth is a *Traverse Table* to every Quarter Point of the Compass, to the 100 part of a League, or Mile.

S. What is the use of this *Traverse Table*?

T. This *Table* shews the difference of Latitude and departure from the Meridian.

S Pray shew me how?

T. Seek the Course Run on the Top of the *Table*, if it be not more than 4 Points from the Meridian (or North and South-Points) and the Miles or Leagues sailed in the Left-hand Column downward, and under the Course and against the Miles or Leagues sailed, (to wit, under *N S*) you will have the difference of Latitude, and under *E. W*, the departure from the Meridian.

Example.

Suppose a Ship Sail North North East 25 Miles, and the difference of Latitude and Departure be required?

North,

North North Eaſt is 2 Points diſtant from the Meridian, wherefore I look for 2 Points at the Top of the Table, and in the Column under it, and againſt 25, I find under *N S* 23.10, (which ſhews that the difference of Latitude is 23 Miles $\frac{1}{100}$ parts, (of 100) and under *E W* 9.57 for the departure from the Meridian. But if the Courſe be above 4 Points from the Meridian, then you muſt ſeek for it at the bottom of the Table, and the Miles or Leagues ſailed in the Right-hand Column upward.

Example.

Suppoſe a Ship ſail North Weſt by Weſt one Quarter-point Weſterly, 40 Leagues, and the difference of Latitude and departure be required?

North Weſt by Weſt one Quarter-point Weſterly, is 5 ¼ Points diſtant from the Meridian, wherefore I ſeek for it at the bottom of the Table, and then look upward in the Right-hand Column for 40, the Leagues ſailed, and over the Courſe, and againſt the Leagues ſailed (over *N S*) is 20.56 the difference of Latitude, and over *E W* is 34.31 the departure from the Meridian.

S. This I very well underſtand, but how if the Leagues or Miles ſailed, be an odd Number above 60, for I ſee that the Table skips from 60 to 70, &c. What muſt I do then?

T. Then you muſt divide your number of Leagues or Miles into two or three parts, *viz.* Hundreds, Tens, and Units.

Example.

Suppoſe a Ship ſail 274 Miles Eaſt North Eaſt, half a Point Eaſterly, and the difference of Latitude and departure be required?

Firſt, find the Courſe (which is 6 ¼ Points) at the bottom of the Table, then ſeek your Miles in the Right-hand Column: Thus,

		Diff. Lat.		Departure.	
Firſt 200	Aginſt which and over the	58	4 and	191	38
Then 70	Courſe (over *N S*) is	20	31 over	66	90
Laſtly 4		01	16 *EW*	03	83

So 274 Miles ſailed upon this Point gives 79 51 ——— 262 11

The Uſe of Table IX. page 29.

T. THE ninth is a *Table* of *Meridional Parts*.

S. When do you make uſe of this *Table* of *Meridional Parts?*

T. I make uſe of it in *Mercator*'s ſailing, for which it chiefly ſerveth.

S. Why do you call it a *Table* of *Meridional Parts?*

T. Becauſe

T. Becaufe it fhews the *Meridional Parts* for every Degree and Minute of Latitude, which *Parts* ferveth to find the *Meridional Miles* or *Minutes,* between two places.

S. What is the firft thing to be known?

T. The firft thing to be known is, that you muft enter the Degrees of Latitude on the Head of the *Table,* (where they are marked or laid down) and the Minutes down the Left-hand Column; then confider whether both places lies on the fame fide of the Equinoctial, or the one on the one fide of the Equinoctial, and the other on the other, or whether one be on the Equinoctial, and the other wide thereof; for the cafe differs according to the Propofition, as you will better underftand by thefe Directions.

1. When both places are in Latitude North, (or in Latitude South) then fubtract the Meridional Parts anfwering to the leffer Latitude, out of thofe for the greater, and the remainder will be the Meridional difference of Latitude.

2. When one place is in Latitude North, and the other in Latitude South, add the Meridional Minutes belonging to each Latitude together, and the fum is the Meridional Minutes between them.

3. When one place lyeth under the Equinoctial, then the Meridional Minutes that are found under the Degrees of Latitude the other place lyeth in, is the Meridional difference of Latitude.

Example 1.

Admit it is required to find the Meridional Parts, or Minutes, between the Latitude 39° 35' North, and 48° 50' North ?

Under 48° and right againft 50' in the Left-hand Column you will find . . . . } 3366.9
Under 39° and againft 35' is 2590.2
The Meridional parts between the Lat. propofed, are 0776.7

Example 2.

To find the Meridional Parts between 18° 20' South Latitude, and 37° 45' North.

Under 37° and againft 45' is 2449.8
Under 18° and againft 20' is 1119.2
The Meridional parts between the two places are 1330.1

Example

Example 3.

To find the Meridional Minutes between the Equinoctial and Latitude 46° 14'.

Under 46° and against 14' is 3135.8, the Meridional Parts required.

Note, That in this *Table* you are to cut off the last Figure of Meridional Parts as in the precedent Examples, because it is but so many 10th. parts of a Minute, the reason that it must be neglected when it is under 5, (¼ Min.) but if above that Number you may add a Minute for it, and so you will find the Meridional parts of 46° 14' should be rather 3136 then 3135, because of the ₁₀⁸ of a Minute remaining.

The Use of Table X. *page* 57.

T. THE tenth is a *Table* of the Miles East or West, that Answer the Degrees of Longitude in the fourth Rumb.

S. For what use is this *Table* ?

T. Its use is to turn Miles of Easting or Westing into Degrees of Longitude, and Degrees of Longitude into Miles of Easting or Westing.

S. If it be so, it must needs be very necessary in *Navigation*, therefore pray shew me well its use that I may not mistake in the Practice of it ?

T. Your request is very just, and I hope to satisfie you, since two or three Examples will make it easie and intelligible as you can desire.

Admit then, that a Ship sail from the Equinoctial on some Point between the North and the East to 15 Degrees of Latitude, and then find she hath made 600 Miles departure East. To find the Degrees of Longitude answering to this 600 Miles, look in your Table for 15 Degrees Latitude, and right against it you will find 900 Miles (that is to say, the Miles East the Ship would have been if she had sailed on the fourth Rumb, either *N E, N W, S E,* or *S W.*) and 15 Degrees 10 Minutes. Then say by the Rule of Three,

As 900 Miles is to 15 Degrees 10 Minutes,

So is 600 Miles to a fourth Number, which being divided by 60 will give 10 Degrees 6 Minutes of Longitude; but if you work your *Traverse* by Leagues, you must first reduce the Leagues proceeding from your *Traverse-Table* into Miles, and then work as before.

Practice.

If 900 Miles give 15 Deg. 10 Min. What will 600 Miles give?

$$
\begin{array}{r}
60 \\
\hline
910 \\
600 \\
\hline
546000
\end{array}
$$

$$990) \underline{556000} (606$$
$$600$$

Example.

Admit a Ship sail from the Parallel of 25 Degrees of Latitude North, (on some Point between the South and the West) to 15 Degrees of Latitude (also) North, and then find she hath made 400 Miles departure West, the Degrees of Longitude answering to this 400 Miles is required?

In this case, you must first subtract the Number, which in your *Table* answers 15 Degrees of Latitude North, from that of 25 Degrees; that is to say, (the lesser Number from the greater, or) 900 Miles from 1500 Miles, and 15 Degrees 10 Minutes from 25 Degrees 50 Minutes, and there will remain 600 Miles, and 10 Degrees 40 Minutes; then say by the *Rule of Three*: If 600 Miles give 10 Degrees 40 Minutes; What will 400 Miles give? (and the Quotient of the Divisor being divided by 60) there will come 7 Degrees 6 Minutes.

Practice.

If 600 Miles give 10 Degr. 40 Min. What will 400 Miles give?

$$
\begin{array}{r}
60 \\
\hline
640 \\
400 \\
\hline
256000
\end{array}
$$

$$600) \underline{256000} (426$$
$$400$$

S. What must I do, if by the difference of Longitude I would find the Miles East or West, answering thereunto?
T. You must then Reverse the *Rule of Three*, and do the contrary of what you have done before.

As

As for Example.

Admit that in the Parallel of 15 Degrees the difference of Longitude is of 10 Degree 6 Min. $\frac{600}{900}$, and would find the Miles East (or West) answering thereunto.

Look in the said Table against the Latitude of 15 Degrees, and there you will find 900 Miles and 15 Degrees 10 Minutes; therefore say by the *Rule of Three:* If 15 Degrees 10 Minutes give 900 Miles; What will 10 Degrees 6 Min. $\frac{600}{900}$ give? and there will come 599 Miles, and 310 remaining, which with the 600 over and above the 10 Degrees 6 Min. makes a Minute more, since the Divisor is but 910; and so the sum is 600 Miles.

Practice.

If 15 Deg. 10 Min. give 900 Miles; What will 10 Deg. 6 Min. $\frac{600}{900}$ give?

$$\begin{array}{ccc}
60 & & 60 \\
\hline
910 & 600 & 606 \\
& 310 & 900 \\
& \overline{910} & \overline{545400}
\end{array}$$

$$91)\ 54540\ (599$$
$$31$$

The Use of Table XI. page 65.

T. THE eleventh is a *Table* to change Degrees and Minutes of any Parallel into Miles.

S. What must I do to change Degrees of a Parallel into Miles?

T. You must look the Latitude in the left Column, and the Degrees and Minutes of Longitude on the head of the *Table*, and right against the Latitude, and under the Degrees of Longitude you will have the Miles and Parts sought.

If the Longitude given consists of Degrees and Minutes, look the Miles and Parts answering the Degrees first, and then those for the Minutes; and if either the Degrees and Minutes be above 10, you must enter several Times, as you will see by these Examples.

Example 1.

Admit I have altered 2 Degrees 45 Minutes of Longitude under the Parallel of 46 Degrees Latitude, and it be required to find the Miles and Parts answering thereto?

I enter

I enter the *Table* as is directed with $\begin{cases} \text{2 Degrees which gives 83.4 Miles.} \\ \text{40 Minutes which gives 27.28} \\ \text{5 Minutes which gives 03.5} \end{cases}$

In all . . 114.7 Miles.

Example 2.

To find the Miles anfwering to 13 Degrees 27 Minutes in the Parallel of 44 Degrees of Latitude.

10 Degrees gives	: . . .	431.6
3 Degrees gives	. . . .	129.5
20 Minutes gives	. . . .	014.4
7 Minutes gives	. . . .	005.0
		580.5 Miles.

S. Why do you cut off the laft Figure from the reft?

T. Becaufe it fheweth but the Parts of a Minute.

S. How many of thofe Parts do you carry to a Minute?

T. I carry Ten, as you fee by the precedent Examples; by which you may underftand that the laft five of your Addition fignifie only $\frac{1}{10}$ of a Mile.

The Ufe of Table XII. page 73.

T. THE twelfth is a *Table* to reduce Miles Eaft and Weft into Degrees of Longitude.

S. Is the ufe of this *Table* as eafie as the laft?

T. Yes, for who underftands the laft muft needs underftand this, there being no more in it then to enter the Miles on the head, and the Degrees of Latitude in the Left-hand Column.

Example.

Admit a Ship fail 218 Miles Eaft or Weft in the Parallel of 48 Degrees Latitude, and it is demanded how many Degrees of Longitude are altered?

200 Miles gives	4°	58′	54″
10 Miles gives	0	14	57
8 Miles gives	0	11	57
In all	5	28	48

The

The Use of Table *XIII. page* 87.

T. THE thirteenth is a *Table* of *Rumbs* with the difference of Latitude and Longitude.

S. I believe this *Table* is very neceſſary in *Navigation*, therefore pray give me ſome Examples that may make it very eaſie to me?

T. It is what I did deſign before, in hopes that you will mind thoſe that followeth, ſince it is what you ſo earneſtly deſire.

Prop. 1.

The Latitudes of two places, and their difference of Longitude being given to find the Courſe and Diſtance.

Admit a Ship ſet from Latitude 49 Degrees 38 Minutes North, and ſail on ſome Rumb between the South and the Weſt, 'till ſhe fall in Latitude 36 Degrees 20 Minutes North, and have altered her Longitude 27 Degrees 50 Minutes; What Courſe has ſhe kept, and what diſtance has ſhe run? I ſeek under the fourth Rumb (in this *Table*) againſt Latitude 49 Degrees 38 Minutes, and find the Longitude anſwering thereto 57 Degrees 23 Minutes, and againſt 36 Degrees 20 Minutes, 39 Degrees 3 Minutes; the difference of theſe two Longitudes 57 Degrees 23 Minutes, and 39 Degrees 3 Minutes, is 18 Degrees 20 Minutes which ſhould be 27 Degrees 50 Minutes, therefore the fourth was not the Rumb. Again, againſt theſe Latitudes I take out the Longitude anſwering to the fifth Rumb, and find them to differ 27 Degrees 28 Minutes, which being only a few Minutes under the true difference of Longitude 27 Degrees 50 Minutes, I conclude her to have ſailed on the fifth Rumb.

To find the diſtance, ſubtract the given Latitudes one from the other, and the difference 13 Degrees 18 Minutes, look in the Left-hand Column of the *Table*, (under the Title *Latitude*) and under the fifth Rumb you find 1437 Miles of 60 to a Degree, the diſtance required?

If the Latitudes had been, the one North, and the other South, you muſt have added them and taken the ſum for the difference.

Prop. 2.

The Courſe and Diſtance being given with the Latitude departed, to find the Latitude ſhe is in, and the difference of Longitude.

Admit a Ship ſail from Latitude 48 Degrees North, on the fourth Rumb; to wit, *S W* 860 Miles: What is the Latitude ſhe is in, and her difference of Longitude? Looking under the fourth Rumb for the diſtance 860, I find 863 the neareſt to it, and right againſt it (in the Column

Colum of Latitude) 10 Degree 10 Minutes, but making a proportionable allowance for the 3 Miles difference, the Latitude is 10 Degrees 8 Minutes, which being subtracted from 48 Degrees, there will remain 37 Degrees 52 Minutes the Latitude she is arrived in. Lastly, against these Latitudes (48 Degrees and 37 Degrees 52 Minutes) under the Rumb look the Correspondent Longitudes, which are 54 Degrees 52 Minutes, and 40 Degrees 58 Minutes, and their difference 13 Degrees 54 Minutes, is the true difference of Longitude.

Prop. 3.

Both the Latitudes and the Course being given, to find the distance sailed and the difference of Longitude.

A Ship departs from an Island in 40 Degrees Latitude North, and sails North-East by North, 'till she be in Latitude 46 Degrees 10 Minutes North, her distance sailed and difference of Longitude is required?

Against the Latitude 40 Degrees, and under the third Rumb you'l find the Longitude to be 29 Degrees 12 Minutes, and against the Latitude 46 Degrees 10 Minutes, and under the same Rumb you'l find 34 Degrees 52 Minutes, then from 34 Degrees 52 Minutes subtract 29 Degrees 12 Minutes, the remainder 5 Degrees 40 Minutes, is the difference of Longitude required?

Secondly, take the Latitudes one from the other, and look the remainder 6 Degrees 10 Minutes, in the Column of Latitude and under the third Rumb you'l have 445 Miles the Distance.

Prop. 4.

The Latitudes of two places and the distance being given, to find the Course and difference of Longitude.

A Ship sails from Latitude 50 Degrees North, 1265 Miles, and then arrives in Latitude 38 Degrees 20 Minutes; What Course has she steered, and what is the difference of Longitude ? Find the difference of Latitude 11 Degrees 40 Minutes, in the Column of Latitude and right against it, searching under all the Rumbs for the distance 1265 Miles, I find it nearly under the fifth Rumb, the Rumb sailed on.

Then against Latitude 50°, under the fifth Rumb is 88° 44' Longit.
Against 38° 20' Latitude, under the fifth Rumb is 62 12.

Remains the difference of Longitude 24 32

Prop. 6.

Prop. 5.

The difference of Longitude, Distance, and one Latitude being given, to find the other Latitude and the Course.

S. How is this to be done ?

T. You must take the Rumb which you judge fittest, and examine what difference of Latitude will answer to the distance given, with which Latitude search all the Rumbs for a difference of Longitude like that given, and where you find it, that is the Rumb sought.

As for Example.

Admit a Ship sail from Latitude 32 Degrees 10 Minutes North, 'till she alter her Longitude 7 Degrees 23 Minutes, and make her Distance sailed 637 Miles.

I chuse the third Rumb, and search under it 'till I find 637, it is against 8 Degrees 50 Minutes Latitude; which added to 32 Degrees 10 Minutes, gives 41 Degrees the second Latitude; the Longitude that answer to these two Latitudes under the Third Rumb, are 30 Degrees 5 Minutes, and 22 Degrees 42 Minutes, whose difference is 7 Degrees 23 Minutes, just agreeing to my difference of Longitude given; whence I conclude the third Rumb to be that sailed on.

Prop. 6.

The difference of Longitude, the Course and one Latitude being given, to find the other Latitude, and the Distance.

Admit a Ship sail from Latitude 43 Degrees 30 Minutes North, on the sixth Rumb or *ENE*, 'till she alter her Longitude 17 Degrees 25 Minutes, to find the other Latitude and Distance.

Look under the sixth Rumb for the Longitude answering to 43 Degrees 30 Minutes Latitude, and you'l find 116 Degrees 51 Minutes, to which add 17 Degrees 25 Minutes, the difference of Longitude given, and see for the sum 134 Degrees 16 Minutes, under the same Rumb and the Title Longitude, and you'l find 48 Degrees 30 Minutes, the other Latitud.

To find the distance, subtract the distance answering to Latitude 43 Degrees 30 Minutes, from the distance answering to 48 Degrees 30 Minutes Latitude; and the difference 684 Miles, is the distance required?

When the Latitudes are of different Denomination (or one is North and the other South) add them, and the sum will be the difference of Latitude.

The

The Use of Table XIV. page 113.

T. THE fourteenth is a *Table* of the Latitude and Longitude of Places.

S. From what Meridian are the Longitudes accounted?

T. The Longitudes are accounted here from (the Meridian of) *Pico Tenerif* Eafterly, as in moft of the Chards.

S. What fignifie the Letters *N* and *S* in the Column of Latitude?

T. The *N* fignifies that the Latitude before it is North, but the *S* fignifies Latitude South.

Example 1.

What is the Latitude and Longitude of Barbadoes, one of the Weft-India *Iflands?*

Againft *Barbadoes* you will find 13 Degrees 12 Minutes North Latitude, and 319 Degrees 40 Minutes of Longitude.

Example 2.

What is the Latitude and Longitude of Bantam (*in the* Eaft-India?)

Againft *Bantam* you will find 6 Degrees 15 Minutes South Latitude, and 126 Degrees 30 Minutes of Longitude.

S. How fhall I find the difference of Longitude between any two places?

T. You muft take the Longitude of the two places, and fubtract the leffer Longitude out of the greater, and if the remainder be lefs than 180 Degrees; that is, the difference of Longitude, but if the remainder be more than 180, fubtract it from 360, the laft remainder is the difference of Longitude.

Example 1.

What is the difference of Longitude between Barbadoes and the Weft-end of Tercera, one of the Weftern *Iflands?*

The Longitude of the *Weft-end* of *Tercera*, is . .	349°	10'
The Longitude of *Barbadoes*	319	40
Difference of Longitude . .	029	30

Example 2.

Example 2.

What is the difference of Longitude between Barbadoes *and the* Lizard?

The Longitude of *Barbadoes*	319°	40'
Longitude of the *Lizard*	12	37
The remainder being greater then 180°. .	307	03
I fubtract from	360	00
Remains the difference of Longitude . .	052	57

e.Mixon

Mizen-top-maſt and running Rigging.

a. The Mizen-top-maſt.
b. The Mizen-top-ſail-brace.
c. The Mizen-top-ſail-clew-line.
d. The Mizen-top-ſail-ſheet.
e. The Mizen-top-ſail-lifts.
f. The Mizen-crow-foot.
g. Hoiſting-line for a penant.
h. The Mizen-ſheet.

Main-Maſt.

a. The Main-top-gallant-maſt.
b. The Main-top-gallant-leefts.
c. The Main-top-gallant-yard.
d. The Main-top-gallant-braces.
e. The Main-top-maſt.
f. The Main-top-maſt-back-ſtay
g. The Main-top-ſail-lifts.
h. The Main-top-ſail-braces.
k. The Main-top-ſail-clew-lines.
l. The Main-top-ſail-leath-lines.
n. The Main-top-ſail-bunt-lines.
m. The Main-lifts.
o. The Main-yard.
r. The Main-braces.
s. The Main-ſheets.
t. The Main-tacks.
v. The Main-ſhrowds.

Fore-Maſt.

a. The Fore-top-gallant-maſt.
b. The Fore-top-gallant-lifts.
c. The Fore-top-gallant-yard.
d. The Fore-top-gallant-braces.
e. The Fore-top-maſt.
f. The Fore-top-maſt-back-ſtay.
g. The Fore-top-ſail-lifts.
h. The Fore-top-ſail-braces.
k. The Fore-top-ſail-clew-lines.
l. The Fore-top-ſail-leach-lines.
n. The Fore-top-ſail-bunt-lines.
m. The Fore-lifts.
o. The Fore-yard.
p. The Fore-leath-lines.
q. The Fore-bunt-lines.
r. The Fore-braces.
s. The Fore-ſheets.
t. The Fore-tacks.
v. The Fore-ſhrowds.
x. The Fore-clew-garnet.

The Bow-Sprit.

a. The Sprit-ſail-top-maſt.
b. The Sprit-ſail-top-ſail-lifts.
c. The Sprit-top-ſail-yard.
d. The Sprit-ſail-top-maſt-ſhrowds.
e. The Sprit-ſail-top-ſail-braces.
f. The Sprit-ſail-top-ſail-crow-foot.
g. The Sprit-ſail-top-ſail-ſheets.
k. The Horſe on the Bow-ſprit.
l. Standing lifts for Sprit-ſail-yard.
m. The Sprit-ſail-yard.
n. The Sprit-ſail-ſheets.
o. The Sprit-ſail-clew-lines.
r. The crean-line.

FINIS.

ource UK Ltd.
UK
90319
UK00005B/361/P